Abbreviations for Units

A	ampere	H	henry	nm	nanometer (10^{-9} m)
Å	angstrom (10^{-10} m)	h	hour	pt	pint
atm	atmosphere	Hz	hertz	qt	quart
Btu	British thermal unit	in	inch	rev	revolution
Bq	becquerel	J	joule	R	roentgen
C	coulomb	K	kelvin	Sv	seivert
°C	degree Celsius	kg	kilogram	s	second
cal	calorie	km	kilometer	T	tesla
Ci	curie	keV	kilo-electron volt	u	unified mass unit
cm	centimeter	lb	pound	V	volt
dyn	dyne	L	liter	W	watt
eV	electron volt	m	meter	Wb	weber
°F	degree Fahrenheit	MeV	mega-electron volt	y	year
fm	femtometer, fermi (10^{-15} m)	Mm	megameter (10^6 m)	yd	yard
ft	foot	mi	mile	μm	micrometer (10^{-6} m)
Gm	gigameter (10^9 m)	min	minute	μs	microsecond
G	gauss	mm	millimeter	μC	microcoulomb
Gy	gray	ms	millisecond	Ω	ohm
g	gram	N	newton		

Some Conversion Factors

Length

1 m = 39.37 in = 3.281 ft = 1.094 yd

1 m = 10^{15} fm = 10^{10} Å = 10^9 nm

1 km = 0.6215 mi

1 mi = 5280 ft = 1.609 km

1 lightyear = 1 $c{\cdot}y$ = 9.461 $\times$ 10^{15} m

1 in = 2.540 cm

Volume

1 L = 10^3 cm^3 = 10^{-3} m^3 = 1.057 qt

Time

1 h = 3600 s = 3.6 ks

1 y = 365.24 d = 3.156 $\times$ 10^7 s

Speed

1 km/h = 0.278 m/s = 0.6215 mi/h

1 ft/s = 0.3048 m/s = 0.6818 mi/h

Angle–angular speed

1 rev = 2π rad = 360°

1 rad = 57.30°

1 rev/min = 0.1047 rad/s

Force–pressure

1 N = 10^5 dyn = 0.2248 lb

1 lb = 4.448 N

1 atm = 101.3 kPa = 1.013 bar = 76.00 cmHg = 14.70 lb/in^2

Mass

1 u = [(10^{-3} mol^{-1})/N_A] kg = 1.661 $\times$ 10^{-27} kg

1 tonne = 10^3 kg = 1 Mg

1 slug = 14.59 kg

1 kg weighs about 2.205 lb

Energy–power

1 J = 10^7 erg = 0.7373 ft·lb = 9.869 $\times$ 10^{-3} L·atm

1 kW·h = 3.6 MJ

1 cal = 4.184 J = 4.129 $\times$ 10^{-2} L·atm

1 L·atm = 101.325 J = 24.22 cal

1 eV = 1.602 $\times$ 10^{-19} J

1 Btu = 778 ft·lb = 252 cal = 1054 J

1 horsepower = 550 ft·lb/s = 746 W

Thermal conductivity

1 W/(m·K) = 6.938 Btu·in/(h·ft^2·F°)

Magnetic field

1 T = 10^4 G

Viscosity

1 Pa·s = 10 poise

fifth edition

PHYSICS

FOR SCIENTISTS AND ENGINEERS

Volume 2C
Elementary Modern Physics

W. H. Freeman and Company
New York

Publisher:	Susan Finnemore Brennan
Senior Development Editors:	Kathleen Civetta/Jennifer Van Hove
Assistant Editors:	Rebecca Pearce/Amanda McCorquodale/Eileen McGinnis
Marketing Manager:	Mark Santee
Project Editors:	Georgia L. Hadler/Cathy Townsend, PreMediaONE, A Black Dot Group Company
Text Designer:	Marsha Cohen
Cover Designer:	Blake Logan
Illustrations:	Network Graphics/PreMediaONE, A Black Dot Group Company
Photo Editors:	Patricia Marx/Dena Betz
Production Manager:	Julia DeRosa
Media and Supplements Editor:	Brian Donnellan
Composition:	PreMediaONE, A Black Dot Group Company
Manufacturing:	RR Donnelley & Sons Company

Cover image: Digital Vision

Library of Congress Cataloging-in-Publication Data

Physics for Scientists and Engineers. - 5th ed.

    p. cm.

    By Paul A. Tipler and Gene Mosca

    Includes index.

    ISBN: 0-7167-0809-4 (Vol. 1 Hardback Ch. 1-20, R)

    ISBN: 0-7167-0900-7 (Vol. 1A Softcover Ch. 1-13, R)

    ISBN: 0-7167-0903-1 (Vol. 1B Softcover Ch. 14-20)

    ISBN: 0-7167-0810-8 (Vol. 2 Hardback Ch. 21-41)

    ISBN: 0-7167-0902-3 (Vol. 2A Softcover Ch. 21-25)

    ISBN: 0-7167-0901-5 (Vol. 2B Softcover Ch. 26-33)

    ISBN: 0-7167-0906-6 (Vol. 2C Softcover Ch. 34-41)

    ISBN: 0-7167-8339-8 (Standard Hardback Ch. 1-33, R)

    ISBN: 0-7167-4389-2 (Extended Hardback Ch. 1-41)

Printed in the United States of America

First printing 2003

CONTENTS IN BRIEF

VOLUME 2

PART V   LIGHT

PART VI   MODERN PHYSICS: QUANTUM MECHANICS, RELATIVITY, AND THE STRUCTURE OF MATTER

APPENDIX

CONTENTS

CHAPTER 10

CONSERVATION OF ANGULAR MOMENTUM / 309

CHAPTER R

SPECIAL RELATIVITY / R-1

CHAPTER 11

GRAVITY / 339

CHAPTER 12 *

STATIC EQUILIBRIUM AND ELASTICITY / 370

PART II   OSCILLATIONS AND WAVES /425

CHAPTER 14

OSCILLATIONS / 425

CHAPTER 15

TRAVELING WAVES / 465

CHAPTER 16

SUPERPOSITION AND STANDING WAVES / 503

VOLUME 2

PART IV   ELECTRICITY AND MAGNETISM/651

CHAPTER 21

THE ELECTRIC FIELD I: DISCRETE CHARGE DISTRIBUTIONS / 651

CHAPTER 22

THE ELECTRIC FIELD II: CONTINUOUS CHARGE DISTRIBUTIONS / 682

CHAPTER 23

ELECTRIC POTENTIAL / 717

CHAPTER 24

ELECTROSTATIC ENERGY AND CAPACITANCE / 748

CHAPTER 25

ELECTRIC CURRENT AND DIRECT-CURRENT CIRCUITS / 786

CHAPTER 26

THE MAGNETIC FIELD / 829

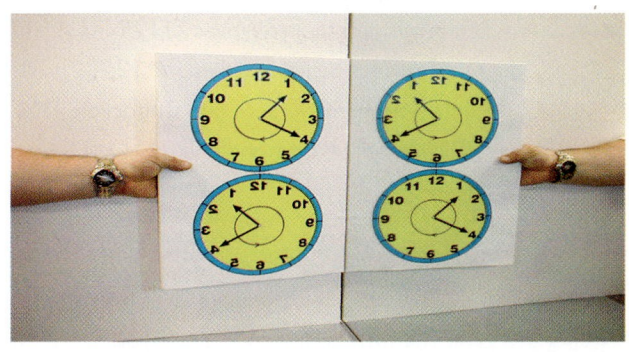

PART VI MODERN PHYSICS: QUANTUM MECHANICS, RELATIVITY, AND THE STRUCTURE OF MATTER

CHAPTER 34

WAVE-PARTICLE DUALITY AND QUANTUM PHYSICS / 1117

CHAPTER 35

APPLICATIONS OF THE SCHRÖDINGER EQUATION / 1149

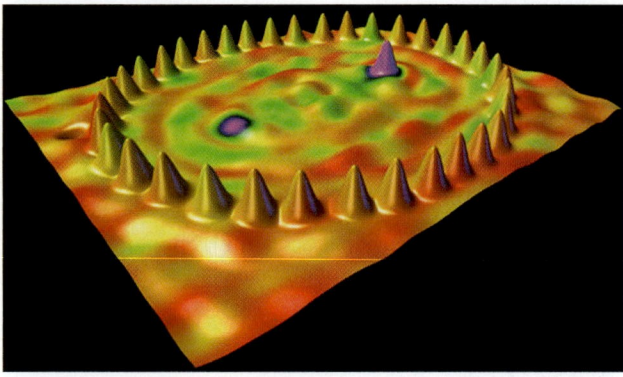

CHAPTER 36

ATOMS / 1171

CHAPTER 37

MOLECULES / 1208

CHAPTER 38

SOLIDS / 1228

CHAPTER 39

RELATIVITY / 1267

PREFACE

We are exceptionally pleased to present the fifth edition of *Physics for Scientists and Engineers*. Over the course of this revision, we have built upon the strengths of the fourth edition so that the new text is an even more reliable, engaging and motivating learning tool for the calculus-based introductory physics course. With the help of reviewers and the many users of the fourth edition we have carefully scrutinized and refined every aspect of the book, with an eye toward improving student comprehension and success. Our goals included helping students to increase their problem-solving ability, making the text more accessible and fun to read, and keeping the text flexible for the instructor.

Examples

One of the most important ways we've addressed our goals was to add some new features to the side-by-side worked examples that were introduced in the fourth edition. These examples juxtapose the problem-solving steps with the necessary equations so that it's easier for students to watch the problem unfold.

The side-by-side format for the worked examples came from a student suggestion; we've just added a few finishing touches:

• After each problem statement, students are asked to *Picture the Problem*. Here, the problem is analyzed both conceptually and visually, with students frequently directed to draw a free-body diagram. Each step of the solution is then presented with a written statement in the left-hand column and the corresponding mathematical equations in the right-hand column.

• *Remarks* at the end of the example point out the importance or relevance of the example, or suggest a different way to approach it.

• NEW *Plausibility Checks* remind students to check their results for mathematical accuracy, and for reasonableness as well.

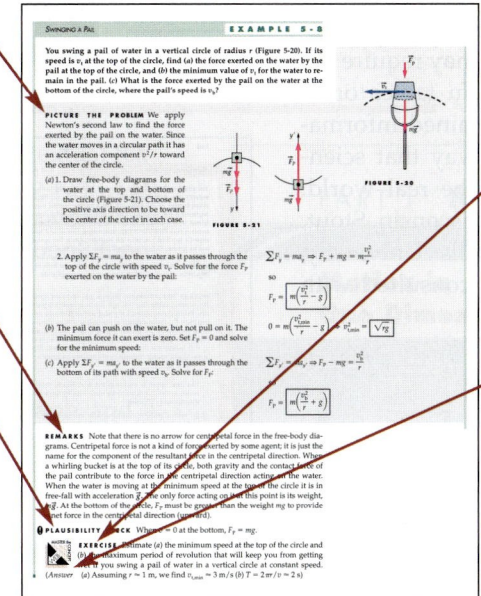

• An *Exercise* often follows the solution of the example, allowing students to check their understanding by solving a similar problem without help. Answers are included with the Exercise to provide immediate feedback and alternative solutions.

• NEW *Master the Concept Exercises* appear at least once in each chapter and help build students' problem-solving skills online.

Content improvements

Chapter R, an optional "mini" chapter in Volume 1, brief enough to be covered in a lecture or two, allows instructors to include this popular modern topic early in the course. The chapter avoids the abstraction associated with the Lorentz transformations and focuses on the basic concepts of length contraction, time dilation, and simultaneity, using thought experiments involving meter sticks and light clocks. The relation between relativistic momentum and relativistic energy is also developed.

Quantum Theory: Chapters 17, "Wave-Particle Duality and Quantum Physics," and 27, "The Microscopic Theory of Electrical Conduction" of the fourth edition have been moved to their more traditional location in Volume II of the fifth edition as Chapters 37 and 38. Should instructors wish to include these chapters earlier in the course, both chapters are available on the web at www.whfreeman.com/tipler5e.

Changes in Approach: Dozens of smaller, yet significant improvements in content have been made throughout the book. For example:

- Motion-diagrams are introduced in Section 3-3 and used to estimate the direction of the acceleration vector using the definition of acceleration.

- In Section 4-4, frictional forces are now introduced qualitatively, allowing for free-body diagrams that include frictional forces. A quantitative treatment of frictional forces appears in Section 5-1.

- Section 4-7 introduces problems with two or more objects. Selecting a separate set of coordinate axes for each object is a robust problem-solving practice when using Newton's laws with systems consisting of two or more objects. The value of this practice is revealed in the example where Steve is sliding down the glacier while Paul has already fallen over its edge.

- In Section 8-8, "Systems With Variable Mass," the basic equation of motion for an object with continuously varying mass (the rocket equation) is developed using an object that is acquiring mass—like an open boxcar in the rain—rather than one that is losing mass—like a rocket spewing exhaust gasses. This approach facilitates both the development of the basic equation of motion and the application of it to certain situations.

- In Chapter 9, "Rotation," there is a new section that provides problem-solving guidelines for applying Newton's Second Law to rotation.

- In Section 13-3 the discussion of buoyancy now includes the buoyant force on objects supported by a submerged surface.

- In Chapter 18, work-energy relations are expressed in terms of the work done on the system. The first law of thermodynamics is now expressed in terms of the work done on the system also. (The Educational Testing Service has adopted the convention that the work term in first law of thermodynamics be the work done on the system. This will be adhered to on all Advanced Placement physics exams.)

More engineering and biological applications

Additional applications emphasize the relevance of physics to students' experiences, further studies, and future careers.

New focus on common pitfalls

Topics that commonly cause confusion are identified with a new ❗ icon where the difficulty is addressed. For example, in Section 3-4 the icon is used to identify the discussion pointing out that the horizontal and vertical motions are independent in projectile motion.

For instructor and student convenience, the fifth edition of *Physics for Scientists and Engineers* is available in five paperback volumes—

Vol. 1A Mechanics (Ch. 1-13, plus a mini-chapter on relativity, Ch. R) 0-7167-0900-7

Vol. 1B Oscillations & Waves; Thermodynamics (Ch. 14-20) 0-7167-0903-1

Vol. 2A Electricity (Ch. 21-25) 0-7167-0902-3

Vol. 2B Electrodynamics, Light (Ch. 26-33) 0-7167-0901-5

Vol. 2C Elementary Modern Physics (Ch. 34-41) 0-7167-0906-6

or in four hardcover versions—

Vol. 1 Mechanics, Oscillations and Waves; Thermodynamics (Ch. 1-20, R) 0-7167-0809-4

Vol. 2 Electricity, Magnetism, Light & Modern Physics (Ch. 21-41) 0-7167-0810-8

Standard Version (Vol 1A-2B) 0-7167-8339-8

Extended Version (Vol 1A-2C) 0-7167-4389-2

New design and improved illustrations

The book has a warmer, more colorful look. Each piece of art has been carefully considered and many have been revised to increase clarity. Approximately 245 new figures have been added, including many new free-body diagrams within the worked examples. New photos bring to life the many real-world applications of physics.

Optional sections

The book was designed to allow professors to be flexible by designating certain sections "optional." These sections are marked with an *, and professors who choose to skip this section can do so knowing that their students won't be missing any material they will need in later chapters.

Summary

End of chapter summaries are organized with important topics on the left and relevant remarks and equations on the right. Here the key equations from the chapter appear together for easy reference.

Exploring essays

Students are invited to examine interesting extensions of the chapter concepts in Exploring sections, which are now found on the Web. These short pieces relate the chapter concepts to everything from the weather to transducers.

Media and Print Supplements

The supplements package has been updated and improved in response to reviewer suggestions and those from users of the fourth edition.

For the Student:

Student Solutions Manual: *Vol. 1, 0-7167-8333-9; Vol. 2, 0-7167-8334-7.* The new manual prepared by David Mills of College of the Redwoods, Charles Adler of St. Mary's College of Maryland, Ed Whittaker of Stevens Institute of Technology, George Zober of Yough Senior High School and Patricia Zober of Ringgold High School provides solutions for about twenty-five percent of the problems in the textbook, using the same side-by-side format and level of detail as the textbook's worked examples.

Study Guide: *Vol. 1, 0-7167-8332-0; Vol. 2, 0-7167-8331-2.* Prepared by Gene Mosca of the United States Naval Academy and Todd Ruskell of Colorado School of Mines, the Study Guide describes the key ideas and potential pitfalls of each chapter, and also includes true and false questions that test essential definitions and relations, questions and answers that require qualitative reasoning, and problems and solutions.

Student Web Site: Robin Jordan of Florida Atlantic University has put together a site designed to make studying and testing easier for both students and professors. The Web site includes:

- **On-line quizzing:** Multiple choice quizzes are available for each chapter. Students will receive immediate feedback, and the quiz results are collected for the instructor in a grade book.

- **iSOLVE homework service**: *0-7167-5802-4.* About one-fourth of the book's end-of-chapter problems, 1,100 altogether, are available on-line in W.H. Freeman's iSOLVE homework service. This service will offer each student a different version of every problem similar to CAPA and WebAssign, and the iSOLVE problems will be marked with an icon in the textbook. Homework scores can be collected in a grade book. Students may purchase access to iSOLVE for three semesters at a time.

• **iSOLVE Checkpoint problems:** A third of our iSOLVE questions are Checkpoint problems, which prompt students to describe how they arrived at their answer and to indicate their confidence level. All student responses will be gathered and included in the instructor's grade book report. Rolf Enger of the U.S. Air Force Academy inspired the development of Checkpoints to help professors gauge their student's understanding of the material.

• **Master the Concept exercises:** For each chapter, one or more exercises from the book will be available on-line so students can practice working the problem with randomized variables and step-by-step guidance. The on-line exercise will walk the student slowly through the problem-solving process and use interactive animations, simulations, video, and other graphic aids to help students visualize the problem. Teachers can collect grade book information on their progress. These premium examples are called out in the book with a Master the Concept icon.

Homework services: In addition to the iSOLVE network, there are three other homework services that are compatible with this textbook. End of chapter problems are available in WebAssign as well as CAPA: A Computer-Assisted Personalized Approach. A list of all the fifth edition problems included in WebAssign and CAPA is posted on the instructor's section of the *Physics* Web site. Our text is also compatible with the University of Texas Interactive Homework Service.

> **The iSolve homework service is available at**
> www.whfreeman.com/tipler5e
>
> **For more information about WebAssign, CAPA or UTX homework services, find their Web sites at:**
> http://webassign.net/info
> http://www.pa.msu.edu/educ/CAPA/
> http://hw.utexas.edu/hw.html

For the Instructor:

Instructor's Resource CD-ROM: *0-7167-9839-5*. This multi-faceted resource will give instructors the tools to make their own Web sites and presentations. The CD contains illustrations from the text in .jpg format, Powerpoint Lecture Slides for each chapter of the book, Lab Demonstration Videos, and Applied Physics videos in QuickTime format, and Presentation Manager Pro v.2.0, as well as all of the solutions to the end-of-chapter problems in editable Microsoft Word format.

Instructor's Resource Manual: The updated IRM contains Classroom Demonstrations for each chapter, a film and video guide with suggestions for each chapter, links to valuable Web sites, and links to free sources for Physlets, animations, and other teaching tools. This manual will be available on the book's Web site at www.whfreeman.com/tipler5e.

Instructor's Solutions Manual: *Vol. 1, 0-7167-9640-6; Vol. 2, 0-7167-9639-2*. This guide contains fully worked solutions for all of the problems in the textbook, using the side-by-side format wherever possible. It is available in print and is also included in editable Word files on the Instructor's CD-ROM.

Test Bank: *In print, 0-7167-9652-X; CD-ROM, 0-7167-9653-8*. Prepared by Mark Riley of Florida State University and David Mills of College of the Redwoods, this set of more than 4,000 multiple choice questions is available both in print and on a CD-ROM for Windows and Macintosh users. All questions refer to specific sections in the book. The CD-ROM version of the Test Bank makes it easy to add, edit and re-sequence questions to suit your needs.

Transparencies: *0-7167-9664-3*. Approximately 150 full color acetates of figures and tables from the text are included, with type enlarged for projection.

Acknowledgments

We are grateful to the many instructors, students, colleagues, and friends who have contributed to this, and to earlier editions.

Charles Adler of St. Mary's College of Maryland authored the excellent new problems. David Mills of the College of the Redwoods extensively revised the solutions manual. Robin Jordan of Florida Atlantic University created the innovative Master the Concept exercises and iSOLVE Checkpoint problems. Laura McCullough of the University of Wisconsin, Stout, and Thomas Foster of Southern Illinois University, Edwardsville, drawing from their background in Physics Education Research, were instrumental in providing context-rich examples in every chapter as well as our new Estimation and Approximation problems. We received invaluable help in accuracy checking of text and problems from professors:

Karamjeet Arya,
San Jose State University

Michael Crivello,
San Diego Mesa College

David Faust,
Mt. Hood Community College

Jerome Licini,
Lehigh University

Dan Lucas,
University of Wisconsin

Jeannette Myers,
Clemson University

Marian Peters,
Appalachian State University

Paul Quinn,
Kutztown University

Michael G. Strauss,
University of Oklahoma

George Zober,
Yough Senior High School

Patricia Zober,
Ringgold High School

Many instructors and students have provided extensive and helpful reviews of one or more chapters. They have each made a fundamental contribution to the quality of this revision, and deserve our gratitude. We would like to thank the following reviewers:

Edward Adelson,
The Ohio State University

Todd Averett,
The College of William and Mary

Yildirim M. Aktas,
University of North Carolina at Charlotte

Karamjeet Arya,
San Jose State University

Alison Baski,
Virginia Commonwealth University

Gary Stephen Blanpied,
University of South Carolina

Ronald Brown,
California Polytechnic State University

Robert Coakley,
University of Southern Maine

Robert Coleman,
Emory University

Andrew Cornelius,
University of Nevada at Las Vegas

Peter P. Crooker,
University of Hawaii

N. John DiNardo,
Drexel University

William Ellis,
University of Technology - Sydney

John W. Farley,
University of Nevada at Las Vegas

David Flammer,
Colorado School of Mines

Tom Furtak,
Colorado School of Mines

Patrick C. Gibbons,
Washington University

John B. Gruber,
San Jose State University

Christopher Gould,
University of Southern California

Phuoc Ha,
Creighton University

Theresa Peggy Hartsell,
Clark College

James W. Johnson,
Tallahassee Community College

Thomas O. Krause,
Towson University

Donald C. Larson,
Drexel University

Paul L. Lee,
California State University, Northridge

Peter M. Levy,
New York University

Jerome Licini,
Lehigh University

Edward McCliment,
University of Iowa

Robert R. Marchini,
The University of Memphis

Pete E.C. Markowitz,
Florida International University

Fred Lipschultz,
University of Connecticut

Graeme Luke,
Columbia University

Howard McAllister,
University of Hawaii

M. Howard Miles,
Washington State University

Matthew Moelter,
University of Puget Sound

Eugene Mosca,
United States Naval Academy

Aileen O'Donughue,
St. Lawrence University

Jack Ord,
University of Waterloo

Richard Packard,
University of California

George W. Parker,
North Carolina State University

Edward Pollack,
University of Connecticut

John M. Pratte,
Clayton College and State University

Brooke Pridmore,
Clayton State College

David Roberts,
Brandeis University

Lyle D. Roelofs,
Haverford College

Larry Rowan,
University of North Carolina
 at Chapel Hill

Lewis H. Ryder,
University of Kent, Canterbury

Bernd Schuttler,
University of Georgia

Cindy Schwarz,
Vassar College

Murray Scureman,
Amdahl Corporation

Scott Sinawi,
Columbia University

Wesley H. Smith,
University of Wisconsin

Kevork Spartalian,
University of Vermont

Kaare Stegavik,
University of Trondheim, Norway

Jay D. Strieb,
Villanova University

Martin Tiersten,
City College of New York

Oscar Vilches,
University of Washington

Fred Watts,
College of Charleston

John Weinstein,
University of Mississippi

David Gordon,
Wilson, MIT

David Winter,
Columbia University

Frank L.H. Wolfe,
University of Rochester

Roy C. Wood,
New Mexico State University

Yuriy Zhestkov,
Columbia University

Of course, our work is never done. We hope to continue to receive comments and suggestions from our readers so that we can improve the text and correct any errors. If you believe you have found an error, or have any other comments, suggestions, or questions, send us a note at asktipler@whfreeman.com. We will incorporate corrections into the text during subsequent reprinting.

Finally, we would like to thank our friends at W. H. Freeman and Company for their help and encouragement. Susan Brennan, Kathleen Civetta, Georgia Lee Hadler, Julia DeRosa, Margaret Comaskey, Dena Betz, Rebecca Pearce, Brian Donnellan, Jennifer Van Hove, Patricia Marx, and Mark Santee were extremely generous with their creativity and hard work at every stage of the process. We are also grateful for the contributions of Cathy Townsend and Denise Kadlubowski at PreMediaONE and the help of our colleagues Larry Tankersley, John Ertel, Steve Montgomery, and Don Treacy.

Paul Tipler
Alameda, California

Gene Mosca
Annapolis, Maryland

ABOUT THE AUTHORS

PAUL A TIPLER

Paul Tipler was born in the small farming town of Antigo, Wisconsin, in 1933. He graduated from high school in Oshkosh, Wisconsin, where his father was superintendent of the Public Schools. He received his B.S. from Purdue University in 1955 and his Ph.D. at the University of Illinois in 1962, where he studied the structure of nuclei. He taught for one year at Wesleyan University in Connecticut while writing his thesis, then moved to Oakland University in Michigan, where he was one of the original members of the Physics department, playing a major role in developing the physics curriculum. During the next 20 years, he taught nearly all the physics courses and wrote the first and second editions of his widely used textbooks *Modern Physics* (1969, 1978) and *Physics* (1976, 1982). In 1982, he moved to Berkeley, California, where he now resides, and where he wrote *College Physics* (1987) and the third edition of *Physics* (1991). In addition to physics, his interests include music, hiking, and camping, and he is an accomplished jazz pianist and poker player.

GENE MOSCA

Gene Mosca was born in New York City and grew up on Shelter Island, New York. His undergraduate studies were at Villanova University and his graduate studies were at the University of Michigan and the University of Vermont, where he received his Ph.D. in 1974. He taught at Southampton High School, the University of South Dakota, and Emporia State University. Since 1986 Gene has been teaching at the U.S. Naval Academy. There he coordinated the core physics course for 16 semesters, and instituted numerous enhancements to both the laboratory and classroom. Proclaimed by Paul Tipler as, "the best reviewer I ever had," Mosca authored the popular Study Guide for the third and fourth editions of the text.

PART VI MODERN PHYSICS: Quantum Mechanics, Relativity, and the Structure of Matter

Wave–Particle Duality and Quantum Physics

CHAPTER 34

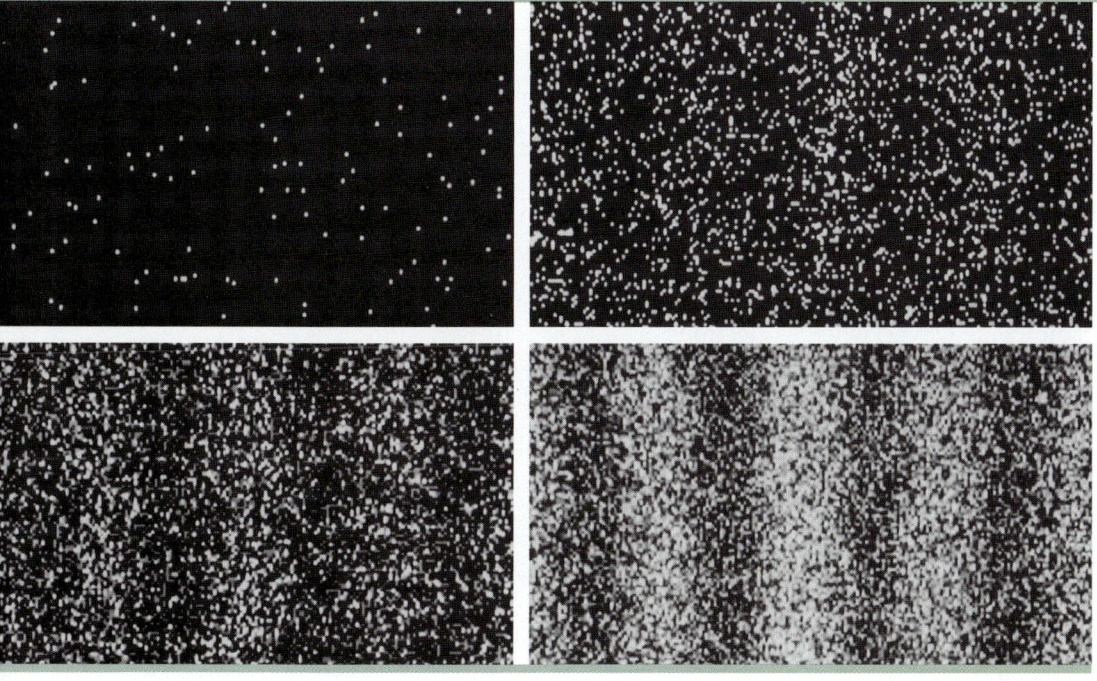

ELECTRON INTERFERENCE PATTERN PRODUCED BY ELECTRONS INCIDENT ON A BARRIER CONTAINING TWO SLITS: (A) 10 ELECTRONS, (B) 100 ELECTRONS, (C) 3000 ELECTRONS, AND (D) 70,000 ELECTRONS. THE MAXIMA AND MINIMA DEMONSTRATE THE WAVE NATURE OF THE ELECTRON AS IT TRAVERSES THE SLITS. INDIVIDUAL DOTS ON THE SCREEN INDICATE THE PARTICLE NATURE OF THE ELECTRON AS IT EXCHANGES ENERGY WITH THE DETECTOR. THE PATTERN IS THE SAME WHETHER ELECTRONS OR PHOTONS (PARTICLES OF LIGHT) ARE USED.

How do you calculate the wavelength of an electron? This is revealed in Example 34-4.

We have seen that the propagation of waves through space is quite different from the propagation of particles. Waves bend around corners (diffraction) and interfere with one another, producing interference patterns. If a wave encounters a small aperture, the wave spreads out on the other side as if the aperture were a point source. If two coherent waves of equal intensity I_0 meet in space, the result can be a wave of intensity $4I_0$ (constructive interference), an intensity of zero (destructive interference), or a wave of intensity between zero and $4I_0$, depending on the phase difference between the waves at their meeting point.

The propagation of particles is quite unlike the propagation of waves. Particles travel in straight lines until they collide with something, after which the particles again travel in straight lines. If two particle beams meet in space, they never produce an interference pattern.

Particles and waves also exchange energy differently. Particles exchange energy in collisions that occur at specific points in space and in time. The energy of waves, on the other hand, is spread out in space and deposited continuously as the wave fronts interact with matter.

Sometimes the propagation of a wave cannot be distinguished from the propagation of a beam of particles. If the wavelength λ is very small compared to distances from the edges of objects, diffraction effects are negligible and the wave travels in straight lines. Also, interference maxima and minima are so close together in space as to be unobservable. The result is that the wave interacts with a detector, like a beam of numerous small particles each exchanging a small amount of energy; the exchange cannot distinguish particles from waves. For example, you do not observe the individual air molecules bouncing off your face if the wind blows on it. Instead, the interaction of billions of particles appears to be continuous, like that of a wave.

At the beginning of the twentieth century, it was thought that sound, light, and other electromagnetic radiation (e.g., radio) were waves; whereas electrons, protons, atoms, and similar constituents of nature were understood to be particles. The first 30 years of the century revealed startling developments in theoretical and experimental physics, such as the finding that light, thought to be a wave, actually exchanges energy in discrete lumps or quanta, just like particles; and the finding that an electron, thought to be a particle, exhibits diffraction and interference as it propagates through space, just like a wave.

The fact that light exchanges energy like a particle implies that light energy is not continuous but is *quantized*. Similarly, the wave nature of the electron, along with the fact that the standing wave condition requires a discrete set of frequencies, implies that the energy of an electron in a confined region of space is not continuous, but is quantized to a discrete set of values.

➤ In this chapter, we begin by discussing some basic properties of light and electrons, examining their wave and particle characteristics. We then consider some of the detailed properties of matter waves, showing, in particular, how standing waves imply the quantization of energy. Finally, we discuss some of the important features of the theory of quantum physics, which was developed in the 1920s and which has been extremely successful in describing nature. Quantum physics is now the basis of our understanding of both the microscopic and very low temperature worlds.

34-1 Light

The question of whether light consists of a beam of particles or waves in motion is one of the most interesting in the history of science. Isaac Newton used a particle theory of light to explain the laws of reflection and refraction; however, for refraction, Newton needed to assume that light travels faster in water or glass than in air, an assumption later shown to be false. The chief early proponents of the wave theory were Robert Hooke and Christian Huygens, who explained refraction by assuming that light travels more slowly in glass or water than it does in air. Newton rejected the wave theory because, in his time, light was believed to travel through a medium only in straight lines—diffraction had not yet been observed.

Because of Newton's great reputation and authority, his particle theory of light was accepted for more than a century. Then, in 1801, Thomas Young demonstrated the wave nature of light in a famous experiment in which two coherent light sources are produced by illuminating a pair of narrow, parallel slits with a single source (Figure 34-1). In Chapter 33, we saw that when light encounters a small opening, the opening acts as a point source of waves (Figure 33-7). In Young's experiment, each slit acts as a line source, which is equivalent to a point source in two dimensions. The interference pattern is observed on a screen placed behind the slits. Interference maxima occur at angles so that the path difference is an integral number of wavelengths. Similarly, interference minima occur if the path difference is one-half wavelength or any odd number of half wavelengths. Figure 34-1b shows the intensity pattern as seen on the screen. Young's experiment and many other experiments demonstrate that light propagates like a wave.

In the early nineteenth century, the French physicist Augustin Fresnel (1788–1827) performed extensive experiments on interference and diffraction and put the wave theory on a rigorous mathematical basis. Fresnel showed that the observed straight-line propagation of light is a result of the very short wavelengths of visible light.

The classical wave theory of light culminated in 1860 when James Clerk Maxwell published his mathematical theory of electromagnetism. This theory yielded a wave equation that predicted the existence of electromagnetic waves that propagate with a speed that can be calculated from the laws of electricity and magnetism. The fact that the result of this calculation was $c \approx 3 \times 10^8$ m/s, the same as the speed of light, suggested to Maxwell that light is an electromagnetic wave. The eye is sensitive to electromagnetic waves with wavelengths in the range from approximately 400 nm (1 nm = 10^{-9} m) to approximately 700 nm. This range is called *visible light*. Other electromagnetic waves (e.g., microwaves, radio, television, and X rays) differ from light only in wavelength and in frequency.

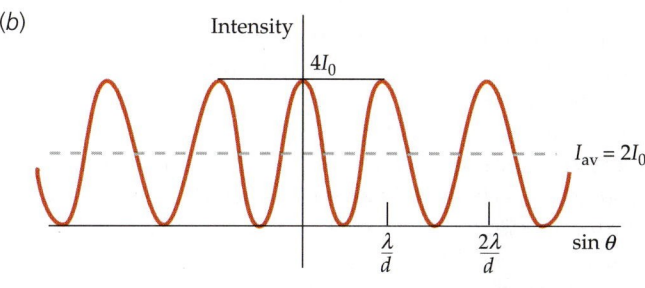

FIGURE 34-1 (*a*) Two slits act as coherent sources of light for the observation of interference in Young's experiment. Cylindrical waves from the slits overlap and produce an interference pattern on a screen far away. (*b*) The intensity pattern produced in Figure 34-1*a*. The intensity is maximum at points where the path difference is an integral number of wavelengths, and the intensity is zero where the path difference is an odd number of half wavelengths.

34-2　The Particle Nature of Light: Photons

The diffraction of light and the existence of an interference pattern in the two-slit experiment give clear evidence that light has wave properties. However, early in the twentieth century, it was found that light energy comes in discrete amounts.

The Photoelectric Effect

The quantum nature of light and the quantization of energy were suggested by Albert Einstein in 1905 in his explanation of the photoelectric effect. Einstein's work marked the beginning of quantum theory, and for his work, Einstein received the Nobel Prize for physics. Figure 34-2 shows a schematic diagram of the basic apparatus for studying the photoelectric effect. Light of a single frequency enters an evacuated chamber and falls on a clean metal surface C (C for cathode), causing electrons to be emitted. Some of these electrons strike the second metal plate A (A for anode), constituting an electric current between the plates. Plate A is negatively charged, so the electrons are repelled by it, with only the most energetic electrons reaching the plate. The maximum kinetic energy of the emitted electrons is measured by slowly increasing the voltage until the current becomes zero. Experiments give the surprising result that the maximum kinetic energy of the emitted electrons is *independent of the intensity* of the incident light. Classically, we would expect that increasing the rate at which light energy falls on the metal surface would increase the energy absorbed by individual electrons and, therefore, would increase the maximum kinetic energy of the electrons emitted. Experimentally, this is not what happens. The maximum kinetic energy of the emitted electrons is the same for a given wavelength of incident light, no matter how intense the light. Einstein demonstrated that this experimental result can be explained if light energy is quantized in small bundles called **photons.** The energy E of each photon is given by

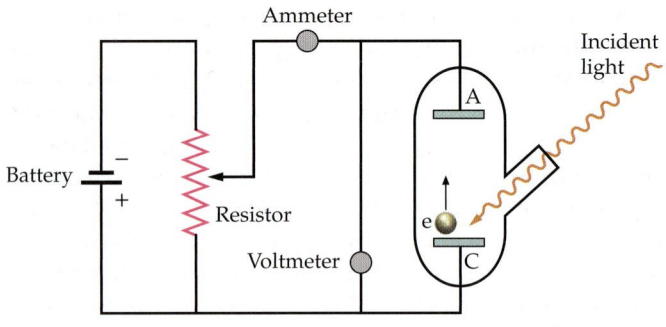

FIGURE 34-2 A schematic drawing of the apparatus for studying the photoelectric effect. Light of a single frequency enters an evacuated chamber and strikes the cathode C, which then ejects electrons. The current in the ammeter measures the number of these electrons that reach the anode A per unit time. The anode is made electrically negative with respect to the cathode to repel the electrons. Only those electrons with enough initial kinetic energy to overcome the repulsion can reach the anode. The voltage between the two plates is slowly increased until the current becomes zero, which happens when even the most energetic electrons do not make it to plate A.

$$E = hf = \frac{hc}{\lambda} \qquad\qquad 34\text{-}1$$

EINSTEIN EQUATION FOR PHOTON ENERGY

where f is the frequency, and h is a constant now known as **Planck's constant.**[†] The measured value of this constant is

$$h = 6.626 \times 10^{-34} \text{ J·s} = 4.136 \times 10^{-15} \text{ eV·s} \qquad\qquad 34\text{-}2$$

PLANCK'S CONSTANT

Equation 34-1 is sometimes called the **Einstein equation.**

At the fundamental level, a light beam consists of a beam of particles—photons—each with energy hf. The intensity (power per unit area) of a monochromatic light beam is the number of photons per unit area per unit of time, times the energy per photon. The interaction of the light beam with the metal surface consists of collisions between photons and electrons. In these collisions, the photons disappear, with each photon giving all its energy to an electron, and the electron emitted from the surface thus receives its energy from a single photon. If the intensity of light is increased, more photons fall on the surface per unit time, and more electrons are emitted. However, each photon still has the same energy hf, so the energy absorbed by each electron is unchanged.

If ϕ is the minimum energy necessary to remove an electron from a metal surface, the maximum kinetic energy of the electrons emitted is given by

$$K_{max} = (\tfrac{1}{2}mv^2)_{max} = hf - \phi \qquad\qquad 34\text{-}3$$

EINSTEIN'S PHOTOELECTRIC EQUATION

The quantity ϕ, called the **work function,** is a characteristic of the particular metal. (Some electrons will have kinetic energies less than $hf - \phi$, because of the loss of energy from traveling through the metal.)

† In 1900, the German physicist Max Planck introduced this constant to explain discrepancies between the theoretical curves and experimental data on the spectrum of blackbody radiation. Planck also assumed that the radiation was emitted and absorbed by a blackbody in quanta of energy hf, but he considered his assumption to be just a calculational device rather than a fundamental property of electromagnetic radiation. Blackbody radiation was discussed in Chapter 20.

According to Einstein's photoelectric equation, a plot of K_{max} versus frequency f should be a straight line with the slope h. This was a bold prediction, because, at the time, there was no evidence that Planck's constant had any application outside of blackbody radiation. In addition, there was no experimental data on K_{max} versus frequency f, because no one before had even suspected that the frequency of the light was related to K_{max}. This prediction was difficult to verify experimentally, but careful experiments by R. A. Millikan approximately 10 years later showed that Einstein's equation was correct. Figure 34-3 shows a plot of Millikan's data.

Photons with frequencies less than a **threshold frequency** f_t, and therefore with wavelengths greater than a **threshold wavelength** $\lambda_t = c/f_t$, do not have enough energy to eject an electron from a particular metal. The threshold frequency and the corresponding threshold wavelength can be related to the work function ϕ by setting the maximum kinetic energy of the electrons equal to zero in Equation 34-3. Then

$$\phi = hf_t = \frac{hc}{\lambda_t} \qquad\qquad 34\text{-}4$$

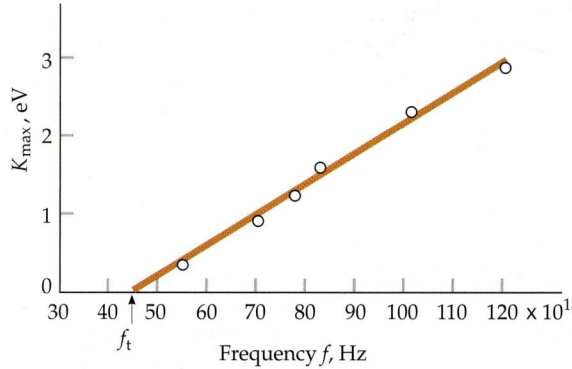

FIGURE 34-3 Millikan's data for the maximum kinetic energy K_{max} versus frequency f for the photoelectric effect. The data fall on a straight line that has a slope h, as predicted by Einstein approximately a decade before the experiment was performed.

Work functions for metals are typically a few electron volts. Since wavelengths are usually given in nanometers and energies in electron volts, it is useful to have the value of hc in electron volt–nanometers:

$$hc = (4.1357 \times 10^{-15}\ eV\cdot s)(2.9979 \times 10^{8}\ m/s) = 1.240 \times 10^{-6}\ eV\cdot m$$

or

$$hc = 1240\ eV\cdot nm \qquad\qquad 34\text{-}5$$

PHOTON ENERGIES FOR VISIBLE LIGHT   **EXAMPLE 34-1**

Calculate the photon energies for light of wavelengths 400 nm (violet) and 700 nm (red). (These are the approximate wavelengths at the two extremes of the visible spectrum.)

1. The energy is related to the wavelength by Equation 34-1:

$$E = hf = \frac{hc}{\lambda}$$

2. For $\lambda = 400$ nm, the energy is:

$$E = \frac{hc}{\lambda} = \frac{1240\ eV\cdot nm}{400\ nm} = \boxed{3.10\ eV}$$

3. For $\lambda = 700$ nm, the energy is:

$$E = \frac{hc}{\lambda} = \frac{1240\ eV\cdot nm}{700\ nm} = \boxed{1.77\ eV}$$

REMARKS We can see from these calculations that visible light contains photons with energies that range from approximately 1.8 eV to 3.1 eV. X rays, which have much shorter wavelengths, contain photons with energies of the order of keV. Gamma rays emitted by nuclei have even shorter wavelengths and photons of energy of the order of MeV.

EXERCISE Find the energy of a photon corresponding to electromagnetic radiation in the FM radio band of wavelength 3 m. (*Answer* 4.13×10^{-7} eV)

EXERCISE Find the wavelength of a photon whose energy is (*a*) 0.1 eV, (*b*) 1 keV, and (*c*) 1 MeV. (*Answer*   (*a*) 12.4 μm, (*b*) 1.24 nm, (*c*) 1.24 pm)

**The intensity of sunlight at the earth's surface is approximately 1400 W/m².
Assuming the average photon energy is 2 eV (corresponding to a wavelength
of approximately 600 nm), calculate the number of photons that strike an area
of 1 cm² each second.**

PICTURE THE PROBLEM   Intensity (power per unit area) is given as is the area.
From this, we can calculate the power, which is the energy per unit time.

Cover the column to the right and try these on your own before looking at the answers.

Steps	Answers
1. The energy ΔE is related to the number N of photons and the energy per photon $hf = 2$ eV.	$\Delta E = Nhf$
2. The intensity I (power per unit area) and the area A are given, so we can find the power.	$I = \dfrac{P}{A}$
3. Knowing the power (energy per unit time) and the time, we can find the energy.	$\Delta E = P\Delta t$
4. Substitute the results from steps 1–3 and solve for N. Take care to get the units correct.	$N = \dfrac{IA\Delta t}{hf} = \boxed{4.38 \times 10^{17}}$

REMARKS   This is an enormous number of photons. In most everyday situations,
the number of photons is so great that the quantization of light is not noticeable.

EXERCISE   Calculate the photon density (in photons per cubic centimeter) of the
sunlight in Example 34-2. The number arriving on an area of 1 cm² in one second
is the number in a column whose cross section is 1 cm² and whose height is the
distance light travels in one second. (*Answer* 1.46×10^7 cm^{-3})

Compton Scattering

The first use of the photon concept was to explain the results of photoelectric-
effect experiments. The photon concept was used by Arthur H. Compton to
explain the results of his measurements of the scattering of X rays by free
electrons in 1923. According to classical theory, if an electromagnetic wave of
frequency f_1 is incident on material containing free charges, the charges will
oscillate with this frequency and reradiate electromagnetic waves of the same
frequency. Compton considered these reradiated waves as scattered photons,
and he pointed out that if the scattering process were a collision between a
photon and an electron (Figure 34-4), the electron would recoil and thus absorb
energy. The scattered photon would then have less energy, and therefore a lower
frequency and longer wavelength, than the incident photon.

According to classical electromagnetic theory (see Section 30-3), the energy
and momentum of an electromagnetic wave are related by

$$E = pc \tag{34-6}$$

The momentum of a photon is thus related to its wavelength λ by $p = E/c = hf/c = h/\lambda$.

$$p = \frac{h}{\lambda} \tag{34-7}$$

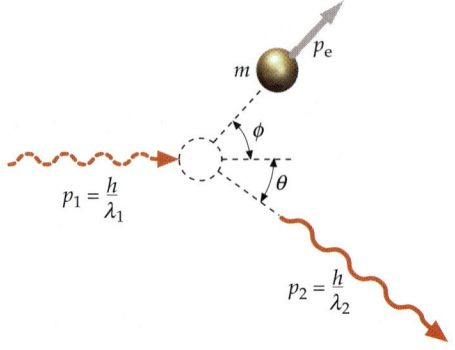

FIGURE 34-4   The scattering of light
by an electron is considered as a collision
of a photon of momentum h/λ_1 and a
stationary electron. The scattered photon
has less energy and therefore has a
greater wavelength than does the
incident electron.

Compton applied the laws of conservation of momentum and energy to the collision of a photon and an electron to calculate the momentum p_2 and thus the wavelength $\lambda_2 = h/p_2$ of the scattered photon (see Figure 34-4). Applying conservation of momentum to the collision gives

$$\vec{p}_1 = \vec{p}_2 + \vec{p}_e \tag{34-8}$$

where $\vec{p}_1$ is the momentum of the incident photon and $\vec{p}_e$ is the momentum of the electron after the collision. The initial momentum of the electron is zero. Rearranging Equation 34-8, we have $\vec{p}_e = \vec{p}_1 - \vec{p}_2$. Taking the dot product of each side with itself gives

$$p_e^2 = p_1^2 + p_2^2 - 2p_1 p_2 \cos \theta \tag{34-9}$$

where θ is the angle the scattered photon makes with the direction of the incident photon. Because the kinetic energy of the electron after the collision can be a significant fraction of the rest energy of an electron, the relativistic expression relating the energy E of the electron to its momentum is used. This expression (Equation R-17) is

$$E = \sqrt{p_e^2 c^2 + (m_e c^2)^2}$$

where m_e is the rest mass of the electron. Applying conservation of energy to the collision gives

$$p_1 c + m_e c^2 = p_2 c + \sqrt{p_e^2 c^2 + (m_e c^2)^2} \tag{34-10}$$

where pc (Equation 34-6) has been used to express the energies of the photons. Eliminating p_e^2 from Equation 34-9 and Equation 34-10 gives

$$\frac{1}{p_2} - \frac{1}{p_1} = \frac{1}{m_e c} (1 - \cos \theta)$$

and substituting for p_1 and p_2, using Equation 34-7 gives

$$\lambda_2 - \lambda_1 = \frac{h}{m_e c} (1 - \cos \theta) \tag{34-11}$$

COMPTON EQUATION

The increase in wavelengths is independent of the wavelength λ_1 of the incident photon. The quantity $h/m_e c$ has dimensions of length and is called the Compton wavelength. Its value is

$$\lambda_C = \frac{h}{m_e c} = \frac{hc}{m_e c^2} = \frac{1240 \text{ eV·nm}}{5.11 \times 10^5 \text{ eV}} = 2.43 \times 10^{-12} \text{ m} = 2.43 \text{ pm} \tag{34-12}$$

Because $\lambda_2 - \lambda_1$ is small, it is difficult to observe unless λ_1 is so small that the fractional change $(\lambda_2 - \lambda_1)/\lambda_1$ is appreciable. Compton used X rays of wavelength 71.1 pm (1 pm $= 10^{-12}$ m $= 10^{-3}$ nm). The energy of a photon of this wavelength is $E = hc/\lambda = (1240 \text{ eV·nm})/(0.0711 \text{ nm}) = 17.4$ keV. Since this is much greater than the binding energy of the valence electrons in most atoms (which is of the order of a few eV), these electrons can be considered to be essentially free. Compton's measurements of $\lambda_2 - \lambda_1$ as a function of scattering angle θ agreed with Equation 34-11, thereby confirming the correctness of the photon concept (i.e., of the particle nature of light).

E X A M P L E   3 4 - 3

An X-ray photon of wavelength 6 pm makes a head-on collision with an electron, so that the scattered photon goes in a direction opposite to that of the incident photon. The electron is initially at rest. (*a*) How much longer is the wavelength of the scattered photon than that of the incident photon? (*b*) What is the kinetic energy of the recoiling electron?

FIGURE 34-5

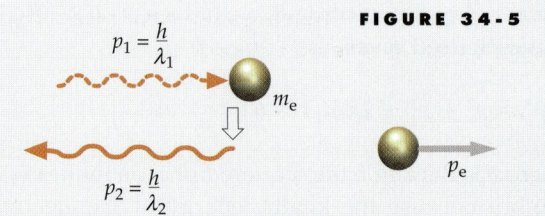

PICTURE THE PROBLEM We can calculate the increase in wavelength, and thus the new wavelength, from Equation 34-11. We then use the new wavelength to find the energy of the scattered photon and then to find the kinetic energy of the recoiling electron from conservation of energy (Figure 34-5).

(*a*) Use Equation 34-11 to calculate the increase in wavelength:

$$\Delta \lambda = \lambda_2 - \lambda_1$$

$$= \frac{h}{m_e c}(1 - \cos \theta)$$

$$= (2.43 \text{ pm})(1 - \cos 180°) = \boxed{4.86 \text{ pm}}$$

(*b*) 1. The kinetic energy of the recoiling electron equals the energy of the incident photon E_1 minus the energy of the scattered photon E_2:

$$K_e = E_1 - E_2 = \frac{hc}{\lambda_1} - \frac{hc}{\lambda_2}$$

2. Calculate λ_2 from the given wavelength of the incident photon and the change found in Part (*a*):

$$\lambda_2 = \lambda_1 + \Delta \lambda = 6 \text{ pm} + 4.86 \text{ pm}$$
$$= 10.86 \text{ pm}$$

3. Substitute the calculated values of E_1 and E_2 to find the energy of the recoiling electron:

$$K_e = \frac{hc}{\lambda_1} - \frac{hc}{\lambda_2}$$

$$= \frac{1240 \text{ eV·nm}}{6.0 \text{ pm}} - \frac{1240 \text{ eV·nm}}{10.86 \text{ pm}}$$

$$= \frac{1.24 \text{ keV·nm}}{6.0 \times 10^{-3} \text{ nm}} - \frac{1.24 \text{ keV·nm}}{10.86 \times 10^{-3} \text{ nm}}$$

$$= 207 \text{ keV} - 114 \text{ keV}$$

$$= \boxed{93 \text{ keV}}$$

REMARKS The kinetic energy of the scattered electron is 93 keV and the rest energy of an electron is 511 keV, so the kinetic energy is 18 percent of the rest energy. Thus, the nonrelativistic formula for the kinetic energy ($\frac{1}{2}m_e v^2$) is not valid.

EXERCISE What is the speed of the scattered electron given by the nonrelativistic formula for the kinetic energy ($\frac{1}{2}m_e v^2$)? (*Answer* 0.6*c*)

34-3 Energy Quantization in Atoms

Ordinary white light has a continuous spectrum; that is, it contains *all* the wavelengths in the visible spectrum. But if atoms in a gas at low pressure are excited by an electric discharge, they emit light of specific wavelengths that are characteristic of the element or the compound. Since the energy of a photon is related to its wavelength by $E = hf = hc/\lambda$, a discrete set of wavelengths implies a discrete set of energies. Conservation of energy then implies that if an atom absorbs or emits a photon, its internal energy changes by a discrete amount, which is ± the energy of the photon. In 1913, this led Niels Bohr to postulate that the internal energy of an atom can have only a discrete set of values. That is, the internal energy of an atom is **quantized.** If an atom radiates light of frequency f, the atom makes a transition from one allowed level to another level that is lower in energy by $\Delta E = hf$. Bohr was able to construct a semiclassical model of the hydrogen atom that had a discrete set of energy levels consistent with the observed spectrum of emitted light.[†] However, the *reason* for the quantization of energy levels in atoms and other systems remained a mystery until the wave nature of electrons was discovered a decade later.

34-4 Electrons and Matter Waves

In 1897, J. J. Thomson showed that the rays of a cathode-ray tube (Figure 34-6) can be deflected by electric and magnetic fields and therefore must consist of electrically charged particles. By measuring the deflections of these particles, Thomson showed that all the particles have the same charge-to-mass ratio q/m. He also showed that particles with this charge-to-mass ratio can be obtained using any material for the cathode, which means that these particles, now called **electrons,** are a fundamental constituent of all matter.

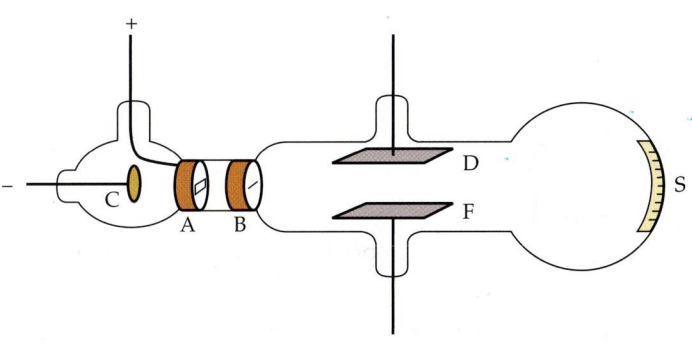

The de Broglie Hypothesis

Since light seems to have both wave and particle properties, it is natural to ask whether matter (e.g., electrons, protons) might also have both wave and particle characteristics. In 1924, a French physics student, Louis de Broglie, suggested this idea in his doctoral dissertation. de Broglie's work was highly speculative, because there was no evidence at that time of any wave aspects of matter.

For the wavelength of electron waves, de Broglie chose

FIGURE 34-6 Schematic diagram of the cathode-ray tube Thomson used to measure q/m for the particles that comprise cathode rays (electrons). Electrons from the cathode C pass through the slits at A and B and strike a phosphorescent screen S. The beam can be deflected by an electric field between plates D and F or by a magnetic field (not shown).

$$\lambda = \frac{h}{p} \qquad\qquad 34\text{-}13$$

DE BROGLIE RELATION FOR THE WAVELENGTH OF ELECTRON WAVES

where p is the momentum of the electron. Note that this is the same as Equation 34-7 for a photon. For the frequency of electron waves, de Broglie chose the Einstein equation relating the frequency and energy of a photon.

$$f = \frac{E}{h} \qquad\qquad 34\text{-}14$$

DE BROGLIE RELATION FOR THE FREQUENCY OF ELECTRON WAVES

These equations are thought to apply to all matter. However, for macroscopic objects, the wavelengths calculated from Equation 34-13 are so small that it is impossible to observe the usual wave properties of interference or diffraction. Even a dust particle with a mass as small as 1 μg is much too massive for any wave characteristics to be noticed, as we see in the following example.

---

† We will study the Bohr model in Chapter 36.

THE DE BROGLIE WAVELENGTH　　　　　　**EXAMPLE 34-4** **Try It Yourself**

Find the de Broglie wavelength of a 10^{-6} g particle moving with a speed of 10^{-6} m/s.

Cover the column to the right and try this on your own before looking at the answers.

Steps

Write the definition of the de Broglie wavelength and substitute the given data.

Answers

$$\lambda = \frac{h}{p} = \frac{h}{mv} = \frac{6.63 \times 10^{-34}\,\text{J·s}}{(10^{-9}\,\text{kg})(10^{-6}\,\text{m/s})}$$

$$= \boxed{6.63 \times 10^{-19}\,\text{m}}$$

REMARKS This wavelength is several orders of magnitude smaller than the diameter of an atomic nucleus, which is about 10^{-15} m.

Since the wavelength found in Example 34-4 is so small, much smaller than any possible apertures or obstacles, diffraction or interference of such waves cannot be observed. In fact, the propagation of waves of very small wavelengths is indistinguishable from the propagation of particles. The momentum of the particle in Example 34-4 was only 10^{-15} kg·m/s. A macroscopic particle with a greater momentum would have an even smaller de Broglie wavelength. We therefore do not observe the wave properties of such macroscopic objects as baseballs and billiard balls.

EXERCISE Find the de Broglie wavelength of a baseball of mass 0.17 kg moving at 100 km/h. (*Answer* 1.4×10^{-34} m)

The situation is different for low-energy electrons and other microscopic particles. Consider a particle with kinetic energy K. Its momentum is found from

$$K = \frac{p^2}{2m}$$

or

$$p = \sqrt{2mK}$$

Its wavelength is then

$$\lambda = \frac{h}{p} = \frac{h}{\sqrt{2mK}}$$

If we multiply the numerator and the denominator by c, we obtain

$$\lambda = \frac{hc}{\sqrt{2mc^2K}} = \frac{1240\,\text{eV·nm}}{\sqrt{2mc^2K}} \qquad\qquad 34\text{-}15$$

WAVELENGTH ASSOCIATED WITH A PARTICLE OF MASS M

where we have used $hc = 1240$ eV·nm. For electrons, $mc^2 = 0.511$ MeV. Then,

$$\lambda = \frac{1240\,\text{eV·nm}}{\sqrt{2mc^2K}} = \frac{1240\,\text{eV·nm}}{\sqrt{2(0.511 \times 10^6\,\text{eV})K}}$$

or

$$\lambda = \frac{1.226}{\sqrt{K}} \text{ nm}, \quad K \text{ in electron volts} \qquad 34\text{-}16$$

ELECTRON WAVELENGTH

Equation 34-15 and Equation 34-16 do not hold for relativistic particles whose kinetic energies are a significant fraction of their rest energies mc^2. (Rest energies were discussed in Chapter 7 and in Chapter R.)

EXERCISE Find the wavelength of an electron whose kinetic energy is 10 eV. (*Answer* 0.388 nm. From this result, we see that a 10-eV electron has a de Broglie wavelength of about 0.4 nm. This is on the same order of magnitude as the size of the atom and the spacing of atoms in a crystal.)

Electron Interference and Diffraction

The observation of diffraction and interference of electron waves would provide the crucial test of the existence of wave properties of electrons. This was first discovered serendipitously in 1927 by C. J. Davisson and L. H. Germer as they were studying electron scattering from a nickel target at the Bell Telephone Laboratories. After heating the target to remove an oxide coating that had accumulated during an accidental break in the vacuum system, they found that the scattered electron intensity as a function of the scattering angle showed maxima and minima. Their target had crystallized, and by accident they had observed electron diffraction. Davisson and Germer then prepared a target consisting of a single crystal of nickel and investigated this phenomenon extensively. Figure 34-7a illustrates their experiment. Electrons from an electron gun are directed at a crystal and detected at some angle ϕ that can be varied. Figure 34-7b shows a typical pattern observed. There is a strong scattering maximum at an angle of 50°. The angle for maximum scattering of waves from a crystal depends on the wavelength of the waves and the spacing of the atoms in the crystal. Using the known spacing of atoms in their crystal, Davisson and Germer calculated the wavelength that could produce such a maximum and found that it agreed with the de Broglie equation (Equation 34-16) for the electron energy they were using. By varying the energy of the incident electrons, they could vary the electron wavelengths and produce maxima and minima at different locations in the diffraction patterns. In all cases, the measured wavelengths agreed with de Broglie's hypothesis.

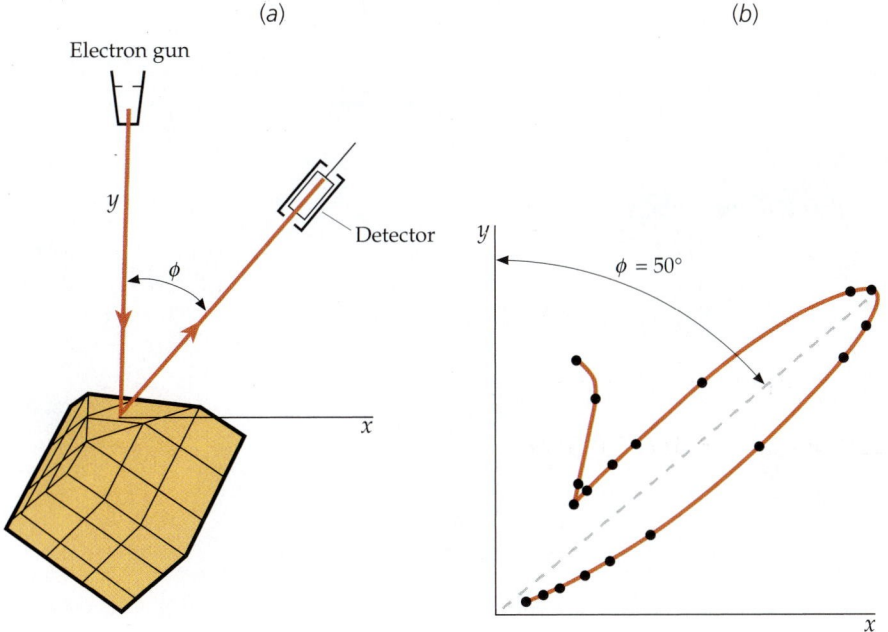

(a) (b)

FIGURE 34-7 The Davisson–Germer experiment. (*a*) Electrons are scattered from a nickel crystal into a detector. (*b*) Intensity of scattered electrons versus scattering angle. The maximum is at the angle predicted by the diffraction of waves of wavelength λ given by the de Broglie formula.

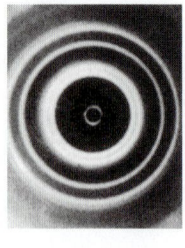  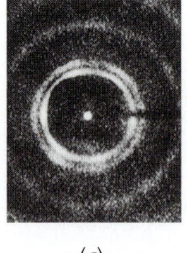 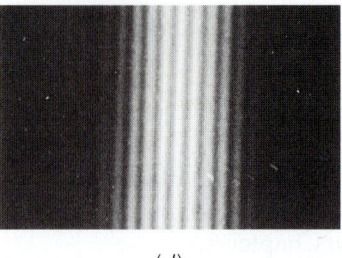

(a) (b) (c) (d)

FIGURE 34-8 (*a*) The diffraction pattern produced by X rays of wavelength 0.071 nm on an aluminum foil target. (*b*) The diffraction pattern produced by 600-eV electrons ($\lambda = 0.050$ nm) on an aluminum foil target. (*c*) The diffraction of 0.0568 eV neutrons ($\lambda = 0.12$ nm) incident on a copper foil. (*d*) A two-slit electron diffraction-interference pattern.

Another demonstration of the wave nature of electrons was provided in the same year by G. P. Thomson (son of J. J. Thomson) who observed electron diffraction in the transmission of electrons through thin metal foils. A metal foil consists of tiny, randomly oriented crystals. The diffraction pattern resulting from such a foil is a set of concentric circles. Figure 34-8*a* and Figure 34-8*b* show the diffraction pattern observed using X rays and electrons on an aluminum foil target. Figure 34-8*c* shows the diffraction patterns of neutrons on a copper foil target. Note the similarity of the patterns. The diffraction of hydrogen and helium atoms was observed in 1930. In all cases, the measured wavelengths agree with the de Broglie predictions. Figure 34-8*d* shows a diffraction pattern produced by electrons incident on two narrow slits. This experiment is equivalent to Young's famous double-slit experiment with light. The pattern is identical to the pattern observed with photons of the same wavelength. (Compare with Figure 34-1*b*.)

Shortly after the wave properties of the electron were demonstrated, it was suggested that electrons rather than light might be used to *see* small objects. As discussed in Chapter 33, reflected waves or transmitted waves can resolve details of objects only if the details are larger than the wavelength of the reflected wave. Beams of electrons, which can be focused electrically, can have very small wavelengths—much shorter than visible light. Today, the electron microscope (Figure 34-9) is an important research tool used to visualize specimens at scales far smaller than those possible with a light microscope.

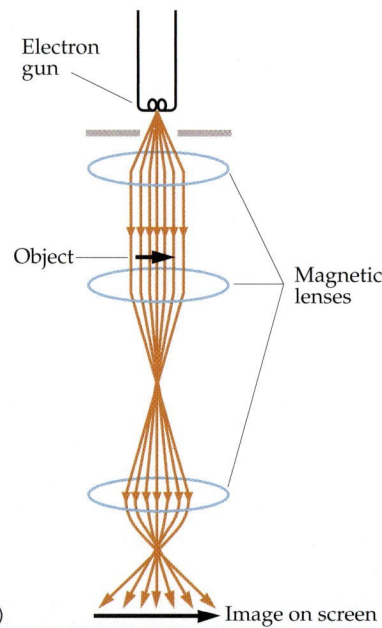

Electron gun

Object

Magnetic lenses

Image on screen

(a)

FIGURE 34-9
(*a*) An electron microscope. Electrons from a heated filament (the electron gun) are accelerated by a large potential difference. The electron beam is made parallel by a magnetic focusing lens. The electrons strike a thin target and are then focused by a second magnetic lens. The third magnetic lens projects the electron beam onto a fluorescent screen to produce the image.
(*b*) An electron micrograph of a DNA molecule.

(b)

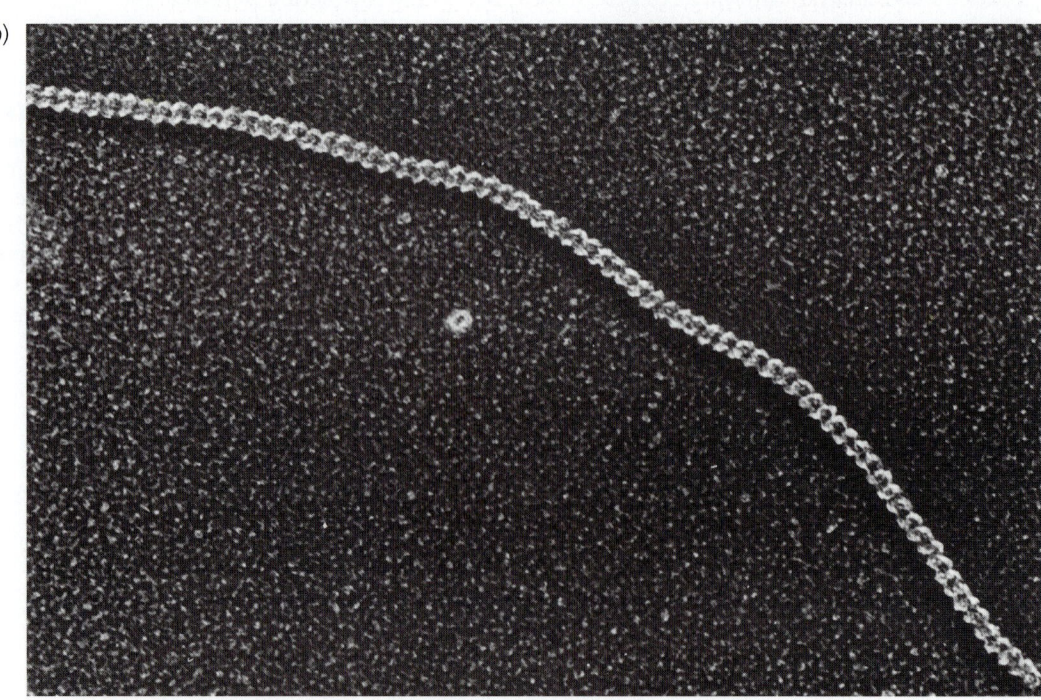

Standing Waves and Energy Quantization

Given that electrons have wave-like properties, it should be possible to produce standing electron waves. If energy is associated with the frequency of a standing wave, as in $E = hf$ (Equation 34-14), then standing waves imply quantized energies.

The idea that the discrete energy states in atoms could be explained by standing waves led to the development by Erwin Schrödinger and others in 1926 of a detailed mathematical theory known as quantum theory, quantum mechanics, or wave mechanics. In this theory, the electron is described by a wave function ψ that obeys a wave equation called the Schrödinger equation. The form of the Schrödinger equation of a particular system depends on the forces acting on the particle, which are described by the potential energy functions associated with those forces. In Chapter 35 we discuss this equation, which is somewhat similar to the classical wave equations for sound or for light. Schrödinger solved the standing wave problem for the hydrogen atom, the simple harmonic oscillator, and other systems of interest. He found that the allowed frequencies, combined with $E = hf$, resulted in the set of energy levels found experimentally for the hydrogen atom, thereby demonstrating that quantum theory provides a general method of finding the quantized energy levels for a given system. Quantum theory is the basis for our understanding of the modern world, from the inner workings of the atomic nucleus to the radiation spectra of distant galaxies.

34-5 The Interpretation of the Wave Function

The wave function for waves on a string is the string displacement y. The wave function for sound waves can be either the displacement of the air molecules s, or the pressure P. The wave function for electromagnetic waves is the electric field $\vec{E}$ and the magnetic field $\vec{B}$. What is the wave function for electron waves? The symbol we use for this wave function is ψ (the Greek letter psi). When Schrödinger published his wave equation, neither he nor anyone else knew just how to interpret the wave function ψ. We can get a hint about how to interpret ψ by considering the quantization of light waves. For classical waves, such as sound or light, the energy per unit volume in the wave is proportional to the square of the wave function. Since the energy of a light wave is quantized, the energy per unit volume is proportional to the number of photons per unit volume. We might therefore expect the square of the photon's wave function to be proportional to the number of photons per unit volume in a light wave. But suppose we have a very low-energy source of light that emits just one photon at a time. In any unit volume, there is either one photon or none. The square of the wave function must then describe the *probability* of finding a photon in some unit volume.

The Schrödinger equation describes a single particle. The square of the wave function for a particle must then describe the *probability density*, which is the probability per unit volume, of finding the particle at a location. The probability of finding the particle in some volume element must also be proportional to the size of the volume element dV. Thus, in one dimension, the probability of finding a particle in a region dx at the position x is $\psi^2(x)\, dx$. If we call this probability $P(x)\, dx$, where $P(x)$ is the **probability density,** we have

$$P(x) = \psi^2(x)$$

34-17

PROBABILITY DENSITY

Generally the wave function depends on time as well as position, and is written $\psi(x,t)$. However, for standing waves, the probability density is independent of time. Since we will be concerned mostly with standing waves in this chapter, we omit the time dependence of the wave function and write it $\psi(x)$ or just ψ.

The probability of finding the particle in dx at point x_1 or at point x_2 is the sum of the separate probabilities $P(x_1)\,dx + P(x_2)\,dx$. If we have a particle at all, the probability of finding the particle somewhere must be 1. Then the sum of the probabilities over all the possible values of x must equal 1. That is,

$$\int_{-\infty}^{\infty} \psi^2\,dx = 1 \qquad\qquad 34\text{-}18$$

NORMALIZATION CONDITION

$$\int_{-\infty}^{\infty} \sin^2\left(\frac{n\pi x}{L}\right)dx = 1$$

$$\frac{1}{2} - \frac{1}{2}\sin\left(\frac{2n\pi x}{L}\right)$$

$$\frac{1}{2}x + \frac{L}{n4\pi}\cos\left(\frac{2\pi n x}{L}\right)\Big|_{-\infty}^{\infty} = 1$$

Equation 34-18 is called the **normalization condition.** If ψ is to satisfy the normalization condition, it must approach zero as x approaches infinity. This places a restriction on the possible solutions of the Schrödinger equation. There are solutions to the Schrödinger equation that do not approach zero as x approaches infinity. However, these are not acceptable as wave functions.

PROBABILITY CALCULATION FOR A CLASSICAL PARTICLE   **EXAMPLE 34-5**

It is known that a classical point particle moves back and forth with constant speed between two walls at $x = 0$ and $x = 8$ cm (Figure 34-10). No additional information about the location of the particle is known. (*a*) What is the probability density $P(x)$? (*b*) What is the probability of finding the particle at $x = 2$ cm? (*c*) What is the probability of finding the particle between $x = 3.0$ cm and $x = 3.4$ cm?

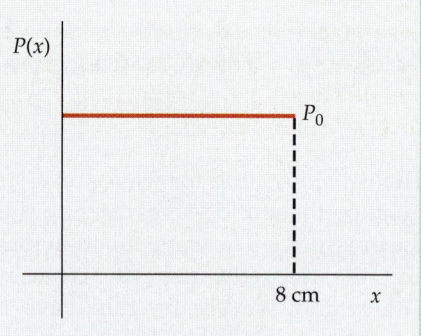

FIGURE 34-10 The probability function $P(x)$.

PICTURE THE PROBLEM We do not know the initial position of the particle. Since the particle moves with constant speed, it is equally likely to be anywhere in the region $0 < x < 8$ cm. The probability density $P(x)$ is therefore independent of x, for $0 < x < 8$ cm, and zero outside of this range. We can find $P(x)$, for $0 < x < 8$ cm, by normalization, that is, by requiring that the probability that the particle is somewhere between $x = 0$ and $x = 8$ cm is 1.

(*a*) 1. The probability density $P(x)$ is uniform between the walls and zero elsewhere:

$$P(x) = P_0, \quad 0 < x < 8\text{ cm}$$
$$P(x) = 0, \quad x < 0 \quad\text{or}\quad x > 8\text{ cm}$$

2. Apply the normalization condition:

$$\int_{-\infty}^{+\infty} P(x)\,dx = \int_{0}^{8\text{ cm}} P_0\,dx = P_0\,(8\text{ cm}) = 1$$

3. Solve for P_0:

$$P_0 = \boxed{\frac{1}{8\text{ cm}}}$$

(*b*) The probability of finding the particle in some range Δx is proportional to $P_0\Delta x = \Delta x/(8\text{ cm})$. Since it is given that $\Delta x = 0$, the probability of finding the particle at the point $x = 2$ cm is 0. Alternatively, since there is an infinite number of points between $x = 0$ and $x = 8$ cm, and the particle is equally likely to be at any point, the chance that the particle will be at any one particular point must be zero.

> The probability of finding the particle at the point $x = 2$ cm is 0.

(c) Since the probability density is constant, the probability of a particle being in some range Δx in the region $0 < x < 8$ cm is $P_0 \Delta x$. The probability of the particle being in the region 3.0 cm $< x <$ 3.4 cm is thus:

$$P_0 \Delta x = \left(\frac{1}{8 \text{ cm}} \right) 0.4 \text{ cm} = \boxed{0.05}$$

REMARKS Note in step 2 of Part (a) that we need only integrate from 0 to 8 cm, because $P(x)$ is zero outside this range.

34-6 Wave–Particle Duality

We have seen that light, which we ordinarily think of as a wave, exhibits particle properties when it interacts with matter, as in the photoelectric effect or in Compton scattering. Electrons, which we usually think of as particles, exhibit the wave properties of interference and diffraction when they pass near the edges of obstacles. All carriers of momentum and energy (e.g., electrons, atoms, light, or sound), exhibit both wave and particle characteristics. It might be tempting to say that an electron, for example, is both a wave and a particle, but what does this mean? In classical physics, the concepts of waves and particles are mutually exclusive. A **classical particle** behaves like a piece of shot; it can be localized and scattered, it exchanges energy suddenly at a point in space, and it obeys the laws of conservation of energy and momentum in collisions. It does *not* exhibit interference or diffraction. A **classical wave,** on the other hand, behaves like a water wave; it exhibits diffraction and interference, and its energy is spread out continuously in space and time. They are mutually exclusive. Nothing can be both a classical particle and a classical wave at the same time.

After Thomas Young observed the two-slit interference pattern with light in 1801, light was thought to be a classical wave. On the other hand, the electrons discovered by J. J. Thomson were thought to be classical particles. We now know that these classical concepts of waves and particles do not adequately describe the complete behavior of any phenomenon.

Everything propagates like a wave and exchanges energy like a particle.

Often the concepts of the classical particle and the classical wave give the same results. If the wavelength is very small, diffraction effects are negligible, so the waves travel in straight lines like classical particles. Also, interference is not seen for waves of very short wavelength, because the interference fringes are too closely spaced to be observed. It then makes no difference which concept we use. If diffraction is negligible, we can think of light as a wave propagating along rays, as in geometrical optics, or as a beam of photon particles. Similarly, we can think of an electron as a wave propagating in straight lines along rays or, more commonly, as a particle.

We can also use either the wave or particle concept to describe exchanges of energy if we have a large number of particles and we are interested only in the average values of energy and momentum exchanges.

The Two-Slit Experiment Revisited

The wave–particle duality of nature is illustrated by the analysis of the experiment in which a single electron is incident on a barrier with two slits. The analysis is the same whether we use an electron or a photon (light). To describe the propagation of an electron, we must use wave theory. Let us assume the source is a point source, such as a needlepoint, so we have spherical waves spreading out from the source. After passing through the two slits, the wavefronts spread out—as if each slit was a source of wavefronts. The wave function ψ at a point on a screen or film far from the slits depends on the difference in path lengths from the source to the point, one path through one slit, and the other path through the other slit. At points on the screen for which the difference in path lengths is either zero or an integral number of wavelengths, the amplitude of the wave function is a maximum. Since the probability of detecting the electron is proportional to ψ^2, the electron is very likely to arrive near these points. At points for which the path difference is an odd number of half wavelengths, the wave function ψ is zero, so the electron is very unlikely to arrive near these points. The chapter opening photos show the interference pattern produced by 10 electrons, 100 electrons, 3000 electrons, and 70,000 electrons. Note that, although the electron propagates through the slits like a wave, the electron interacts with the screen at a single point like a particle.

The Uncertainty Principle

An important principle consistent with the wave–particle duality of nature is the uncertainty principle. It states that, in principle, it is impossible to simultaneously measure both the position and the momentum of a particle with unlimited precision. A common way to measure the position of an object is to look at the object with light. If we do this, we scatter light from the object and determine the position by the direction of the scattered light. If we use light of wavelength λ, we can measure the position x only to an uncertainty Δx of the order of λ because of diffraction effects.

$$\Delta x \sim \lambda$$

To reduce the uncertainty in position, we therefore use light of very short wavelength, perhaps even X rays. In principle, there is no limit to the accuracy of such a position measurement, because there is no limit on how small the wavelength λ can be.

We can determine the momentum p_x of the object if we know the mass and can determine its velocity. The momentum of the object can be found by measuring the object's position at two nearby times and computing its velocity. If we use light of wavelength λ, the photons carry momentum h/p_x. If these photons are scattered by the object we are looking at, the scattering changes the momentum of the object in an uncontrollable way. Each photon carries momentum h/λ, so the uncertainty in the momentum Δp_x of the object, introduced by looking at it, is of the order of h/λ:

$$\Delta p_x \sim \frac{h}{\lambda}$$

If the wavelength of the radiation is small, the momentum of each photon will be large and the momentum measurement will have a large uncertainty. Reducing the intensity of light cannot eliminate this uncertainty; such a reduction merely reduces the number of photons in the beam. To *see* the object, we must scatter at least one photon. Therefore, the uncertainty in the momentum measurement of the object will be large if λ is small, and the uncertainty in the position measurement of the object will be large if λ is large.

Of course we could always *look at* the objects by scattering electrons instead of photons, but the same difficulty remains. If we use low-momentum electrons to reduce the uncertainty in the momentum measurement, we have a large uncertainty in the position measurement because of diffraction of the electrons. The relation between the wavelength and momentum $\lambda = h/p_x$ is the same for electrons as it is for photons.

The product of the intrinsic uncertainties in position and momentum is

$$\Delta x\, \Delta p_x \sim \lambda \times \frac{h}{\lambda} = h$$

If we define precisely what we mean by uncertainties in measurement, we can give a precise statement of the uncertainty principle. If Δx and Δp are defined to be the standard deviations in the measurements of position and momentum, it can be shown that their product must be greater than or equal to $\hbar/2$.

$$\Delta x\, \Delta p_x \geq \tfrac{1}{2}\hbar \qquad\qquad\qquad 34\text{-}19$$

where $\hbar = h/2\pi$.[†]

Equation 34-19 provides a statement of the uncertainty principle first enunciated by Werner Heisenberg in 1927. In practice, the experimental uncertainties are usually much greater than the intrinsic lower limit that results from wave–particle duality.

34-7 A Particle in a Box

We can illustrate many of the important features of quantum physics without solving the Schrödinger equation by considering a simple problem of a particle of mass m confined to a one-dimensional box of length L, like the particle in Example 34-5. This can be considered a crude description of an electron confined within an atom or a proton confined within a nucleus. If a classical particle bounces back and forth between the walls of the box, the particle's energy and momentum can have any values. However, according to quantum theory, the particle is described by a wave function ψ, whose square describes the probability of finding the particle in some region. Since we are assuming that the particle is indeed inside the box, the wave function must be zero everywhere outside the box. If the box is between $x = 0$ and $x = L$, we have

$$\psi = 0, \text{ for } x \leq 0 \text{ and for } x \geq L$$

In particular, if we assume the wave function to be continuous everywhere, it must be zero at the end points of the box $x = 0$ and $x = L$. This is the same condition as the condition for standing waves on a string fixed at $x = 0$ and $x = L$, and the results are the same. The allowed wavelengths for a particle in the box are those where the length L equals an integral number of half wavelengths (Figure 34-11).

$$L = n\frac{\lambda_n}{2}, \qquad n = 1, 2, 3, \dots \qquad\qquad 34\text{-}20$$

STANDING-WAVE CONDITION FOR A PARTICLE IN A BOX OF LENGTH L

The total energy of the particle is its kinetic energy

$$E = \frac{1}{2}mv^2 = \frac{p^2}{2m}$$

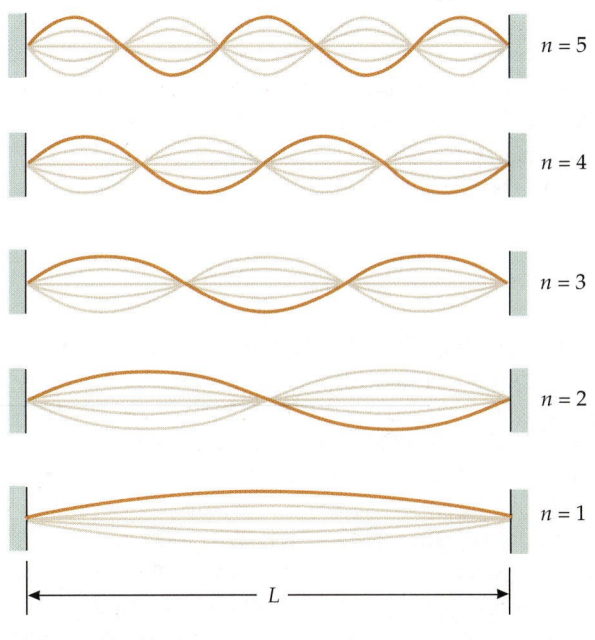

$n = 5$

$n = 4$

$n = 3$

$n = 2$

$n = 1$

L

FIGURE 34-11 Standing waves on a string fixed at both ends. The standing-wave condition is the same as for standing electron waves in a box.

[†] The combination $h/2\pi$ occurs so often it is given a special symbol, somewhat analogous to giving the special symbol ω for $2\pi f$, which occurs often in oscillations.

Substituting the de Broglie relation $p_n = h/\lambda_n$,

$$E_n = \frac{p_n^2}{2m} = \frac{(h/\lambda_n)^2}{2m} = \frac{h^2}{2m\lambda_n^2}$$

Then the standing-wave condition $\lambda_n = 2L/n$ gives the allowed energies.

$$E_n = n^2 \frac{h^2}{8mL^2} = n^2 E_1 \qquad\qquad 34\text{-}21$$

ALLOWED ENERGIES FOR A PARTICLE IN A BOX

where

$$E_1 = \frac{h^2}{8mL^2} \qquad\qquad 34\text{-}22$$

GROUND-STATE ENERGY FOR A PARTICLE IN A BOX

is the energy of the lowest state, which is the ground state.

The condition $\psi = 0$ at $x = 0$ and $x = L$ is called a **boundary condition.** Boundary conditions in quantum theory lead to energy quantization. Figure 34-12 shows the energy-level diagram for a particle in a box. Note that the lowest energy is not zero. This result is a general feature of quantum theory. If a particle is confined to some region of space, the particle has a minimum kinetic energy, which is called the **zero-point energy.** The smaller the region of space the particle is confined to, the greater its zero-point energy. In Equation 34-22, this is indicated by the fact that E_1 varies as $1/L^2$.

If an electron is confined (i.e., bound to an atom) in some energy state E_i, the electron can make a transition to another energy state E_f with the emission of a photon (if $E_f < E_i$; if E_f is greater than E_i, the system absorbs a photon). The transition from state 3 to the ground state is indicated in Figure 34-12 by the vertical arrow. The frequency of the emitted photon is found from the conservation of energy[†]

$$hf = E_i - E_f \qquad\qquad 34\text{-}23$$

The wavelength of the photon is then

$$\lambda = \frac{c}{f} = \frac{hc}{E_i - E_f} \qquad\qquad 34\text{-}24$$

Standing-Wave Functions

The amplitude of a vibrating string fixed at $x = 0$ and $x = L$ is given by Equation 16-15:

$$y_n = A_n \sin k_n x$$

where A_n is a constant and $k_n = 2\pi/\lambda_n$ is the wave number. The wave functions for a particle in a box (which can be obtained by solving the Schrödinger equation, as we will see in Chapter 35) are the same

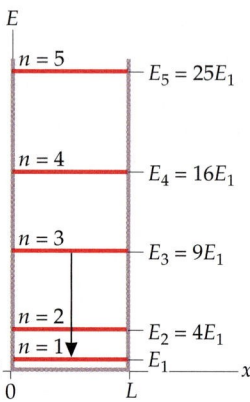

FIGURE 34-12 Energy-level diagram for a particle in a box. Classically, a particle can have any energy value. Quantum mechanically, only those energy values given by Equation 34-22 are allowed. A transition between the state $n = 3$ and the ground state $n = 1$ is indicated by the vertical arrow.

[†] This equation was first proposed by Niels Bohr in his semiclassical model of the hydrogen atom in 1913, about 10 years before de Broglie's suggestion that electrons have wave properties. We will study the Bohr model in Chapter 36.

$$\psi_n(x) = A_n \sin k_n x$$

where $k_n = 2\pi/\lambda_n$. Using $\lambda_n = 2L/n$, we have

$$k_n = \frac{2\pi}{\lambda_n} = \frac{2\pi}{2L/n} = \frac{n\pi}{L}$$

The wave functions can thus be written

$$\psi_n(x) = A_n \sin\left(n\pi\frac{x}{L}\right)$$

The constant A_n is determined by the normalization condition (Equation 34-18):

$$\int_{-\infty}^{\infty} \psi^2\, dx = \int_0^L A_n^2 \sin^2\left(n\pi\frac{x}{L}\right) dx = 1$$

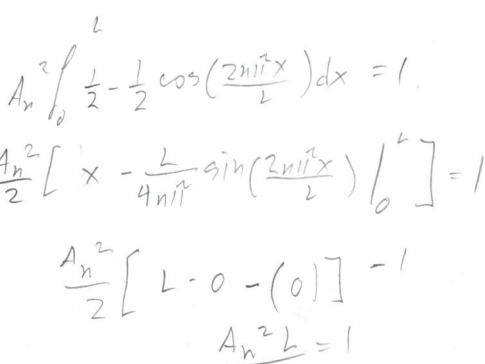

Note that we need integrate only from $x = 0$ to $x = L$ because $\psi(x)$ is zero everywhere else. The result of evaluating the integral and solving for A_n is

$$A_n = \sqrt{\frac{2}{L}}$$

independent of n. The normalized wave functions for a particle in a box are thus

$$\psi_n(x) = \sqrt{\frac{2}{L}} \sin\left(n\pi\frac{x}{L}\right) \qquad\qquad 34\text{-}25$$

WAVE FUNCTIONS FOR A PARTICLE IN A BOX

These standing-wave functions for $n = 1$, $n = 2$, and $n = 3$ are shown in Figure 34-13.

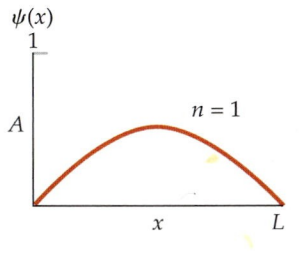
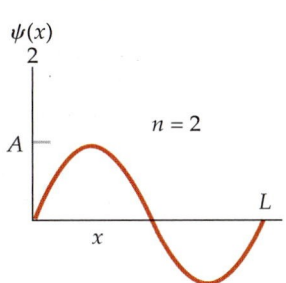
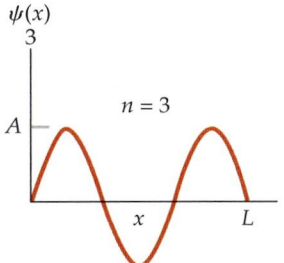

FIGURE 34-13 Standing-wave functions for $n = 1$, $n = 2$, and $n = 3$.

The number n is called a **quantum number.** It characterizes the wave function for a particular state and for the energy of that state. In our one-dimensional problem, a quantum number arises from the boundary condition on the wave function that it must be zero at $x = 0$ and $x = L$. In three-dimensional problems, three quantum numbers arise, one associated with a boundary condition in each dimension.

Figure 34-14 shows plots of ψ^2 for the ground state $n = 1$, the first excited state $n = 2$, the second excited state $n = 3$, and the state $n = 10$. In the ground state, the particle is most likely to be found near the center of the box, as indicated by the maximum value of ψ^2 at $x = L/2$. In the first excited state, the particle is least likely to be found near the center of the box because ψ^2 is small near $x = L/2$.

For very large values of n, the maxima and minima of ψ^2 are very close together, as illustrated for $n = 10$. The average value of ψ^2 is indicated in this figure by the dashed line. For very large values of n, the maxima are so closely spaced that ψ^2 cannot be distinguished from its average value. The fact that $(\psi^2)_{av}$ is constant across the whole box means that the particle is equally likely to be found anywhere in the box—the same as in the classical result. This is an example of **Bohr's correspondence principle:**

> In the limit of very large quantum numbers, the classical calculation and the quantum calculation must yield the same results.

BOHR'S CORRESPONDENCE PRINCIPLE

The region of very large quantum numbers is also the region of very large energies. For large energies, the percentage change in energy between adjacent quantum states is very small, so energy quantization is not important (see Problem 83).

We are so accustomed to thinking of the electron as a classical particle that we tend to think of an electron in a box as a particle bouncing back and forth between the walls. But the probability distributions shown in Figure 34-14 are stationary; that is, they do not depend on time. A better picture for an electron in a bound state is a cloud of charge with the charge density proportional to ψ^2. Figure 34-14 can then be thought of as plots of the charge density versus x for the various states. In the ground state, $n = 1$, the electron cloud is centered in the middle of the box and is spread out over most of the box, as indicated in Figure 34-14a. In the first excited state, $n = 2$, the charge density of the electron cloud has two maxima, as indicated in Figure 34-14b. For very large values of n, there are many closely spaced maxima and minima in the charge density resulting in an average charge density that is approximately uniform throughout the box. This electron-cloud picture of an electron is very useful in understanding the structure of atoms and molecules. However, it should be noted that whenever an electron is observed to interact with matter or radiation, it is always observed as a whole unit charge.

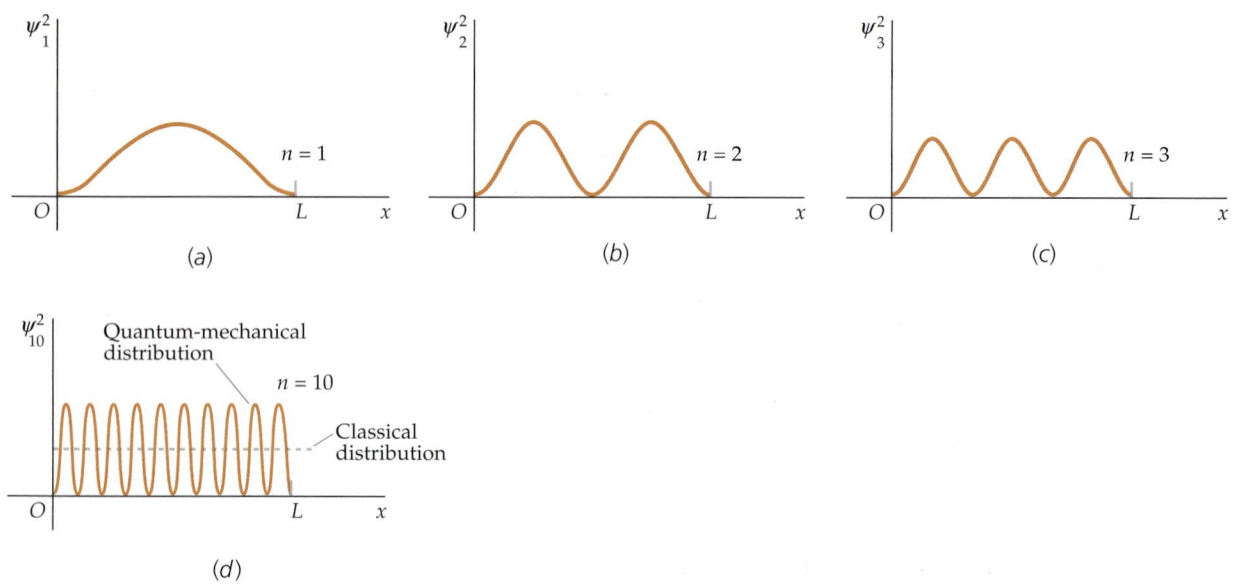

FIGURE 34-14 ψ^2 versus x for a particle in a box of length L for (a) the ground state, $n = 1$; (b) the first excited state, $n = 2$; (c) the second excited state, $n = 3$; and (d) the state $n = 10$. For $n = 10$, the maxima and minima of ψ^2 are so close together that individual maxima may be hard to distinguish. The dashed line indicates the average value of ψ^2. It gives the classical prediction that the particle is equally likely to be found near any point in the box.

PHOTON EMISSION BY A PARTICLE IN A BOX **EXAMPLE 34-6**

An electron is in a one-dimensional box of length 0.1 nm. (*a*) Find the ground-state energy. (*b*) Find the energy in electron volts of the five lowest states, and then sketch an energy-level diagram. (*c*) Find the wavelength of the photon emitted for each transition from the state $n = 3$ to a lower-energy state.

PICTURE THE PROBLEM For Part (*a*) and Part (*b*), the energies are given by $E_n = n^2 E_1$ (Equation 34-21), where the ground-state energy $E_1 = h^2/8mL^2$ (Equation 34-22). For Part (*c*), the photon wavelengths are given by $\lambda = hc/(E_i - E_f)$ (Equation 34-24).

(*a*) Use $hc = 1240$ eV·nm and $mc^2 = 5.11 \times 10^5$ eV to calculate E_1:

$$E_1 = \frac{(hc)^2}{8(mc^2)L^2}$$

$$= \frac{(1240 \text{ eV·nm})^2}{8(5.11 \times 10^5 \text{ eV})(0.1 \text{ nm})^2} = \boxed{37.6 \text{ eV}}$$

(*b*) Calculate $E_n = n^2 E_1$ for $n = 2, 3, 4,$ and 5:

$$E_2 = (2)^2(37.6 \text{ eV}) = \boxed{150 \text{ eV}}$$

$$E_3 = (3)^2(37.6 \text{ eV}) = \boxed{338 \text{ eV}}$$

$$E_4 = (4)^2(37.6 \text{ eV}) = \boxed{602 \text{ eV}}$$

$$E_5 = (5)^2(37.6 \text{ eV}) = \boxed{940 \text{ eV}}$$

(*c*) 1. Use the energies found in Part (*b*) to calculate the wavelength for a transition from state 3 to state 2:

$$\lambda = \frac{hc}{E_3 - E_2}$$

$$= \frac{1240 \text{ eV·nm}}{338 \text{ eV} - 150 \text{ eV}} = \boxed{6.60 \text{ nm}}$$

2. Then use the energies in Part (*a*) and Part (*b*) to calculate the wavelength for a transition from state 3 to state 1:

$$\lambda = \frac{hc}{E_3 - E_1}$$

$$= \frac{1240 \text{ eV·nm}}{338 \text{ eV} - 37.6 \text{ eV}} = \boxed{4.13 \text{ nm}}$$

REMARKS The energy-level diagram is shown in Figure 34-15. The transitions from $n = 3$ to $n = 2$ and from $n = 3$ to $n = 1$ are indicated by the vertical arrows. The ground-state energy of 37.6 eV is on the same order of magnitude as the kinetic energy of the electron in the ground state of the hydrogen atom, which is 13.6 eV. In the hydrogen atom, the electron also has potential energy of -27.2 eV in the ground state, giving it a total ground-state energy of -13.6 eV.

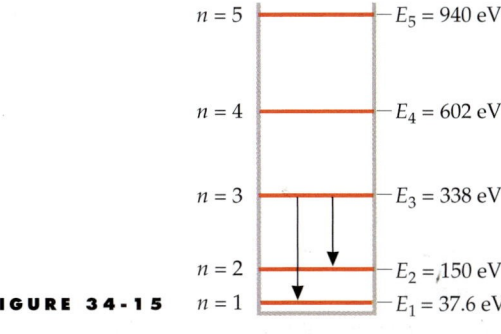

$n = 5$		$E_5 = 940$ eV
$n = 4$		$E_4 = 602$ eV
$n = 3$		$E_3 = 338$ eV
$n = 2$		$E_2 = 150$ eV
FIGURE 34-15 $n = 1$		$E_1 = 37.6$ eV

EXERCISE Calculate the wavelength of the photon emitted if the electron in the box makes a transition from $n = 4$ to $n = 3$. (*Answer* 4.69 nm)

34-8 Expectation Values

The solution of a classical mechanics problem is typically specified by giving the position of a particle as a function of time. But the wave nature of matter prevents us from doing this for microscopic systems. The most that we can know is the probability of measuring a certain value of the position x. If we measure the position for a large number of identical systems, we get a range of values corresponding to the probability distribution. The average value of x obtained from such measurements is called the **expectation value** and is written $<x>$. The expectation value of x is the same as the average value of x that we would expect to obtain from a measurement of the positions of a large number of particles with the same wave function $\psi(x)$.

Since $\psi^2(x)\,dx$ is the probability of finding a particle in the region dx, the expectation value of x is

$$< x > = \int_{-\infty}^{+\infty} x\psi^2(x)\,dx \qquad\qquad 34\text{-}26$$

EXPECTATION VALUE OF X DEFINED

The expectation value of any function $f(x)$ is given by

$$< f(x) > = \int_{-\infty}^{+\infty} f(x)\psi^2(x)\,dx \qquad\qquad 34\text{-}27$$

EXPECTATION VALUE OF $F(X)$ DEFINED

*Calculating Probabilities and Expectation Values

The problem of a particle in a box allows us to illustrate the calculation of the probability of finding the particle in various regions of the box and the expectation values for various energy states. We give two examples, using the wave functions given by Equation 34-25.

THE PROBABILITY OF THE PARTICLE BEING
FOUND IN A SPECIFIED REGION OF A BOX

E X A M P L E   3 4 - 7

A particle in a one-dimensional box of length L is in the ground state. Find the probability of finding the particle (a) anywhere in the region of length $\Delta x = 0.01L$, centered at $x = \frac{1}{2}L$ and (b) in the region $0 < x < \frac{1}{4}L$.

PICTURE THE PROBLEM The probability P of finding the particle in some infinitesimal range dx is $\psi^2\,dx$. For a particle in the ground state, ψ is given by Equation 34-25, with $n = 1$; ψ^2 is illustrated in Figure 34-14. The probability of finding x in some region is just the area under this curve for the region. For Part (a), the region is $\Delta x = 0.01L$, centered at $x = L/2$, and the area under the ψ^2 versus x curve is shown in Figure 34-16a. This area is $\approx \psi^2\,\Delta x$. For Part (b), the region is $0 < x < L/4$, and the area under the curve is shown in Figure 34-16b. To calculate this area, we must integrate ψ^2 from $x = 0$ to $x = L/4$.

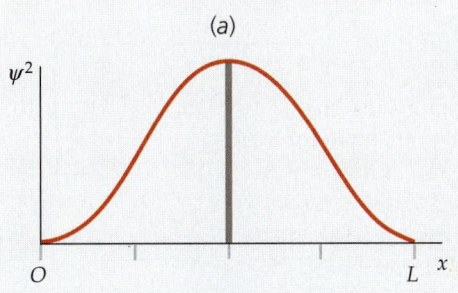
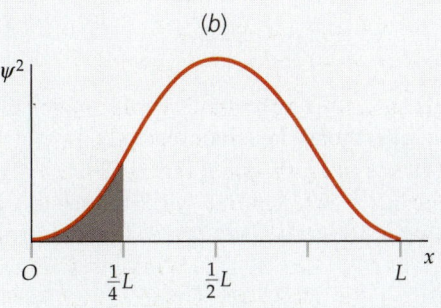

FIGURE 34-16

(a) 1. The probability of finding the particle is the area under the curve shown in Figure 34-16a. To calculate this area, we need to calculate the height of curve at $x = L/2$:

$$\psi(x) = \sqrt{\frac{2}{L}} \sin\left(\pi\frac{x}{L}\right)$$

so

$$\psi^2(L/2) = \frac{2}{L}\sin^2\frac{\pi}{2} = \frac{2}{L}$$

2. The area is the height times the width, and the width is $\Delta x = 0.01L$:

$$P = \psi^2(L/2)\Delta x = \frac{2}{L}\times 0.01L = \boxed{0.02}$$

(b) 1. The probability of finding the particle is the area under the curve shown in Figure 34-16b. To calculate this area, we need to integrate from $x = 0$ to $x = L/4$:

$$P = \int_0^{L/4}\psi^2(x)\,dx = \int_0^{L/4}\frac{2}{L}\sin^2\frac{\pi x}{L}\,dx$$

2. The integral can be evaluated a number of ways. If a table of integrals is used, a change in the integration variable in required. Changing the integration variable to $\theta = \pi x/L$ gives:

$$P = \frac{2}{\pi}\int_0^{\pi/4}\sin^2\theta\,d\theta$$

3. The integral can be found in tables:

$$\int_0^{\pi/4}\sin^2\theta\,d\theta = \left(\frac{\theta}{2} - \frac{\sin^2 2\theta}{4}\right)\Big|_0^{\pi/4} = \left(\frac{\pi}{8} - \frac{1}{4}\right)$$

4. Use the result from Part (b), step 3 to calculate the probability:

$$P = \frac{2}{\pi}\left(\frac{\pi}{8} - \frac{1}{4}\right) = \boxed{0.091}$$

 **REMARKS** An integral was not necessary for Part (a) because the area of interest could be well approximated by a rectangle of height ψ^2 and width x. The chance of finding the particle in the region $\Delta x = 0.01L$ at $x = \frac{1}{2}L$ is approximately 2 percent. The chance of finding the particle in the region $0 < x < L/4$ is about 9.1 percent.

CALCULATING EXPECTATION VALUES   **E X A M P L E   3 4 - 8**

Find (a) $<x>$ and (b) $<x^2>$ for a particle in its ground state in a box of length L.

PICTURE THE PROBLEM We use $<f(x)> = \int f(x)\psi^2(x)\,dx$, with

$$\psi_n(x) = \sqrt{\frac{2}{L}}\sin\frac{n\pi x}{L}.$$

(a) 1. Write $<x>$ using the ground-state wave function given by Equation 34-25, with $n = 1$:

$$<x> = \int_{-\infty}^{+\infty} x\psi^2(x)\,dx = \frac{2}{L}\int_0^L x\sin^2\left(\frac{\pi x}{L}\right)dx$$

2. To evaluate this integral by using a table of integrals, first change the integration variable to $\theta = \pi x/L$:

$$<x> = \frac{2}{L}\left(\frac{L}{\pi}\right)^2\int_0^\pi \theta\sin^2\theta\,d\theta$$

$$= \frac{2L}{\pi^2}\int_0^\pi \theta\sin^2\theta\,d\theta$$

3. The table of integrals gives:

$$\int_0^\pi \theta \sin^2 \theta \, d\theta = \left[\frac{\theta^2}{4} - \frac{\theta \sin 2\theta}{4} - \frac{\cos 2\theta}{8} \right]_0^\pi = \frac{\pi^2}{4}$$

4. Substitute this value into the expression in step 2:

$$<x> = \frac{2L}{\pi^2} \int_0^\pi \theta \sin^2 \theta \, d\theta = \frac{2L}{\pi^2} \frac{\pi^2}{4} = \boxed{\frac{L}{2}}$$

(b) 1. Repeat step 1 and step 2 of Part (a) for $<x^2>$:

$$<x^2> = \int_{-\infty}^{+\infty} x^2 \psi^2(x) \, dx = \int_0^L x^2 \frac{2}{L} \sin^2(\pi x/L) \, dx$$

$$= \frac{2}{L}\left(\frac{L}{\pi}\right)^3 \int_0^\pi \theta^2 \sin^2 \theta \, d\theta = \frac{2L^2}{\pi^3} \int_0^\pi \theta^2 \sin^2 \theta \, d\theta$$

2. Evaluating the integral using a table of integrals gives:

$$\int_0^\pi \theta^2 \sin^2 \theta \, d\theta = \left[\frac{\theta^3}{6} - \left(\frac{\theta^2}{4} - \frac{1}{8}\right) \sin 2\theta - \frac{\theta \cos 2\theta}{4} \right]_0^\pi$$

$$= \frac{\pi^3}{6} - \frac{\pi}{4}$$

3. Substitute this value into the expression in step 1 of Part (b):

$$<x^2> = \frac{2L^2}{\pi^3}\left(\frac{\pi^3}{6} - \frac{\pi}{4}\right) = \left(\frac{1}{3} - \frac{1}{2\pi^2}\right)L^2 = \boxed{0.283L^2}$$

REMARKS The expectation value of x is $L/2$, as we would expect, because the probability distribution is symmetric about the midpoint of the box. Note that $<x^2>$ is not equal to $<x>^2$.

34-9 Energy Quantization in Other Systems

The quantized energies of a system are generally determined by solving the Schrödinger equation for that system. The form of the Schrödinger equation depends on the potential energy of the particle. The potential energy for a one-dimensional box from $x = 0$ to $x = L$ is shown in Figure 34-17. This potential energy function is called an **infinite square-well potential**, and it is described mathematically by

$$U(x) = 0, \quad 0 < x < L$$

$$U(x) = \infty, \quad x < 0 \quad \text{or} \quad x > L \qquad\qquad 34\text{-}28$$

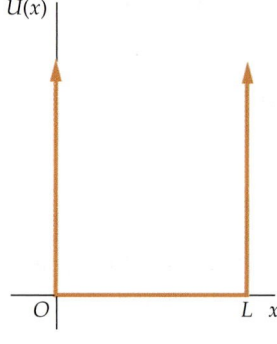

FIGURE 34-17 The infinite square-well potential energy. For $x < 0$ and $x > L$, the potential energy $U(x)$ is infinite. The particle is confined to the region in the well $0 < x < L$.

Inside the box the particle moves freely, so the potential energy is uniform. For convenience, we choose the value of this potential energy to be zero. Outside the box the potential energy is infinite, so the particle cannot exist outside the box no matter what its energy. We did not need to solve the Schrödinger equation for this potential because the wave functions and quantized frequencies are the same as for a string fixed at both ends, which we studied in Chapter 16. Although this problem seems artificial, actually it is useful for some physical problems, such as a neutron inside a nucleus.

The Harmonic Oscillator

More realistic than the particle in a box is the harmonic oscillator, which applies to an object of mass m on a spring of force constant k or to any system undergoing small oscillations about a stable equilibrium. Figure 34-18 shows the potential energy function

$$U(x) = \tfrac{1}{2}kx^2 = \tfrac{1}{2}m\omega_0^2 x^2$$

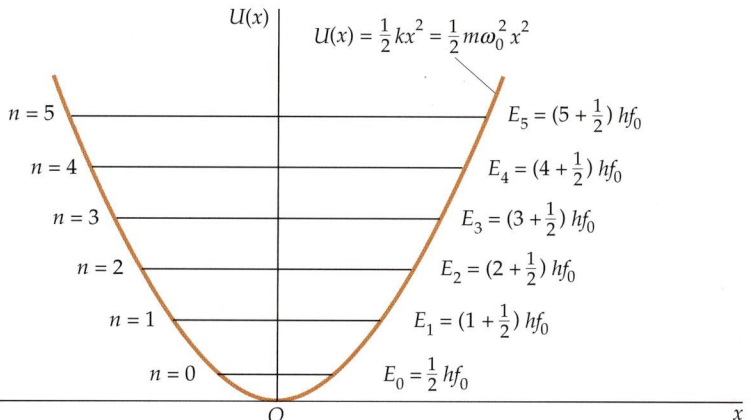

FIGURE 34-18 Harmonic oscillator potential energy function. The allowed energy levels are indicated by the equally spaced horizontal lines. Also, $\omega_0 = 2\pi f_0$.

where $\omega_0 = \sqrt{k/m}$ is the natural frequency of the oscillator. Classically, the object oscillates between $x = +A$ and $x = -A$. Its total energy is $E = \frac{1}{2}m\omega_0^2 A^2$, which can have any nonnegative value, including zero.

In quantum theory, the particle is represented by the wave function $\psi(x)$, which is determined by solving the Schrödinger equation for this potential. Normalizable wave functions $\psi_n(x)$ occur only for discrete values of the energy E_n given by

$$E_n = (n + \tfrac{1}{2})hf_0, \quad n = 0, 1, 2, 3, \ldots \qquad \text{34-29}$$

where $f_0 = \omega_0/2\pi$ is the classical frequency of the oscillator. Note that the energy levels of a harmonic oscillator are evenly spaced with separation hf, as indicated in Figure 34-18. Compare this with the uneven spacing of the energy levels for the particle in a box, as shown in Figure 34-12. If a harmonic oscillator makes a transition from energy level n to the next lowest energy level $n - 1$, the frequency f of the photon emitted is given by $hf = E_i - E_f$ (Equation 34-23). Applying this equation gives

$$hf = E_n - E_{n-1} = (n + \tfrac{1}{2})hf_0 - (n - 1 + \tfrac{1}{2})hf_0 = hf_0$$

The frequency f of the emitted photon is therefore equal to the classical frequency f_0 of the oscillator.

The Hydrogen Atom

In the hydrogen atom, an electron is bound to a proton by the electrostatic force of attraction (discussed in Chapter 21). This force varies inversely as the square of the separation distance (exactly like the gravitational attraction of the earth and sun). The potential energy of the electron–proton system therefore varies inversely with separation distance (Equation 23-9). As in the case of gravitational potential energy, the potential energy of the electron–proton system is chosen to be zero if the electron is an infinite distance from the proton. Then for all finite distances, the potential energy is negative. Like the case of an object orbiting the earth, the electron–proton system is a bound system if its total energy is negative. Like the energies of a particle in a box and of a harmonic oscillator, the energies are described by a quantum number n. As we will see in Chapter 36, the allowed energies of the hydrogen atom are given by

$$E_n = -\frac{13.6 \text{ eV}}{n^2}, \quad n = 1, 2, 3, \ldots \qquad \text{34-30}$$

The lowest energy corresponds to $n = 1$. The ground-state energy is thus -13.6 eV. The energy of the first excited state is $-(13.6 \text{ eV})/2^2 = -3.40$ eV. Figure 34-19 shows the energy-level diagram for the hydrogen atom. The vertical arrows indicate transitions from a higher state to a lower state with the emission of electromagnetic radiation. Only those transitions ending at the first excited state ($n = 2$) involve energy differences in the range of visible light of 1.77 eV to 3.10 eV, as calculated in Example 34-1.

Other atoms are more complicated than the hydrogen atom, but their energy levels are similar in many ways to those of hydrogen. Their ground-state energies are of the order of -1 eV to -10 eV, and many transitions involve energies corresponding to photons in the visible range.

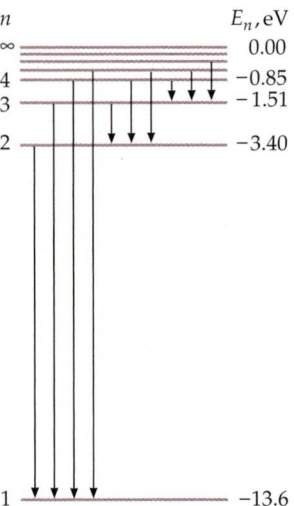

FIGURE 34-19 Energy-level diagram for the hydrogen atom. The energy of the ground state is -13.6 eV. As n approaches ∞ the energy approaches 0, which is the highest energy state for an electron bound to the atom.

SUMMARY

1. All phenomenon propagate like waves and interact like particles.

2. The quantum of light is called a photon and has energy $E = hf$, where h is Planck's constant.

3. The relation between wavelength and momentum of electrons, photons, and other particles is given by the de Broglie relation $\lambda = h/p$.

4. Energy quantization in bound systems arises from standing-wave conditions, which are equivalent to boundary conditions on the wave function.

5. The uncertainty principle is a fundamental law of nature that places theoretical restrictions on the precision of a simultaneous measurement of the position and momentum of a particle. It follows from the general properties of waves.

Topic	Relevant Equations and Remarks	
1. Constants and Values		
Planck's constant	$h = 6.626 \times 10^{-34}$ J·s $= 4.136 \times 10^{-15}$ eV·s	**34-2**
hc	$hc = 1240$ eV·nm	**34-5**
Compton Wavelength	$\lambda_C = \dfrac{h}{m_e c} = 2.43$ pm	**34-12**
2. The Particle Nature of Light: Photons	Energy is quantized.	
Photon energy	$E = hf$	**34-1**
Momentum of a photon	$p = \dfrac{h}{\lambda}$	**34-7**

3. **Frequency–Wavelength (energy–momentum) Relations**

Photons	$E = pc,$ so $\lambda f = c$	**R-17**
Nonrelativistic particles	$K = \dfrac{p^2}{2m},$ so $\lambda = \dfrac{hc}{\sqrt{2mc^2 K}}$	**34-15**
Electrons	$\lambda = \dfrac{1.226}{\sqrt{K}}$ nm, K in electron volts	**34-16**
Photoelectric effect	$K_{max} = (\tfrac{1}{2}mv^2)_{max} = hf - \phi$	**34-3**
	where ϕ is the work function of the cathode.	
Compton scattering	$\lambda_2 - \lambda_1 = \dfrac{h}{m_e c}(1 - \cos\theta) = \lambda_C(1 - \cos\theta)$	**34-11**

4. **Quantum Mechanics**

The state of a particle, such as an electron, is described by its wave function ψ, which is the solution of the Schrödinger wave equation.

Probability density

The probability of finding the particle in some region of space dx is given by

$$P(x) = \psi^2(x)\, dx \tag{34-17}$$

Normalization condition

$$\int_{-\infty}^{\infty} \psi^2 \, dx = 1 \tag{34-18}$$

Quantum number

The wave function for a particular energy state is characterized by a quantum number n. In three dimensions there are three quantum numbers, one associated with a boundary condition in each dimension.

Expectation value

The expectation value of x is the same as the average value of x that we would expect to obtain from a measurement of the positions of a large number of particles with the same wave function $\psi(x)$.

$$<x> = \int_{-\infty}^{+\infty} x\psi^2(x)\, dx \tag{34-26}$$

$$<f(x)> = \int_{-\infty}^{+\infty} f(x)\psi^2(x)\, dx \tag{34-27}$$

5. **Wave–Particle Duality**

Light, electrons, neutrons, and all carriers of momentum and energy exhibit both wave and particle properties. Everything propagates like a classical wave exhibiting diffraction and interference, but exchanges energy in discrete lumps like a classical particle. Because the wavelength of macroscopic objects is so small, diffraction and interference are not observed. Also, if a macroscopic amount of energy is exchanged, so many quanta are involved that the particle nature of the energy is not evident.

6. **Uncertainty Principle**

The wave–particle duality of nature leads to the uncertainty principle, which states that the product of the uncertainty in a measurement of position and the uncertainty in a measurement of momentum must be greater than or equal to $\tfrac{1}{2}\hbar$, where $\hbar$ is Planck's constant divided by 2π.

$$\Delta x \, \Delta p \geq \tfrac{1}{2}\hbar \tag{34-19}$$

- Single-concept, single-step, relatively easy
- •• Intermediate-level, may require synthesis of concepts
- ••• Challenging
- **SSM** Solution is in the *Student Solutions Manual*
- **iSOLVE** Problems available on iSOLVE online homework service
- **iSOLVE**✓ These "Checkpoint" online homework service problems ask students additional questions about their confidence level, and how they arrived at their answer.

In a few problems, you are given more data than you actually need; in a few other problems, you are required to supply data from your general knowledge, outside sources, or informed estimates.

Conceptual Problems

1 • **SSM** The quantized character of electromagnetic radiation is revealed by (*a*) the Young double-slit experiment. (*b*) diffraction of light by a small aperture. (*c*) the photoelectric effect. (*d*) the J. J. Thomson cathode-ray experiment.

2 •• Two monochromatic light sources, A and B, emit the same number of photons per second. The wavelength of A is $\lambda_A = 400$ nm, and the wavelength of B is $\lambda_B = 600$ nm. The power radiated by source B is (*a*) equal to the power of source A. (*b*) less than the power of source A. (*c*) greater than the power of source A. (*d*) cannot be compared to the power from source A using the available data.

3 • True or false:

(*a*) In the photoelectric effect, the current is proportional to the intensity of the incident light.
(*b*) In the photoelectric effect, the work function of a metal depends on the frequency of the incident light.
(*c*) In the photoelectric effect, the maximum kinetic energy of electrons emitted varies linearly with the frequency of the incident light.
(*d*) In the photoelectric effect, the energy of a photon is proportional to its frequency.

4 • In the photoelectric effect, the number of electrons emitted per second is (*a*) independent of the light intensity. (*b*) proportional to the light intensity. (*c*) proportional to the work function of the emitting surface. (*d*) proportional to the frequency of the light.

5 • **SSM** The work function of a surface is ϕ. The threshold wavelength for emission of photoelectrons from the surface is (*a*) hc/ϕ. (*b*) ϕ/hf. (*c*) hf/ϕ. (*d*) none of the answers are correct.

6 •• When light of wavelength λ_1 is incident on a certain photoelectric cathode, no electrons are emitted, no matter how intense the incident light is. Yet, when light of wavelength $\lambda_2 < \lambda_1$ is incident, electrons are emitted, even when the incident light has low intensity. Explain.

7 • True or false:

(*a*) The de Broglie wavelength of an electron varies inversely with its momentum.
(*b*) Electrons can be diffracted.
(*c*) Neutrons can be diffracted.
(*d*) An electron microscope is used to look at electrons.

8 • If the de Broglie wavelength of an electron and a proton are equal, then (*a*) the velocity of the proton is greater than the velocity of the electron. (*b*) the velocity of the proton and the electron are equal. (*c*) the velocity of the proton is less than the velocity of the electron. (*d*) the energy of the proton is greater than the energy of the electron. (*e*) both (*a*) and (*d*) are correct.

9 • A proton and an electron have equal kinetic energies. It follows that the de Broglie wavelength of the proton is (*a*) greater than that of the electron. (*b*) equal to that of the electron. (*c*) less than that of the electron.

10 • Can the expectation value of x ever equal a value for which the probability function $P(x)$ is zero? Give a specific example.

11 • **SSM** Explain why the maximum kinetic energy of electrons emitted in the photoelectric effect does not depend on the intensity of the incident light, but the total number of electrons emitted does depend on the intensity of the incident light.

12 •• A six-sided die has the numeral 1 painted on three sides and the numeral 2 painted on the other three sides. (*a*) What is the probability of a 1 coming up when the die is thrown? (*b*) What is the expectation value of the numeral that comes up when the die is thrown?

13 •• It was once believed that if two identical experiments are done on identical systems under the same conditions, the results must be identical. Explain why this is not true and how it can be modified, so that it is consistent with quantum physics.

Estimation and Approximation

14 •• Students in a physics lab are trying to determine the value of Planck's constant h, using a photoelectric apparatus similar to the one shown in Figure 34-2. For a light source, the students use a helium–neon laser with tunable wavelength. The data that the students obtain for the maximum electron kinetic energy are

λ	544 nm	594 nm	604 nm	612 nm	633 nm
K_{max}	0.360 eV	0.199 eV	0.156 eV	0.117 eV	0.062 eV

(a) Using a spreadsheet program or graphing calculator, convert the wavelengths to frequencies and plot K_{max} versus frequency. (b) Use the graph to estimate the value of Planck's constant implied by the students' data. (*Note:* You may wish to use the linear regression function of your spreadsheet program or graphing calculator.) (c) Compare your result with the accepted value for Planck's constant.

15 • • The cathode that was used by the students in the experiment described in Problem 14 is known to be constructed from one of the following metals, with the given work function

Metals	Tungsten	Silver	Potassium	Cesium
Work function	4.58 eV	2.4 eV	2.1 eV	1.9 eV

Solve this problem using the same data as given in Problem 14. (a) Using a spreadsheet program or graphing calculator, convert the wavelengths to frequencies, and plot K_{max} versus frequency. (b) Use the graph to estimate the value of the work function implied by the students' data. (*Note:* You may wish to use the linear regression function of your spreadsheet program or graphing calculator.) (c) Which metal was most likely used for the cathode in their experiment?

16 • • **SSM** Students in an advanced physics lab use X rays to measure the Compton wavelength, λ_C. The students obtain the following wavelength shifts $\lambda_2 - \lambda_1$ as a function of scattering angle θ

θ	45°	75°	90°	135°	180°
$\lambda_2 - \lambda_1$	0.647 pm	1.67 pm	2.45 pm	3.98 pm	4.95 pm

Use their data to estimate the value for the Compton wavelength. Compare this number with the accepted value.

17 • • **SSM** Baseball, tennis, golf, and soccer are sports that involve placing a ball in play with a certain speed. Estimate which of these sports has a ball with the smallest de Broglie wavelength when the ball is moving with the highest speed typically created by a professional athlete.

The Particle Nature of Light: Photons

18 • **ISOLVE** ✔ Find the photon energy in joules and in electron volts for an electromagnetic wave of frequency (a) 100 MHz in the FM radio band and (b) 900 kHz in the AM radio band.

19 • What are the frequencies of photons that have the following energies? (a) 1 eV, (b) 1 keV, and (c) 1 MeV.

20 • **SSM** Find the photon energy for light of wavelength (a) 450 nm, (b) 550 nm, and (c) 650 nm.

21 • Find the photon energy if the wavelength is (a) 0.1 nm (about 1 atomic diameter) and (b) 1 fm (1 fm = 10^{-15} m, about 1 nuclear diameter).

22 • • **ISOLVE** The wavelength of light emitted by a 3-mW helium–neon laser is 632 nm. If the diameter of the laser beam is 1 mm, what is the density of photons in the beam?

23 • **SSM** Lasers used in the telecommunications network typically have a wavelength near 1.55 μm. How many photons per second are being transmitted if such a laser has an output power of 2.5 mW?

The Photoelectric Effect

24 • The work function for tungsten is 4.58 eV. (a) Find the threshold frequency and wavelength for the photoelectric effect. Find the maximum kinetic energy of the electrons if the wavelength of the incident light is (b) 200 nm and (c) 250 nm.

25 • **ISOLVE** ✔ When light of wavelength 300 nm is incident on potassium, the emitted electrons have maximum kinetic energy of 2.03 eV. (a) What is the energy of an incident photon? (b) What is the work function for potassium? (c) What would be the maximum kinetic energy of the electrons if the incident light had a wavelength of 430 nm? (d) What is the threshold wavelength for the photoelectric effect with potassium?

26 • The threshold wavelength for the photoelectric effect for silver is 262 nm. (a) Find the work function for silver. (b) Find the maximum kinetic energy of the electrons if the incident radiation has a wavelength of 175 nm.

27 • The work function for cesium is 1.9 eV. (a) Find the threshold frequency and threshold wavelength for the photoelectric effect. Find the maximum kinetic energy of the electrons if the wavelength of the incident light is (b) 250 nm and (c) 350 nm.

28 • • **SSM** **ISOLVE** When a surface is illuminated with light of wavelength 780 nm, the maximum kinetic energy of the emitted electrons is 0.37 eV. What is the maximum kinetic energy if the surface is illuminated with light of wavelength 410 nm?

Compton Scattering

29 • Find the shift in wavelength of photons scattered by electrons at $\theta = 60°$.

30 • When photons are scattered by electrons in carbon, the shift in wavelength is 0.33 pm. Find the scattering angle.

31 • The wavelength of Compton-scattered photons is measured at $\theta = 135°$. If $\Delta\lambda/\lambda$ is to be 2.3 percent, what should the wavelength of the incident photons be?

32 • **SSM** **ISOLVE** Compton used photons of wavelength 0.0711 nm. (a) What is the energy of these photons? (b) What is the wavelength of the photon scattered at $\theta = 180°$? (c) What is the energy of the photon scattered at this angle?

33 • For the photons used by Compton (see Problem 32), find the momentum of the incident photon and the momentum of the photon scattered at 180°; use the conservation of momentum to find the momentum of the recoil electron in this case.

34 • • An X-ray photon of wavelength 6 pm that collides with an electron is scattered by an angle of 90°. (a) What is the change in wavelength of the photon? (b) What is the kinetic energy of the scattered electron?

35 • • How many head-on, Compton-scattering events are necessary to double the wavelength of a photon with an initial wavelength of 200 pm?

Electrons and Matter Waves

36 • Use Equation 34-16 to calculate the de Broglie wavelength for an electron of kinetic energy (a) 2.5 eV, (b) 250 eV, (c) 2.5 keV, and (d) 25 keV.

37 • An electron is moving at $v = 2.5 \times 10^5$ m/s. Find the electron's de Broglie wavelength.

38 • **SOLVE** An electron has a wavelength of 200 nm. Find (a) its momentum and (b) its kinetic energy.

39 •• **SSM** An electron, a proton, and an alpha particle (the nucleus of a helium atom) each have a kinetic energy of 150 keV. Find (a) their momenta and (b) their de Broglie wavelengths.

40 • **SOLVE** A neutron in a reactor has kinetic energy of approximately 0.02 eV. Calculate the de Broglie wavelength of this neutron from Equation 34-15, where $mc^2 = 940$ MeV is the rest energy of the neutron.

41 • Use Equation 34-15 to find the de Broglie wavelength of a proton (rest energy $mc^2 = 938$ MeV) that has a kinetic energy of 2 MeV.

42 • **SSM** A proton is moving at $v = 0.003c$, where c is the speed of light. Find the electron's de Broglie wavelength.

43 • What is the kinetic energy of a proton whose de Broglie wavelength is (a) 1 nm and (b) 1 fm?

44 • Which sport has a ball with the longest de Broglie wavelength; baseball, with a ball weighing 5 oz and moving at 95 mi/h, or tennis, with a ball weighing 2 oz and moving at 130 mi/h?

45 • The energy of the electron beam in Davisson and Germer's experiment was 54 eV. Calculate the wavelength for these electrons.

46 • The distance between Li^+ and Cl^+ ions in a LiCl crystal is 0.257 nm. Find the energy of electrons that have a wavelength equal to this spacing.

47 • **SSM** An electron microscope uses electrons of energy 70 keV. Find the wavelength of these electrons.

48 • What is the de Broglie wavelength of a neutron with speed 10^6 m/s?

Wave–Particle Duality

49 • Suppose you have a spherical object of mass 4 g moving at 100 m/s. What size aperture is necessary for the object to show diffraction? Show that no common objects would be small enough to squeeze through such an aperture.

50 • A neutron has a kinetic energy of 10 MeV. What size object is necessary to observe neutron diffraction effects? Is there anything in nature of this size that could serve as a target to demonstrate the wave nature of 10-MeV neutrons?

51 • What is the de Broglie wavelength of an electron of kinetic energy 200 eV? What are some common targets that could demonstrate the wave nature of such an electron?

A Particle in a Box

52 •• **SSM** Use a spreadsheet program or graphing calculator to plot $\psi(x)$ and the probability distribution $\psi^2(x)$ of a particle in a box for the states $n = 1$, $n = 2$, and $n = 3$.

53 •• (a) Find the energy of the ground state ($n = 1$) and the first two excited states of a proton in a one-dimensional box of length $L = 10^{-15}$ m = 1 fm. (These are the order of magnitude of nuclear energies.) Make an energy-level diagram for this system, and calculate the wavelength of electromagnetic radiation emitted when the proton makes a transition from (b) $n = 2$ to $n = 1$, (c) $n = 3$ to $n = 2$, and (d) $n = 3$ to $n = 1$.

54 •• **SOLVE** (a) Find the energy of the ground state ($n = 1$) and the first two excited states of a proton in a one-dimensional box of length 0.2 nm (about the diameter of a H_2 molecule). Calculate the wavelength of electromagnetic radiation emitted when the proton makes a transition from (b) $n = 2$ to $n = 1$, (c) $n = 3$ to $n = 2$, and (d) $n = 3$ to $n = 1$.

*Calculating Probabilities and Expectation Values

55 •• **SOLVE** A particle is in the ground state of a box of length L. Find the probability of finding the particle in the interval $\Delta x = 0.002L$ at (a) $x = L/2$, (b) $x = 2L/3$, and (c) $x = L$. (Since Δx is very small, you need not do any integration because the wave function is slowly varying.)

56 •• **SSM** **SOLVE** Repeat Problem 55 for a particle in the first excited state ($n = 2$).

57 •• (a) Find $<x>$ for the second excited state ($n = 2$) for a particle in a box of length L and (b) find $<x^2>$.

58 •• A particle in a one-dimensional box is in the first excited state ($n = 2$). (a) Sketch $\psi^2(x)$ versus x for this state. (b) What is the expectation value $<x>$ for this state? (c) What is the probability of finding the particle in some small region dx centered at $x = L/2$? (d) Are your answers for Part (b) and Part (c) contradictory? If not, explain.

59 •• A particle of mass m has a wave function given by $\psi(x) = Ae^{-|x|/a}$, where A and a are constants. (a) Find the normalization constant A. (b) Calculate the probability of finding the particle in the region $-a \leq x \leq a$.

60 •• A one-dimensional box is on the x-axis in the region of $0 \leq x \leq L$. A particle in this box is in its ground state. Calculate the probability that the particle will be found in the region (a) $0 < x < \frac{1}{2}L$, (b) $0 < x < L/3$, and (c) $0 < x < 3L/4$.

61 •• Repeat Problem 60 for a particle in the first excited state of the box.

62 •• The classical probability distribution function for a particle in a box of length L is given by $P(x) = 1/L$. Use this to show that $<x> = L/2$ and $<x^2> = L^2/3$ for a classical particle in the box described in Problem 60.

63 •• (a) For the wave functions

$$\psi_n(x) = \sqrt{\frac{2}{L}} \sin \frac{n\pi x}{L}, \quad n = 1, 2, 3, \ldots$$

corresponding to a particle in the nth state of the box described in Problem 60, show that $<x> = L/2$ and $<x^2> = L^2/3 - L^2/2n^2\pi^2$. (b) Compare this result for $n \gg 1$ with your answer for the classical distribution of Problem 62.

64 •• SSM (a) Use a spreadsheet program or graphing calculator to plot the expectation value for position $<x>$ and the square of the position $<x^2>$ as a function of the quantum number n for the particle in the box described in Problem 60, for values of n from 1 to 100. Assume $L = 1$ m for your graph. Refer to Problem 63. (b) Comment on the significance of any asymptotic limits that your graph shows.

65 •• The wave functions for a particle of mass m in a one-dimensional box of length L centered at the origin (so that the ends are at $x = \pm L/2$) are given by

$$\psi(x) = \sqrt{\frac{2}{L}} \cos \frac{n\pi x}{L}, \qquad n = 1, 3, 5, 7, \ldots$$

and

$$\psi(x) = \sqrt{\frac{2}{L}} \sin \frac{n\pi x}{L}, \qquad n = 2, 4, 6, 8, \ldots$$

Calculate $<x>$ and $<x^2>$ for the ground state.

66 •• Calculate $<x>$ and $<x^2>$ for the first excited state of the box described in Problem 65.

General Problems

67 • SSM ISOLVE A light beam of wavelength 400 nm has an intensity of 100 W/m². (a) What is the energy of each photon in the beam? (b) How much energy strikes an area of 1 cm² perpendicular to the beam in 1 s? (c) How many photons strike this area in 1 s?

68 • ISOLVE A 1-μg particle is moving with a speed of approximately 10^{-1} cm/s in a box of length 1 cm. Treating this as a one-dimensional particle in a box, calculate the approximate value of the quantum number n.

69 • (a) For the classical particle of Problem 68, find Δx and Δp, assuming that these uncertainties are given by $\Delta x/L = 0.01$ percent and $\Delta p/p = 0.01$ percent. (b) What is $(\Delta x \Delta p)/\hbar$?

70 • ISOLVE In 1987, a laser at Los Alamos National Laboratory produced a flash that lasted 1×10^{-12} s and that had a power of 5×10^{15} W. Estimate the number of emitted photons if their wavelength was 400 nm.

71 • ISOLVE You cannot "see" anything smaller than the wavelength λ used. What is the minimum energy of an electron needed in an electron microscope to "see" an atom, which has a diameter of about 0.1 nm?

72 • ISOLVE A common flea that has a mass of 0.008 g can jump vertically as high as 20 cm. Estimate the de Broglie wavelength for the flea immediately after takeoff.

73 •• SSM ISOLVE ✓ Suppose that a 100-W source radiates light of wavelength 600 nm uniformly in all directions and that the eye can detect this light if only 20 photons per second enter a dark-adapted eye with a pupil 7 mm in diameter. How far from the source can the light be detected under these rather extreme conditions?

74 •• ISOLVE The diameter of the pupil of the eye under room-light conditions is approximately 5 mm. (It can vary from approximately 1 mm to 8 mm.) Find the intensity of light of wavelength 600 nm so that 1 photon per second passes through the pupil.

75 •• ISOLVE A lightbulb radiates 90 W uniformly in all directions. (a) Find the intensity at a distance of 1.5 m. (b) If the wavelength is 650 nm, find the number of photons per second that strike a surface of area 1 cm², which is oriented so that the line to the bulb is perpendicular to the surface.

76 •• When light of wavelength λ_1 is incident on the cathode of a photoelectric tube, the maximum kinetic energy of the emitted electrons is 1.8 eV. If the wavelength is reduced to $\lambda_1/2$, the maximum kinetic energy of the emitted electrons is 5.5 eV. Find the work function ϕ of the cathode material.

77 •• A photon of energy E undergoes Compton scattering at an angle of θ. Show that the energy E' of the scattered photon is given by

$$E' = \frac{E}{(E/m_e c^2)(1 - \cos \theta) + 1}$$

78 •• ISOLVE ✓ A particle is confined to a one-dimensional box. In making a transition from the state n to the state $n - 1$, radiation of 114.8 nm is emitted; in the transition from the state $n - 1$ to the state $n - 2$, radiation of wavelength 147 nm is emitted. The ground-state energy of the particle is 1.2 eV. Determine n.

79 •• SSM The Pauli exclusion principle states that no more than one electron may occupy a particular quantum state at a time. Therefore, if we wish to model an atom as a collection of electrons trapped in a one-dimensional box, each electron in the box must have a unique value of the quantum number n. Calculate the energy that the most energetic electron would have for the uranium atom with atomic number 92, assuming the box has a width of 0.05 nm. How does this energy compare to the rest-mass energy of the electron?

80 •• ISOLVE A beam of electrons, each with the same kinetic energy, illuminates a pair of slits separated by a distance $d = 54$ nm. The beam forms bright and dark fringes on a screen located a distance $L = 1.5$ m beyond the two slits. The arrangement is otherwise identical to that used in the optical two-slit interference experiment (described in Chapter 33 and in Figure 33-7) and the fringes have the appearance shown in Figure 34-18(d). The bright fringes are found to be separated by a distance of 0.68 mm. What is the kinetic energy of the electrons in the beam?

81 •• ISOLVE When a surface is illuminated with light of wavelength λ, the maximum kinetic energy of the emitted electrons is 1.2 eV. If the wavelength $\lambda' = 0.8\lambda$ is used, the maximum kinetic energy increases to 1.76 eV. For wavelength $\lambda' = 0.6\lambda$, the maximum kinetic energy of the emitted electrons is 2.676 eV. Determine the work function of the surface and the wavelength λ.

82 •• A simple pendulum of length 1 m has a bob of mass 0.3 kg. The energy of this oscillator is quantized to the values $E_n = (n + \frac{1}{2})hf_0$, where n is an integer and f_0 is the frequency of the pendulum. (a) Find n if the angular amplitude is 10°. (b) Find n if the energy changes by 0.01 percent.

83 •• SSM (a) Show that for large n, the fractional difference in energy between state n and state $n + 1$ for a particle in a one-dimensional box is given approximately by

$$(E_{n+1} - E_n)/E_n \approx 2/n$$

(b) What is the approximate percentage energy difference between the states $n_1 = 1000$ and $n_2 = 1001$? (c) Comment on how this result is related to Bohr's correspondence principle.

84 •• A mode-locked, titanium–sapphire laser has a wavelength of 850 nm and produces 100 million pulses of light each second. Each pulse has a duration of 125 femtoseconds (1 fs $= 10^{-15}$ s) and contains 5×10^9 photons. What is the average power produced by the laser?

85 •• i SOLVE This problem is one of estimating the time lag (expected classically but not observed) in the photoelectric effect. Let the intensity of the incident radiation be 0.01 W/m². (a) If the area of the atom is 0.01 nm², find the energy per second falling on an atom. (b) If the work function is 2 eV, how long would it take classically for this much energy to fall on one atom?

Applications of the Schrödinger Equation

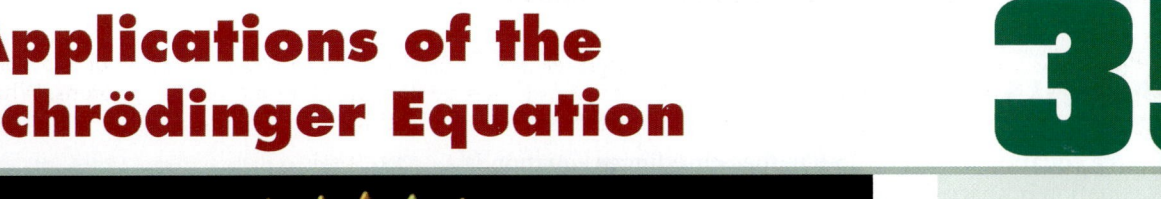

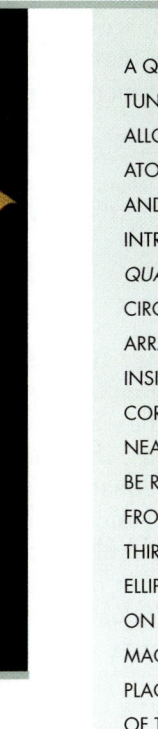

A QUANTUM MIRAGE. THE SCANNING TUNNELING MICROSCOPE (STM) ALLOWS ONE TO PUSH INDIVIDUAL ATOMS AROUND ON A SURFACE AND TO IMAGE THEM. ESPECIALLY INTRIGUING ARE IMAGES OF *QUANTUM CORRALS*, WHICH ARE CIRCULAR OR ELLIPTICAL ARRANGEMENTS ON A SURFACE INSIDE OF WHICH THE WAVES CORRESPONDING TO ELECTRONS NEAR THE SUBSTRATE SURFACE CAN BE REVEALED. THIS IMAGE COMES FROM IBM, WHERE PHYSICISTS PLACED THIRTY-SIX COBALT ATOMS IN AN ELLIPTICAL "STONEHENGE" PATTERN ON A COPPER SURFACE. AN EXTRA MAGNETIC COBALT ATOM WAS PLACED AT ONE OF THE TWO FOCI OF THE ELLIPSE, CAUSING VISIBLE INTERACTIONS WITH THE SURFACE ELECTRON WAVES. BUT THE WAVES ALSO SEEM TO BE INTERACTING WITH A PHANTOM COBALT ATOM AT THE OTHER FOCUS, AN ATOM THAT IS NOT REALLY THERE.

? **Could the phantom cobalt atom described above be caused by reflections of waves from the corral of cobalt atoms? The reflection and transmissions of one-dimensional waves is discussed in Section 35-4.**

I n Chapter 34, we found that electrons and other particles have wave properties and are described by a wave function $\Psi(x, t)$. The probability of finding the particle in some region of space is proportional to the square of the wave function. We mentioned that the wave function is a solution of the Schrödinger equation, and we discussed some solutions qualitatively without reference to the equation itself. In particular, we showed how the standing-wave conditions lead to quantization of energy for a particle confined to a one-dimensional box.

➤ **This chapter is a continuation of the material introduced in Chapter 34. We discuss the Schrödinger equation and apply the equation to the particle in the box problem and to several other situations in which a particle is confined to a region of space to illustrate how boundary conditions lead to energy quantization. We then show how the Schrödinger equation leads to barrier penetration and discuss the extension of the Schrödinger equation to more than one dimension and to more than one particle.**

35-1 The Schrödinger Equation

Like the classical wave equation (Equation 15-9b), the Schrödinger equation is a partial differential equation in space and time. Like Newton's laws of motion, the Schrödinger equation cannot be derived. Its validity, like that of Newton's laws, lies in its agreement with experiment. In one dimension, the Schrödinger equation is[†]

$$-\frac{\hbar^2}{2m}\frac{\partial^2 \Psi(x,t)}{\partial x^2} + U\Psi(x,t) = i\hbar\frac{\partial \Psi(x,t)}{\partial t} \qquad 35\text{-}1$$

TIME-DEPENDENT SCHRÖDINGER EQUATION

where U is the potential energy function. Equation 35-1 is called the **time-dependent Schrödinger equation.** Unlike the classical wave equation, it relates the second space derivative of the wave function to the *first time* derivative of the wave function, and it contains the imaginary number $i = \sqrt{-1}$. The wave functions that are solutions of this equation are not necessarily real. $\Psi(x, t)$ is not a measurable function like the classical wave functions for sound or electromagnetic waves. The probability of finding a particle in some region of space dx is certainly real though, so we must modify slightly the equation for probability density given in Chapter 34 (Equation 34-17). We take for the probability of finding a particle in some region dx

$$P(x, t)\, dx = |\Psi(x, t)|^2\, dx = \Psi^*\Psi\, dx \qquad 35\text{-}2$$

where Ψ^*, the complex conjugate of Ψ, is obtained from Ψ by replacing i by $-i$ wherever it appears.[‡]

In classical mechanics, the standing-wave solutions to the wave equation (Equation 16-16) are of great interest and value. Not surprisingly, standing-wave solutions to the Schrödinger wave equation are also of great interest and value. The wave function for the standing-wave motion of a uniform taut string is $A\sin(kx)\cos(\omega t + \delta)$, and this is representative of all standing waves. A standing wave function can always be expressed as a function of position multiplied by a function of time, where the function of time is one that varies sinusoidally with time. Standing-wave solutions to the one-dimensional Schrödinger wave equation are thus expressed

$$\Psi(x, t) = \psi(x)e^{-i\omega t} \qquad 35\text{-}3$$

where $e^{-i\omega t} = \cos(\omega t) - i\sin(\omega t)$. [In Appendix D, it is shown that $e^{-i\omega t} = \cos(\omega t) - i\sin(\omega t)$.] The right side of Equation 35-1 is then

$$i\hbar\frac{\partial \Psi(x,t)}{\partial t} = i\hbar(-i\omega)\psi(x)e^{-i\omega t} = \hbar\omega\psi(x)e^{-i\omega t} = E\psi(x)e^{-i\omega t}$$

where $E = \hbar\omega$ is the energy of the particle.

The Schrödinger wave equation has standing-wave solutions only if the potential energy function depends upon position alone. Substituting $\psi(x)e^{-i\omega t}$ into Equation 35-1 and canceling the common factor $e^{-i\omega t}$, we obtain an equation for $\psi(x)$, called the **time-independent Schrödinger equation:**

[†] Although we simply state the Schrödinger equation, Schrödinger himself had a vast knowledge of classical wave theory that led him to this equation.

[‡] Every complex number can be written in the form $z = a + bi$, where a and b are real numbers and $i = \sqrt{-1}$. The complex conjugate of z is $z^* = a - bi$, so $z^*z = (a + bi)(a - bi) = a^2 + b^2 = |z|^2$. Complex numbers are discussed more fully in Appendix D.

$$-\frac{\hbar^2}{2m}\frac{d^2\psi(x)}{dx^2} + U(x)\psi(x) = E\psi(x) \qquad 35\text{-}4$$

TIME-INDEPENDENT SCHRÖDINGER EQUATION

where we have written U as $U(x)$ to emphasize that while U may depend on position, U does not depend on time. The function $U(x)$ represents the environment of the particle being described. It is this potential energy function in the Schrödinger equation that establishes the difference between different problems, just as the expressions for forces acting on a particle play in classical physics.

The calculation of the allowed energy levels in a system involves only the time-independent Schrödinger equation, whereas finding the probabilities of transition between these levels requires the solution of the time-dependent equation. In this book, we will be concerned only with the time-independent Schrödinger equation.

The solution of Equation 35-4 depends on the form of the potential energy function $U(x)$. When $U(x)$ is such that the particle is confined to some region of space, only certain discrete energies E_n give solutions ψ_n that can satisfy the normalization condition (Equation 34-18):

$$\int_{-\infty}^{\infty} |\psi|^2 \, dx = 1$$

The complete time-dependent wave functions are then given, from Equation 35-3, by

$$\Psi_n(x, t) = \psi_n(x)e^{-i\omega_n t} = \psi_n(x)e^{-i(E_n/\hbar)t} \qquad 35\text{-}5$$

A Particle in an Infinite Square-Well Potential

We will illustrate the use of the time-independent Schrödinger equation by solving it for the problem of a particle in a box. The potential energy for a one-dimensional box from $x = 0$ to $x = L$ is shown in Figure 35-1. It is called an **infinite square-well potential** and is described mathematically by

$$U(x) = 0, \quad 0 < x < L$$

$$U(x) = \infty, \quad x < 0 \text{ or } x > L \qquad 35\text{-}6$$

Inside the box, the potential energy is zero, whereas outside the box it is infinite. Since we require the particle to be in the box, we have $\psi(x) = 0$ everywhere outside the box. We then need to solve the Schrödinger equation inside the box subject to the condition that $\psi(x)$ must be zero at $x = 0$ and at $x = L$.

Inside the box $U(x) = 0$, so the Schrödinger equation is written

$$-\frac{\hbar^2}{2m}\frac{d^2\psi(x)}{dx^2} = E\psi(x)$$

or

$$\frac{d^2\psi(x)}{dx^2} + k^2\psi(x) = 0 \qquad 35\text{-}7$$

where

$$k^2 = \frac{2mE}{\hbar^2} \qquad 35\text{-}8$$

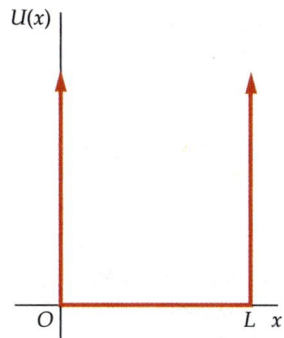

FIGURE 35-1 The infinite square-well potential energy. For $x < 0$ and $x > L$, the potential energy $U(x)$ is infinite. The particle is confined to the region in the well $0 < x < L$.

The general solution of Equation 35-7 can be written as

$$\psi(x) = A \sin kx + B \cos kx \qquad\qquad 35\text{-}9$$

where A and B are constants. At $x = 0$, we have

$$\psi(0) = A \sin(k0) + B \cos(k0) = 0 + B$$

The boundary condition $\psi(x) = 0$ at $x = 0$ thus gives $B = 0$, and Equation 35-9 becomes

$$\psi(x) = A \sin kx \qquad\qquad 35\text{-}10$$

The wave function is thus a sine wave with the wavelength λ related to the wave number k in the usual way, $\lambda = 2\pi/k$. The boundary condition $\psi(x) = 0$ at $x = L$ restricts the possible values of k and therefore the values of the wavelength λ, and (from Equation 35-8) the energy $E = \hbar^2 k^2/2m$. We have

$$\psi(L) = A \sin kL = 0 \qquad\qquad 35\text{-}11$$

This condition is satisfied if kL is π or any integer times π, that is, if k is restricted to the values k_n given by

$$k_n = n\frac{\pi}{L}, \quad n = 1, 2, 3, \ldots \qquad\qquad 35\text{-}12$$

The condition (Equation 35-11) is also satisfied for $n = 0$. The function $\psi(x) = A \sin 0 = 0$ for all values of x is a solution to the wave equation. However, this solution is rejected as a wave function on physical grounds. It cannot be normalized and cannot be a wave function for a particle. Substituting this result into Equation 35-8 and solving for E gives us the allowed energy values:

$$E_n = \frac{\hbar^2 k_n^2}{2m} = \frac{\hbar^2}{2m}\left(n\frac{\pi}{L}\right)^2 = n^2\left(\frac{h^2}{8mL^2}\right) = n^2 E_1 \qquad\qquad 35\text{-}13$$

where

$$E_1 = \frac{h^2}{8mL^2} \qquad\qquad 35\text{-}14$$

Equation 35-14 is the same as Equation 34-22, which we obtained by fitting an integral number of half-wavelengths into the box.

For each value of n, there is wave function $\psi_n(x)$ given by

$$\psi_n(x) = A_n \sin\frac{n\pi x}{L} \qquad\qquad 35\text{-}15$$

which is the same as Equation 34-25 with the constant $A_n = \sqrt{2/L}$ determined by normalization.[†]

35-2 A Particle in a Finite Square Well

The quantization of energy that we found for a particle in an infinite square well is a result that follows from the general solution of the Schrödinger equation for any particle confined to some region of space. We will illustrate this by considering the qualitative behavior of the wave function for a slightly more general potential energy function, the finite square well, which is shown in Figure 35-2.

---

† See Equation 34-18.

This potential energy function is described mathematically by

$$U(x) = U_0, \quad x < 0$$

$$U(x) = 0, \quad 0 < x < L \qquad\qquad 35\text{-}16$$

$$U(x) = U_0, \quad x > L$$

This potential energy function is discontinuous at $x = 0$ and $x = L$, but it is finite everywhere. The solutions of the Schrödinger equation for this type of potential energy function depend on whether the total energy E is greater or less than U_0. We will not discuss the case of $E > U_0$, except to remark that in that case the particle is not confined and any value of the energy is allowed; that is, there is no energy quantization. Here we assume that $0 \le E < U_0$.

Inside the well, $U(x) = 0$, and the time-independent Schrödinger equation is the same as for the infinite well (Equation 35-7):

$$-\frac{\hbar^2}{2m}\frac{d^2\psi(x)}{dx^2} = E\psi(x)$$

or

$$\frac{d^2\psi(x)}{dx^2} + k^2\psi(x) = 0$$

where $k^2 = 2mE/\hbar^2$. The general solution is of the form

$$\psi(x) = A \sin kx + B \cos kx$$

In this case, $\psi(x)$ is not required to be zero at $x = 0$, so B is not zero. Outside the well, the time-independent Schrödinger equation is

$$-\frac{\hbar^2}{2m}\frac{d^2\psi(x)}{dx^2} + U_0\psi(x) = E\psi(x)$$

or

$$\frac{d^2\psi(x)}{dx^2} - \alpha^2\psi(x) = 0 \qquad \Rightarrow \quad \frac{d^2\psi(x)}{dx^2} = \alpha^2\psi(x) \qquad\qquad 35\text{-}17$$

where

$$\alpha^2 = \frac{2m}{\hbar^2}(U_0 - E) > 0 \qquad\qquad 35\text{-}18$$

The wave functions and allowed energies for the particle can be found by solving Equation 35-17 for $\psi(x)$ outside the well and then requiring that $\psi(x)$ and $d\psi(x)/dx$ be continuous at the boundaries $x = 0$ and $x = L$. The solution of Equation 35-17 is not difficult (for positive values of x, it is of the form $\psi(x) = \psi(x) = Ce^{-\alpha x}$), but applying the boundary conditions involves much tedious algebra and is not important for our purpose. The important feature of Equation 35-17 is that the second derivative of $\psi(x)$, which is related to the concavity of the wave function, has the same sign as the wave function ψ. If ψ is positive, $d^2\psi/dx^2$ is also positive and the wave function curves away from the axis, as shown in Figure 35-3a. Similarly, if ψ is negative, $d^2\psi/dx^2$ is negative and ψ again curves away from the axis, as shown in Figure 35-3b. This behavior is

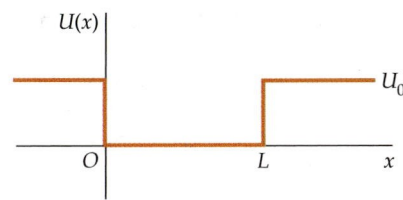

FIGURE 35-2 The finite square-well potential energy.

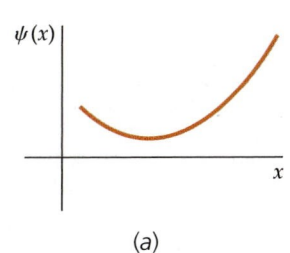

(a)

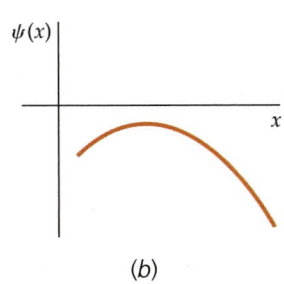

(b)

FIGURE 35-3 (a) A positive function with positive concavity. (b) A negative function with negative concavity.

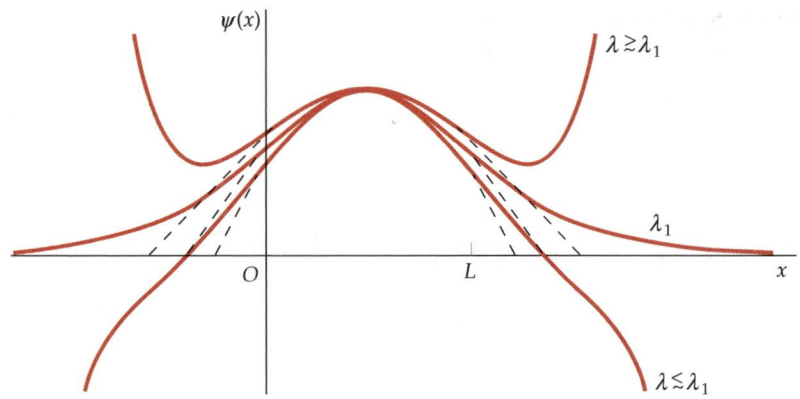

FIGURE 35-4 Functions satisfying the Schrödinger equation with wavelengths near the wavelength λ_1 corresponding to the ground-state energy $E_1 = \hbar^2/2m\lambda_1^2$ in the finite well. If λ is slightly greater than λ_1, the function approaches infinity, like the function in Figure 35-3a. At the critical wavelength λ_1, the function and its slope approach zero together. If λ is slightly less than λ_1, the function crosses the x axis while the slope is still negative. The slope then becomes more negative because its rate of change $d^2\psi/dx^2$ is now negative. This function approaches negative infinity as x approaches infinity.

very different from the behavior inside the well, where ψ and $d^2\psi/dx^2$ have opposite signs so that ψ always curves toward the axis like a sine or cosine function. Because of this behavior outside the well, for most values of the energy E in Equation 35-17, $\psi(x)$ becomes infinite as x approaches $\pm\infty$; that is, most wave functions $\psi(x)$ are not well behaved outside the well. Though they satisfy the Schrödinger equation, such functions are not proper wave functions because they cannot be normalized. The solutions of the Schrödinger equation are well behaved (i.e., they approach 0 as $|x|$ becomes very large) only for certain values of the energy. These energy values are the allowed energies for the finite square well.

Figure 35-4 shows a well-behaved wave function, with a wavelength λ_1 inside the well corresponding to the ground-state energy. The behavior of the wave functions corresponding to nearby wavelengths and energies is also shown. Figure 35-5 shows the wave functions and probability distributions for the ground state and first two excited states. From this figure, we can see that the wavelengths inside the well are slightly longer than the corresponding wavelengths for the infinite well (Figure 34-14), so the corresponding energies are slightly less than those for the infinite well. Another feature of the finite-well problem is that there are only a finite number of allowed energies. For very small values of U_0, there is only one allowed energy.

Note that the wave function penetrates beyond the edges of the well at $x = L$ and $x = 0$, indicating that there is some small probability of finding the particle in the region in which its total energy E is less than its potential energy U_0. This region is called the *classically forbidden region* because the kinetic energy, $E - U_0$, would be negative when $U_0 > E$. Since negative kinetic energy has no meaning in classical physics, it is interesting to speculate on the result of an attempt to observe the particle in the classically forbidden region. It can be shown from the uncertainty principle that if an attempt is made to localize the particle in the classically forbidden region, such a measurement introduces an uncertainty in the momentum of the particle corresponding to a minimum kinetic energy that is greater than $U_0 - E$. This is just great enough to prevent us from measuring a negative kinetic energy. The penetration of the wave function into a classically forbidden region does have important consequences in barrier penetration, which will be discussed in Section 35-4.

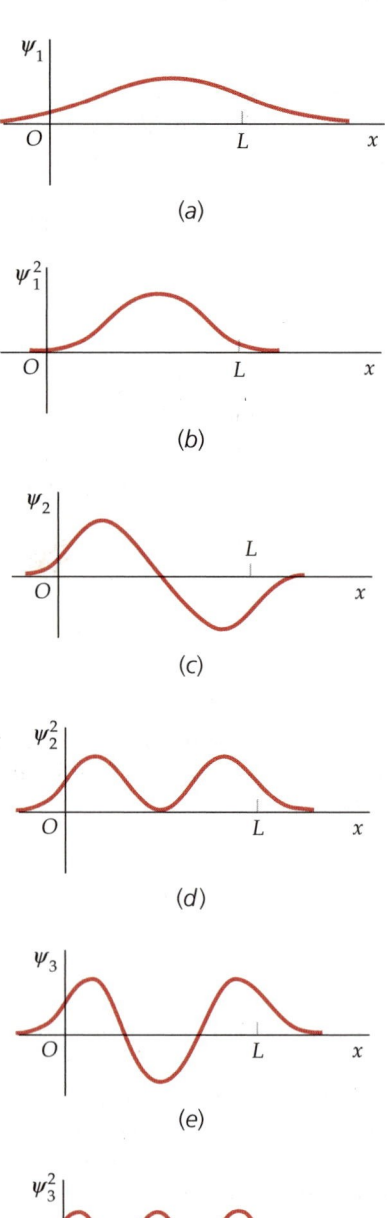

FIGURE 35-5 Graphs of the wave functions $\psi_n(x)$ and probability distributions $\psi_n^2(x)$ for $n = 1$, $n = 2$, and $n = 3$ for the finite square well. Compare these graphs with those of Figure 34-14 for the infinite square well, where the wave functions are zero at $x = 0$ and $x = L$. The wavelengths here are slightly longer than the corresponding wavelengths for the infinite well, so the allowed energies are somewhat smaller.

Much of our discussion of the finite-well problem applies to any problem in which $E > U(x)$ in some region and $E < U(x)$ outside that region, as we see in the next section.

35-3 The Harmonic Oscillator

The potential energy for a particle of mass m attached to a spring of force constant k is

$$U(x) = \tfrac{1}{2}kx^2 = \tfrac{1}{2}m\omega_0^2 x^2 \qquad\qquad 35\text{-}19$$

where $\omega_0 = \sqrt{k/m}$ is the natural frequency of the oscillator. Classically, the object oscillates between $x = +A$ and $x = -A$. The object's total energy is $E = \tfrac{1}{2}m\omega_0^2 A^2$, which can have any positive value or zero.

This potential energy function, shown in Figure 35-6, applies to virtually any system undergoing small oscillations about a position of stable equilibrium. For example, it could apply to the oscillations of the atoms of a diatomic molecule, such as H_2 or HCl, oscillating about their equilibrium separation. Between the classical turning points ($|x| < A$), the total energy is greater than the potential energy, and the Schrödinger equation can be written

$$\frac{d^2\psi(x)}{dx^2} = -k^2\psi(x) \qquad\qquad 35\text{-}20$$

where $k^2 = (2m/\hbar^2)[E - U(x)]$ now depends on x. The solutions of this equation are no longer simple sine or cosine functions because the wave number $k = 2\pi/\lambda$ now varies with x; but since $d^2\psi/dx^2$ and ψ have opposite signs, ψ will always curve toward the axis and the solutions will oscillate.

Outside the classical turning points ($|x| > A$), the potential energy is greater than the total energy and the Schrödinger equation is similar to Equation 35-17:

$$\frac{d^2\psi(x)}{dx^2} = +\alpha^2\psi(x) \qquad\qquad 35\text{-}21$$

except that here $\alpha^2 = (2m/\hbar^2)[U(x) - E] > 0$ depends on x. For $|x| > A$, $d^2\psi/dx^2$ and ψ have the same sign, so ψ will curve away from the axis and there will be only certain values of E for which solutions exist that approach zero as x approaches infinity.

For the harmonic oscillator potential energy function, the Schrödinger equation is

$$-\frac{\hbar^2}{2m}\frac{d^2\psi(x)}{dx^2} + \tfrac{1}{2}m\omega_0^2 x^2\psi(x) = E\psi(x) \qquad\qquad 35\text{-}22$$

[handwritten annotations:]
if $|x| \geq A$ $\tfrac{1}{2}m\omega_0^2 x^2\psi(x) \geq \tfrac{1}{2}m\omega_0^2 A^2\psi(x)$

$\therefore \ \frac{-\hbar^2}{2m}\frac{d^2\psi(x)}{dx^2} + \tfrac{1}{2}m\omega_0 x^2\psi(x) = \tfrac{1}{2}\omega_0^2 A^2\psi(x)$

$\Rightarrow \ \frac{-\hbar^2}{2m}\frac{d^2\psi(x)}{dx^2} = -CE\psi(x)$

$c = $ constant $\quad \frac{d^2\psi(x)}{dx^2}$ and $\psi(x)$ have same sign

Wave Functions and Energy Levels

Rather than pursue a general solution to the Schrödinger equation for this system, we simply present the solution for the ground state and the first excited state.

The ground-state wave function $\psi_0(x)$ is found to be a Gaussian function centered at the origin:

$$\psi_0(x) = A_0 e^{-ax^2} \qquad\qquad 35\text{-}23$$

where A_0 and a are constants. This function and the wave function for the first excited state are shown in Figure 35-7.

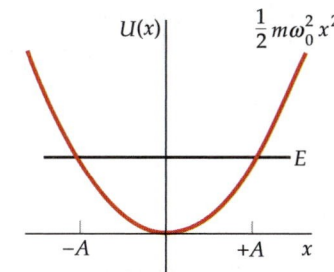

FIGURE 35-6 Harmonic oscillator potential.

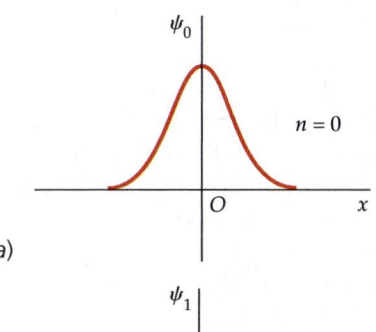

(a)

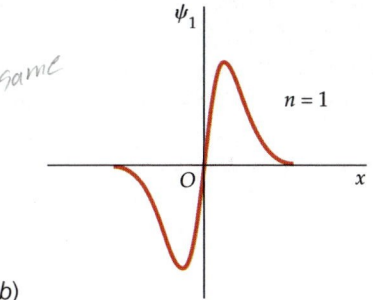

(b)

FIGURE 35-7 (a) The ground-state wave function for the harmonic oscillator potential. (b) The wave function for the first excited state of the harmonic oscillator potential.

VERIFYING THE GROUND-STATE WAVE FUNCTION **EXAMPLE 35-1**

Verify that $\psi_0(x) = A_0 e^{-ax^2}$, where a is a positive constant, is a solution of the Schrödinger equation for the harmonic oscillator.

PICTURE THE PROBLEM We take the first and second derivative of ψ with respect to x and substitute into Equation 35-22. Since this is the ground-state wave function, we write E_0 for the energy E.

1. Compute $d\psi_0/dx$:
$$\frac{d\psi_0(x)}{dx} = \frac{d}{dx}(A_0 e^{-ax^2}) = -2ax A_0 e^{-ax^2}$$

2. Compute $d^2\psi_0/dx^2$:
$$\frac{d^2\psi_0(x)}{dx^2} = -2a A_0 e^{-ax^2} + 4a^2 x^2 A_0 e^{-ax^2}$$

3. Substitute these derivatives into the Schrödinger equation:
$$-\frac{\hbar^2}{2m}\frac{d^2\psi(x)}{dx^2} + \frac{1}{2}m\omega_0^2 x^2 \psi(x) = E\psi(x)$$

$$-\frac{\hbar^2}{2m}(-2a A_0 e^{-ax^2} + 4a^2 x^2 A_0 e^{-ax^2}) + \frac{1}{2}m\omega_0^2 x^2 A_0 e^{-ax^2} = E_0 A_0 e^{-ax^2}$$

4. Cancel the common factor $A_0 e^{-ax^2}$ and show the result in standard polynomial form:
$$-\frac{\hbar^2}{2m}(-2a + 4a^2 x^2) + \frac{1}{2}m\omega_0^2 x^2 = E_0$$

so
$$\left(-\frac{2\hbar^2 a^2}{m} + \frac{1}{2}m\omega_0^2\right)x^2 + \left(\frac{\hbar^2 a}{m} - E_0\right) = 0$$

5. The equation in step 4 must hold for all x. Set $x = 0$ and solve for E_0:
$$0 + \left(\frac{\hbar^2 a}{m} - E_0\right) = 0$$

so
$$E_0 = \frac{\hbar^2 a}{m}$$

6. Substitute this result for E_0 into the equation in step 4 and simplify:
$$\left(-\frac{2\hbar^2 a^2}{m} + \frac{1}{2}m\omega_0^2\right)x^2 + 0 = 0$$

7. It follows that the coefficient of x^2 must equal zero:
$$-\frac{2\hbar^2 a^2}{m} + \frac{1}{2}m\omega_0^2 = 0$$

8. Solve for a:
$$a = \frac{m\omega_0}{2\hbar}$$

9. Substitute this result into the equation for E_0 in step 5:
$$E_0 = \frac{\hbar^2 a}{m} = \frac{1}{2}\hbar\omega_0$$

> We have shown that the given function satisfies the Schrödinger equation for any value of A_0, as long as the energy is given by $E_0 = \frac{1}{2}\hbar\omega_0$.

REMARKS The step 4 equation is a polynomial that is equal to zero. A theorem that would have simplified the solution is: If a polynomial is equal to zero over a continuous range of values of x, then each of the polynomial coefficients is equal to zero. For example, if $Ax^3 + Bx^2 + Cx + D = 0$ on the interval $1 < x < 2$, then $A = B = C = D = 0$. The proof of this result is the topic of Problem 43.

We see from this example that the ground-state energy is given by

$$E_0 = \frac{\hbar^2 a}{m} = \frac{1}{2}\hbar\omega_0 \qquad\qquad 35\text{-}24$$

The first excited state has a node in the center of the potential well, just as with the particle in a box.[†] The wave function $\psi_1(x)$ is

$$\psi_1(x) = A_1 x e^{-ax^2} \qquad\qquad 35\text{-}25$$

where $a = m\omega_0/2\hbar$, as in Example 35-1. This function is also shown in Figure 35-7. Substituting $\psi_1(x)$ into the Schrödinger equation, as was done for $\psi_0(x)$ in Example 35-1, yields the energy of the first excited state,

$$E_1 = \tfrac{3}{2}\hbar\omega_0$$

In general, the energy of the nth excited state of the harmonic oscillator is

$$E_n = (n + \tfrac{1}{2})\hbar\omega_0, \quad n = 0, 1, 2, \ldots \qquad 35\text{-}26$$

as indicated in Figure 35-8. The fact that the energy levels are evenly spaced by the amount $\hbar\omega_0$ is a peculiarity of the harmonic oscillator potential. As we saw in Chapter 34, the energy levels for a particle in a box, or for the hydrogen atom, are not evenly spaced. The precise spacing of energy levels is closely tied to the particular form of the potential energy function.

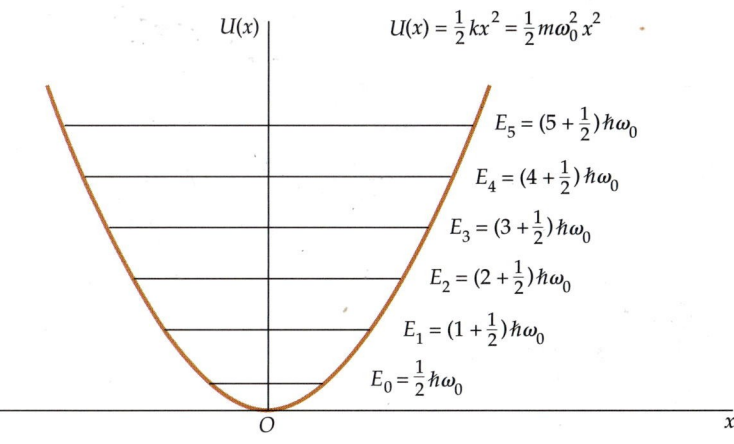

FIGURE 35-8 Energy levels in the harmonic oscillator potential.

35-4 Reflection and Transmission of Electron Waves: Barrier Penetration

In Sections 35-2 and 35-3, we were concerned with bound-state problems in which the potential energy is larger than the total energy for large values of $|x|$. In this section, we consider some simple examples of unbound states for which E is greater than $U(x)$. For these problems, $d^2\psi/dx^2$ and ψ have opposite signs, so $\psi(x)$ curves toward the axis and does not become infinite at large values of $|x|$.

Step Potential

Consider a particle of energy E moving in a region in which the potential energy is the step function

$$U(x) = 0, \quad x < 0$$

$$U(x) = U_0, \quad x > 0$$

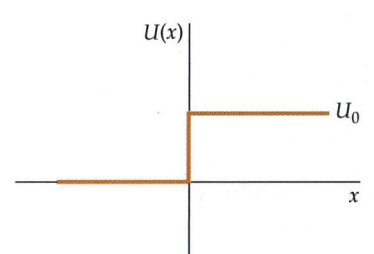

FIGURE 35-9 Step potential. A classical particle incident from the left, with total energy $E > U_0$, is always transmitted. The change in potential energy at $x = 0$ merely provides an impulsive force that reduces the speed of the particle. A wave incident from the left is partially transmitted and partially reflected because the wavelength changes abruptly at $x = 0$.

as shown in Figure 35-9. We are interested in what happens when a particle moving from left to right encounters the step.

The classical answer is simple. To the left of the step, the particle moves with a speed $v = \sqrt{2E/m}$. At $x = 0$, an impulsive force acts on the particle. If the initial energy E is less than U_0, the particle will be turned around and will then move to the left at its original speed; that is, the particle will be reflected by the step. If E is greater than U_0, the particle will continue to move to the right but with reduced speed given by $v = \sqrt{2(E - U_0)/m}$. We can picture this classical problem as a ball rolling along a level surface and coming to a steep hill of height h given by $mgh = U_0$. If the initial kinetic energy of the ball is less than

---

† Each higher-energy state has one additional node in the wave function.

mgh, the ball will roll part way up the hill and then back down and to the left along the lower surface at its original speed. If E is greater than *mgh*, the ball will roll up the hill and proceed to the right at a lesser speed.

The quantum-mechanical result is similar when E is less than U_0. Figure 35-10 shows the wave function for the case $E < U_0$. The wave function does not go to zero at $x = 0$ but rather decays exponentially, like the wave function for the bound state in a finite square-well problem. The wave penetrates slightly into the classically forbidden region $x > 0$, but it is eventually completely reflected. This problem is somewhat similar to that of total internal reflection in optics.

For $E > U_0$, the quantum-mechanical result differs markedly from the classical result. At $x = 0$, the wavelength changes abruptly from $\lambda_1 = h/p_1 = h/\sqrt{2mE}$ to $\lambda_2 = h/p_2 = h/\sqrt{2m(E - U_0)}$. We know from our study of waves that when the wavelength changes suddenly, part of the wave is reflected and part of the wave is transmitted. Since the motion of an electron (or other particle) is governed by a wave equation, the electron sometimes will be transmitted and sometimes will be reflected. The probabilities of reflection and transmission can be calculated by solving the Schrödinger equation in each region of space and comparing the amplitudes of the transmitted waves and reflected waves with that of the incident wave. This calculation and its result are similar to finding the fraction of light reflected from an air–glass interface. If R is the probability of reflection, called the reflection coefficient, this calculation gives

$$R = \frac{(k_1 - k_2)^2}{(k_1 + k_2)^2} \qquad 35\text{-}27$$

where k_1 is the wave number for the incident wave and k_2 is the wave number for the transmitted wave. This result is the same as the result in optics for the reflection of light at normal incidence from the boundary between two media having different indexes of refraction n (Equation 31-11). The probability of transmission T, called the **transmission coefficient,** can be calculated from the reflection coefficient, since the probability of transmission plus the probability of reflection must equal 1:

$$T + R = 1 \qquad 35\text{-}28$$

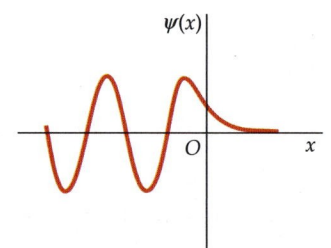

FIGURE 35-10 When the total energy E is less than U_0, the wave function penetrates slightly into the region $x > 0$. However, the probability of reflection for this case is 1, so no energy is transmitted.

REFLECTION AND TRANSMISSION AT A STEP BARRIER **EXAMPLE 35-2**

A particle of energy E_0 traveling in a region in which the potential energy is zero is incident on a potential barrier of height $U_0 = 0.2E_0$. Find the probability that the particle will be reflected.

PICTURE THE PROBLEM We need to calculate the wave numbers k_1 and k_2 and use them to calculate the reflection coefficient R from Equation 35-27. The wave numbers are related to the kinetic energy K by $K = p^2/2m = \hbar^2 k^2/2m$.

1. The probability of reflection is the reflection coefficient:
$$R = \frac{(k_1 - k_2)^2}{(k_1 + k_2)^2}$$

2. Calculate k_1 from the initial kinetic energy E_0:
$$E_0 = \frac{\hbar^2 k_1^2}{2m}$$
$$k_1 = \sqrt{2mE_0/\hbar^2} = 1.41\sqrt{mE_0/\hbar^2}$$

3. Relate k_2 to the final kinetic energy K_2:
$$\frac{\hbar k_2^2}{2m} = K_2 = E_0 - U_0 = E_0 - 0.2E_0 = 0.8E_0$$

4. Solve for k_2:

$$k_2 = \sqrt{2m(0.8E_0)/\hbar^2} = 1.26\sqrt{mE_0/\hbar^2}$$

5. Substitute these values into Equation 35-27 to calculate R:

$$R = \frac{(k_1 - k_2)^2}{(k_1 + k_2)^2} = \frac{(1.41 - 1.26)^2}{(1.41 + 1.26)^2} = \boxed{0.00316}$$

REMARKS The probability of reflection is only 0.3 percent. This probability is small because the barrier height reduces the kinetic energy by only 20 percent. Since k is proportional to the square root of the kinetic energy, the wave number and therefore the wavelength is changed by only 10 percent.

 EXERCISE Express the index of refraction n of light in terms of the wave number k, and show that Equation 31-11 for the reflection of light at normal incidence is the same as Equation 35-27.

In quantum mechanics, a localized particle is represented by a wave packet, which has a maximum at the most probable position of the particle. Figure 35-11 shows a wave packet representing a particle of energy E incident on a step potential of height U_0, which is less than E. After the encounter, there are two wave packets. The relative heights of the transmitted packet and reflected packet indicate the relative probabilities of transmission and reflection. For the situation shown here, E is much greater than U_0, and the probability of transmission is much greater than that of reflection.

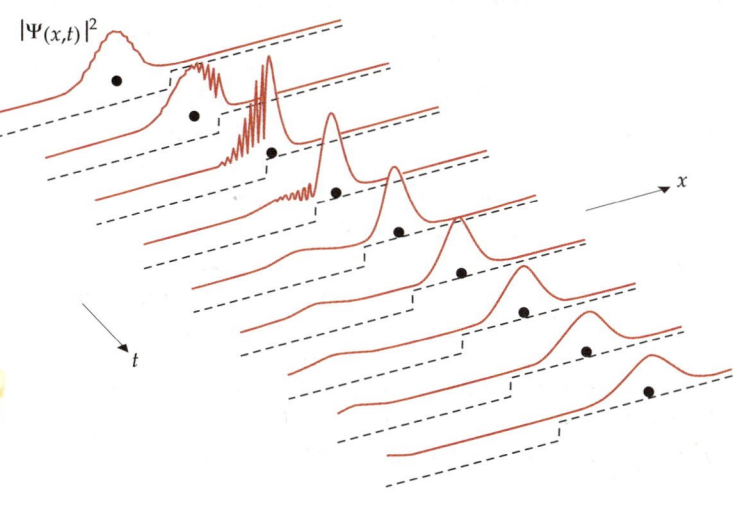

FIGURE 35-11 Time development of a one-dimensional wave packet representing a particle incident on a step potential for $E > U_0$. The position of a classical particle is indicated by the dot. Note that part of the packet is transmitted and part is reflected.

Barrier Penetration

Figure 35-12a shows a rectangular potential barrier of height U_0 and width a given by

$$U(x) = 0, \quad x < 0$$

$$U(x) = U_0, \quad 0 < x < a$$

$$U(x) = 0, \quad x > a$$

We consider a particle of energy E, which is slightly less than U_0, that is incident on the barrier from the left. Classically, the particle would always be reflected. However, a wave incident from the left does not decrease immediately to zero at the barrier, but it will instead decay exponentially in the classically forbidden region $0 < x < a$. Upon reaching the far wall of the barrier ($x = a$), the wave function must join smoothly to a sinusoidal wave function to the right of the barrier, as shown in Figure 35-12b. This implies that there is some probability of the particle (which is represented by the wave function) being found on the far side of the barrier even though, classically, it should never pass through the barrier. For the case in which the quantity $\alpha a = \sqrt{2ma^2(U_0 - E)/\hbar^2}$ is much greater than 1, the transmission coefficient is proportional to $e^{-2\alpha a}$:

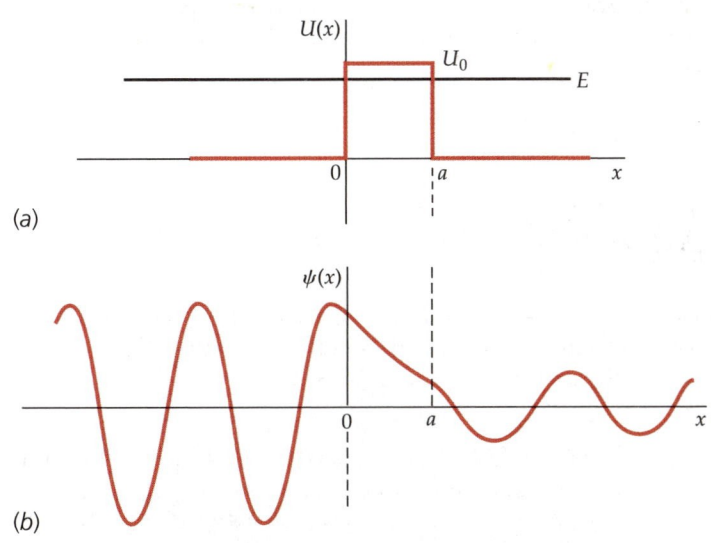

FIGURE 35-12 (a) A rectangular potential barrier. (b) The penetration of the barrier by a wave with total energy less than the barrier energy. Part of the wave is transmitted by the barrier even though, classically, the particle cannot enter the region $0 < x < a$ in which the potential energy is greater than the total energy. To the left of the barrier, there is both an incident and a reflected wave. These waves form a resultant wave so that ψ is a superposition of a standing wave and a traveling wave (traveling toward the barrier). To the right of the barrier is only the transmitted wave that is traveling away from the barrier.

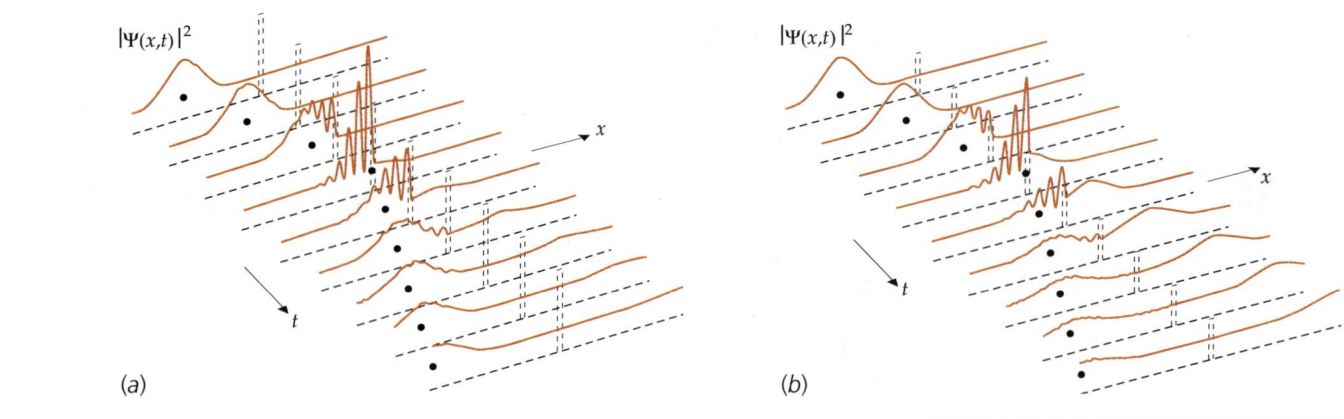

(a) (b)

$$T \propto e^{-2\alpha a} \qquad\qquad 35\text{-}29$$

TRANSMISSION THROUGH A BARRIER

with $\alpha = \sqrt{2m(U_0 - E)/\hbar^2}$. The probability of penetration of the barrier thus decreases exponentially with the barrier thickness a and with the square root of the relative barrier height $(U_0 - E)$.

Figure 35-13a shows a wave packet incident on a potential barrier of height U_0 that is considerably greater than the energy of the particle. The probability of penetration is very small, as indicated by the relative sizes of the reflected and transmitted packets. In Figure 35-13b, the barrier is just slightly greater than the energy of the particle. In this case, the probability of penetration is about the same as the probability of reflection. Figure 35-14 shows a particle incident on two potential barriers of height just slightly greater than the energy of the particle.

As we have mentioned, the penetration of a barrier is not unique to quantum mechanics. When light is totally reflected from a glass–air interface, the light wave can penetrate the air barrier if a second piece of glass is brought within a few wavelengths of the first. This effect can be demonstrated with a laser beam and two 45° prisms (Figure 35-15). Similarly, water waves in a ripple tank can penetrate a gap of deep water (Figure 35-16).

The theory of barrier penetration was used by George Gamow in 1928 to explain the enormous variation in the half-lives for α decay of radioactive nuclei.

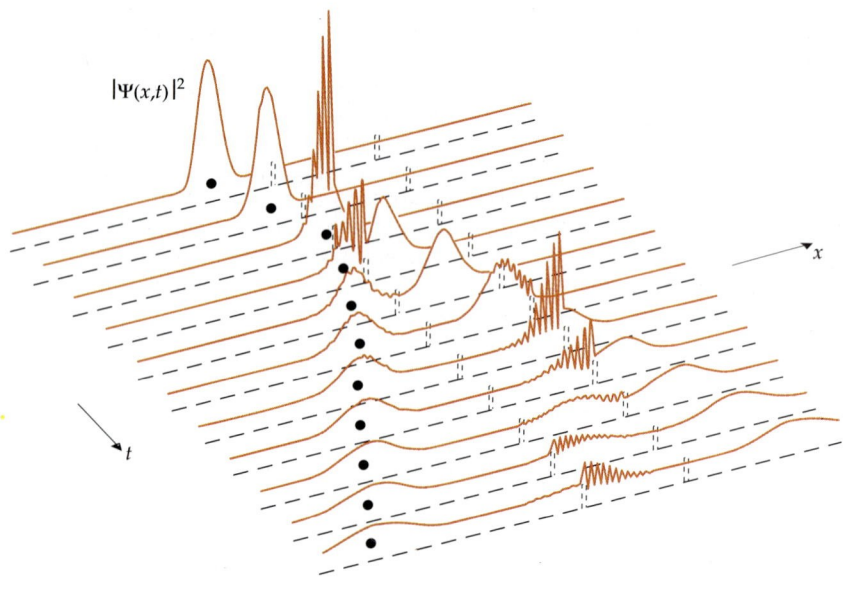

FIGURE 35-14 A wave packet representing a particle incident on two barriers. At each encounter, part of the packet is transmitted and part reflected, resulting in part of the packet being trapped between the barriers for some time.

FIGURE 35-13 Barrier penetration. (a) The same particle incident on a barrier of height much greater than the energy of the particle. A very small part of the packet tunnels through the barrier. In both drawings, the position of a classical particle is indicated by a dot. (b) A wave packet representing a particle incident on a barrier of height just slightly greater than the energy of the particle. For this particular choice of energies, the probability of transmission is approximately equal to the probability of reflection, as indicated by the relative sizes of the transmitted and reflected packets.

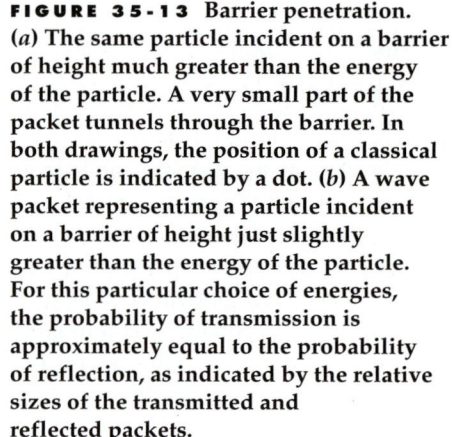

$\therefore$ smaller barrier thickness increases probability of transmission

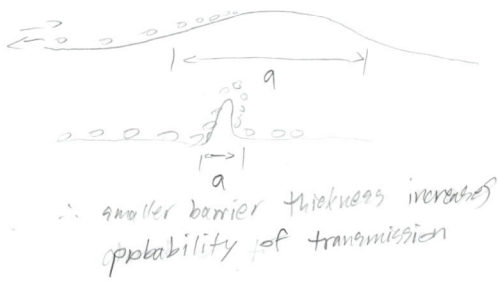

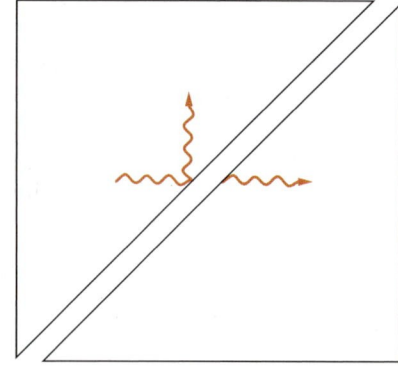

FIGURE 35-15 The penetration of an optical barrier. If the second prism is close enough to the first, part of the wave penetrates the air barrier even when the angle of incidence in the first prism is greater than the critical angle.

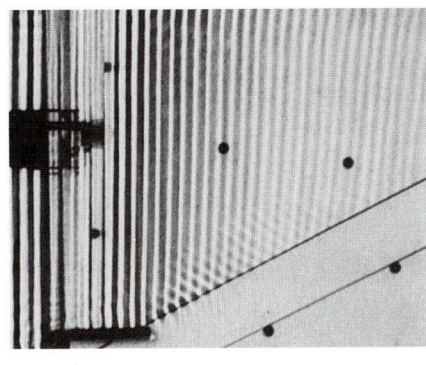
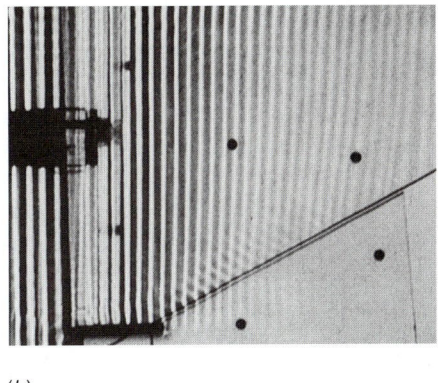

(a)   (b)

FIGURE 35-16 The penetration of a barrier by water waves in a ripple tank. In Figure 35-16a, the waves are totally reflected from a gap of deeper water. When the gap is very narrow, as in Figure 35-16b, a transmitted wave appears. The dark circles are spacers that are used to support the prisms from below.

(Alpha particles are helium nuclei emitted from larger atoms in radioactive decay; they consist of two protons and two neutrons tightly bound together.) In general, the smaller the energy of the emitted α particle, the longer the half-life. The energies of α particles from natural radioactive sources range from approximately 4 MeV to 7 MeV, whereas the half-lives range from approximately 10^{-5} second to 10^{10} years. Gamow represented a radioactive nucleus by a potential well containing an α particle, as shown in Figure 35-17. Without knowing very much about the nuclear force that is exerted on the α particle within the nucleus, Gamow represented it by a square well. Just outside the well, the α particle with its charge of $+2e$ is repelled by the nucleus with its charge $+Ze$, where Ze is the remaining nuclear charge. This force is represented by the Coulomb potential energy $+k(2e)(Ze)/r$. The energy E is the measured kinetic energy of the emitted α particle, because when it is far from the nucleus its potential energy is zero. After the α particle is formed inside the radioactive nucleus, it bounces back and forth inside the nucleus, hitting the barrier at the nuclear radius R. Each time the α particle strikes the barrier, there is some small probability of the particle penetrating the barrier and appearing outside the nucleus. We can see from Figure 35-17 that a small increase in E reduces the relative height of the barrier $U - E$ and also the barrier's thickness. Because the probability of penetration is so sensitive to the barrier thickness and relative height, a small increase in E leads to a large increase in the probability of transmission and therefore to a shorter lifetime. Gamow was able to derive an expression for the half-life as a function of E that is in excellent agreement with experimental results.

In the **scanning tunneling electron microscope** developed in the 1980s, a thin space between a material specimen and a tiny probe acts as a barrier to electrons bound in the specimen. A small voltage applied between the probe and the specimen causes the electrons to *tunnel* through the vacuum separating the two surfaces if the surfaces are close enough together. The tunneling current is extremely sensitive to the size of the gap between the probe and specimen. If a constant tunneling current is maintained as the probe scans the specimen, the surface of the specimen can be mapped out by the motions of the probe. In this way, the surface features of a specimen can be measured with a resolution of the order of the size of an atom.

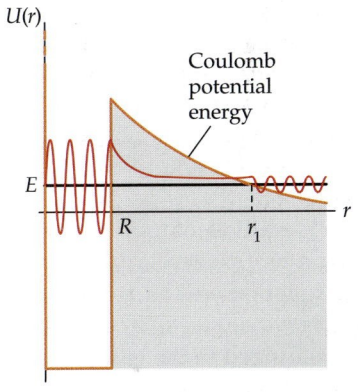

FIGURE 35-17 Model of a potential energy function for an α particle in a radioactive nucleus. The strong attractive nuclear force when r is less than the nuclear radius R can be approximately described by the potential well shown. Outside the nucleus the nuclear force is negligible, and the potential is given by Coulomb's law, $U(r) = +k(2e)(Ze)/r$, where Ze is the nuclear charge and $2e$ is the charge of the α particle. The wave function of the alpha particle, shown in red, is placed on the graph.

35-5 The Schrödinger Equation in Three Dimensions

The one-dimensional time-independent Schrödinger equation is easily extended to three dimensions. In rectangular coordinates, it is

$$-\frac{\hbar^2}{2m}\left(\frac{\partial^2\psi}{\partial x^2} + \frac{\partial^2\psi}{\partial y^2} + \frac{\partial^2\psi}{\partial z^2}\right) + U\psi = E\psi \qquad 35\text{-}30$$

where the wave function ψ and the potential energy U are generally functions of all three coordinates, x, y, and z. To illustrate some of the features of problems in three dimensions, we consider a particle in a three-dimensional infinite square well given by $U(x, y, z) = 0$ for $0 < x < L$, $0 < y < L$, and $0 < z < L$. Outside this cubical region, $U(x, y, z) = \infty$. For this problem, the wave function must be zero at the edges of the well.

There are standard methods in partial differential equations for solving Equation 35-30. We can guess the form of the solution from our knowledge of probability. For a one-dimensional box along the x axis, we have found the probability that a particle is in the region dx at x to be $A_1^2 \sin^2(k_1 x) \, dx$ (from Equation 35-10), where A_1 is a normalization constant and $k_1 = n\pi/L$ is the wave number. Similarly, for a box along the y axis, the probability of a particle being in a region dy at y is $A_2^2 \sin^2(k_2 y) \, dy$. The probability of two independent events occurring is the product of the probabilities of each event occurring.[†] So the probability of a particle being in region dx at x *and* in region dy at y is $A_1^2 \sin^2(k_1 x) \, dx \, A_2^2 \sin^2(k_2 y) \, dy = A_1^2 \sin^2(k_1 x) A_2^2 \sin^2(k_2 y) \, dx \, dy$. The probability of a particle being in the region dx, dy, and dz is $\psi(x, y, z) \, dx \, dy \, dz$, where $\psi(x, y, z)$ is the solution of Equation 35-30. This solution is of the form

$$\psi(x, y, z) = A \sin^2(k_1 x) \sin^2(k_2 y) \sin^2(k_3 z) \, dx \, dy \, dz \qquad \text{35-31}$$

where the constant A is determined by normalization. Inserting this solution into Equation 35-30, we obtain for the energy

$$E = \frac{\hbar^2}{2m}(k_1^2 + k_2^2 + k_3^2)$$

which is equivalent to $E = (p_x^2 + p_y^2 + p_z^2)/(2m)$, with $p_x = \hbar k_1$, and so on. The wave function will be zero at $x = L$ if $k_1 = n_1 \pi/L$, where n_1 is an integer. Similarly, the wave function will be zero at $y = L$ if $k_2 = n_2 \pi/L$, and the wave function will be zero at $z = L$ if $k_3 = n_3 \pi/L$. (It is also zero at $x = 0$, $y = 0$, and $z = 0$.) The energy is thus quantized to the values

$$E_{n_1, n_2, n_3} = \frac{\hbar^2 \pi^2}{2mL^2}(n_1^2 + n_2^2 + n_3^2) = E_1 (n_1^2 + n_2^2 + n_3^2) \qquad \text{35-32}$$

where n_1, n_2, and n_3 are integers and E_1 is the ground-state energy of the one-dimensional well. Note that the energy and wave function are characterized by three quantum numbers, each arising from the boundary conditions for one of the coordinates.

The lowest energy state (the ground state) for the cubical well occurs when $n_1 = n_2 = n_3 = 1$ and has the value

$$E_{1,1,1} = \frac{3\hbar^2 \pi^2}{2mL^2} = 3E_1$$

The first excited energy level can be obtained in three different ways: $n_1 = 2$, $n_2 = n_3 = 1$; $n_2 = 2$, $n_1 = n_3 = 1$; or $n_3 = 2$, $n_1 = n_2 = 1$. Each has a different wave function. For example, the wave function for $n_1 = 2$ and $n_2 = n_3 = 1$ is

$$\psi_{2,1,1} = A \sin\frac{2\pi x}{L} \sin\frac{\pi y}{L} \sin\frac{\pi z}{L} \qquad \text{35-33}$$

(where the value of the normalization constant A is different than the value of the normalization constant in Equation 35-31). There are thus three different quantum states as described by the three different wave functions corresponding to the same energy level. An energy level with which more than one wave function is associated is said to be **degenerate**. In this case, there is threefold degeneracy.

[†] For example, if you throw two dice, the probability of the first die coming up 6 is $1/6$ and the probability of the second die coming up an odd number is $1/2$. The probability of the first die coming up 6 *and* the second die coming up an odd number is $(1/6)(1/2) = 1/12$.

Degeneracy is related to the spatial symmetry of the system. If, for example, we consider a noncubic well, where $U = 0$ for $0 < x < L_1$, $0 < y < L_2$, and $0 < z < L_3$, the boundary conditions at the edges would lead to the quantum conditions $k_1 L_1 = n_1 \pi$, $k_2 L_2 = n_2 \pi$, and $k_3 L_3 = n_3 \pi$, and the total energy would be

$$E_{n_1,n_2,n_3} = \frac{\hbar^2 \pi^2}{2m}\left(\frac{n_1^2}{L_1^2} + \frac{n_2^2}{L_2^2} + \frac{n_3^2}{L_3^2}\right) \qquad \text{35-34}$$

These energy levels are not degenerate if L_1, L_2, and L_3 are all different. Figure 35-18 shows the energy levels for the ground state and first two excited states for an infinite cubic well in which the excited states are degenerate and for a noncubic infinite well in which L_1, L_2, and L_3 are all slightly different so that the excited levels are slightly split apart and the degeneracy is removed. The ground state is the state where the quantum numbers n_1, n_2, and n_3 all equal 1. None of the three quantum numbers can be zero. If any one of n_1, n_2, and n_3 were zero, the corresponding wave number k would also equal zero and the corresponding wave function (Equation 35-31) would equal zero for all values of x, y, and z.

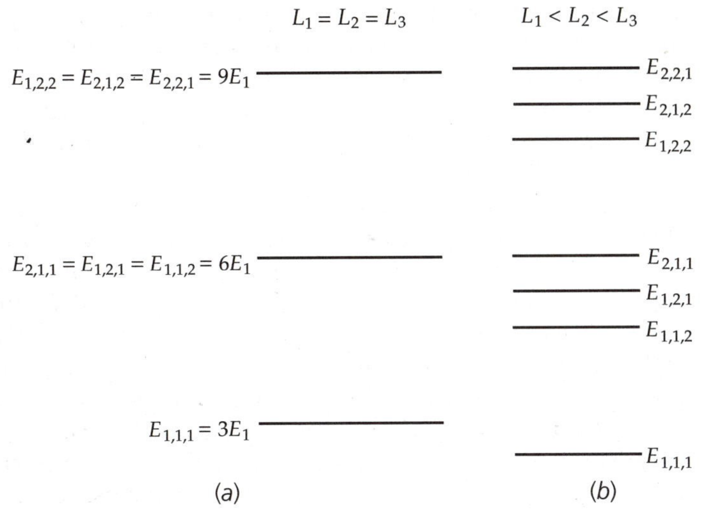

$L_1 = L_2 = L_3$	$L_1 < L_2 < L_3$	
		$E_{2,2,1}$
$E_{1,2,2} = E_{2,1,2} = E_{2,2,1} = 9E_1$		$E_{2,1,2}$
		$E_{1,2,2}$
		$E_{2,1,1}$
$E_{2,1,1} = E_{1,2,1} = E_{1,1,2} = 6E_1$		$E_{1,2,1}$
		$E_{1,1,2}$
$E_{1,1,1} = 3E_1$		
		$E_{1,1,1}$
(a)	(b)	

FIGURE 35-18 Energy-level diagrams for (a) a cubic infinite well and (b) a noncubic infinite well. In Figure 35-18a the energy levels are degenerate; that is, there are two or more wave functions having the same energy. The degeneracy is removed when the symmetry of the potential is removed, as in Figure 35-18b.

ENERGY LEVELS FOR A PARTICLE IN A THREE-DIMENSIONAL BOX

EXAMPLE 35-3

A particle is in a three-dimensional box with $L_3 = L_2 = 2L_1$. Give the quantum numbers n_1, n_2, and n_3 that correspond to the thirteen quantum states of this box that have the lowest energies.

PICTURE THE PROBLEM We can use Equation 35-34 to write the energies in terms of the ratios $L_2/L_1 = 2$ and $L_3/L_1 = 2$, then find by inspection the values of the quantum numbers that give the lowest energies.

1. The energy of a level is given by Equation 35-34:

$$E_{n_1,n_2,n_3} = \frac{\hbar^2 \pi^2}{2m}\left(\frac{n_1^2}{L_1^2} + \frac{n_2^2}{L_2^2} + \frac{n_3^2}{L_3^2}\right)$$

2. Factor out $1/L_1^2$:

$$E_{n_1,n_2,n_3} = \frac{\hbar^2 \pi^2}{2mL_1^2}\left(n_1^2 + n_2^2 \frac{L_1^2}{L_2^2} + n_3^2 \frac{L_1^2}{L_3^2}\right)$$

$$= E_1(n_1^2 + n_2^2/4 + n_3^2/4)$$

3. The lowest energy is $E_{1,1,1}$:

$E_{1,1,1} = E_1(1^2 + 1^2\tfrac{1}{4} + 1^2\tfrac{1}{4}) = 1.5E_1$ (1st)

4. The energy increases the least when we increase n_2 or n_3. Try various values of the quantum numbers:

$E_{1,2,1} = E_{1,1,2} = E_1(1^2 + 2^2\tfrac{1}{4} + 1\tfrac{1}{4}) = 2.25E_1$ (2nd and 3rd)

$E_{1,2,2} = E_1(1^2 + 2^2\tfrac{1}{4} + 2^2\tfrac{1}{4}) = 3.0E_1$ (4th)

$E_{1,3,1} = E_{1,1,3} = E_1(1^2 + 3^2\tfrac{1}{4} + 1^2\tfrac{1}{4}) = 3.50E_1$ (5th and 6th)

$E_{1,3,2} = E_{1,2,3} = E_1(1^2 + 3^2\tfrac{1}{4} + 2^2\tfrac{1}{4}) = 4.25E_1$ (7th and 8th)

$E_{2,1,1} = E_1(2^2 + 1^2\tfrac{1}{4} + 1^2\tfrac{1}{4}) = 4.5E_1$ (9th)

$\left.\begin{array}{l} E_{2,2,1} = E_{2,1,2} = E_1(2^2 + 2^2\tfrac{1}{4} + 1^2\tfrac{1}{4}) = 5.25E_1 \\[6pt] E_{1,4,1} = E_{1,1,4} = E_1(1^2 + 4^2\tfrac{1}{4} + 1^2\tfrac{1}{4}) = 5.25E_1 \end{array}\right\}$ (10th, 11th, 12th, and 13th)

REMARKS Note the degeneracy of the levels.

EXERCISE Find the quantum numbers and energies of the next two energy levels in step 4. (*Answer* $E_{1,3,3} = 5.5E_1$, $E_{1,4,2} = E_{1,2,4} = E_{2,2,2} = 6.0E_1$)

EXAMPLE 35-4 Try It Yourself

Write the degenerate wave functions for the fourth and fifth excited states
(levels 5 and 6) of the results in step 4 of Example 35-3.

PICTURE THE PROBLEM Use Equation 35-33 with $k_i = n_i \pi / L$.

Cover the column to the right and try these on your own before looking at the answers.

Steps   **Answers**

Write the wave functions corresponding to the energies $\psi_{1,3,1} = A \sin \dfrac{\pi x}{L} \sin \dfrac{3\pi y}{L} \sin \dfrac{\pi z}{L}$
$E_{1,3,1}$ and $E_{1,1,3}$.

$\psi_{1,1,3} = A \sin \dfrac{\pi x}{L} \sin \dfrac{\pi y}{L} \sin \dfrac{3\pi z}{L}$

35-6 The Schrödinger Equation for Two Identical Particles

Our discussion of quantum mechanics has thus far been limited to situations
in which a single particle moves in some force field characterized by a potential
energy function U. The most important physical problem of this type is the
hydrogen atom, in which a single electron moves in the Coulomb potential of the
proton nucleus. This problem is actually a two-body problem, since the proton
also moves in the field of the electron. However, the motion of the much more
massive proton requires only a very small correction to the energy of the atom
that is easily made in both classical and quantum mechanics. When we consider
more complicated problems, such as the helium atom, we must apply quantum
mechanics to two or more electrons moving in an external field. Such problems
are complicated by the interaction of the electrons with each other and also by
the fact that the electrons are identical.

The interaction of two electrons with each other is electromagnetic and is
essentially the same as the classical interaction of two charged particles. The
Schrödinger equation for an atom with two or more electrons cannot be solved
exactly, so approximation methods must be used. This is not very different from
the situation in classical problems with three or more particles. However, the
complications arising from the identity of electrons are purely quantum mechanical and have no classical counterpart. They are due to the fact that it is impossible to keep track of which electron is which. Classically, identical particles can be
identified by their positions, which in principle can be determined with unlimited
accuracy. This is impossible quantum mechanically because of the uncertainty
principle. Figure 35-19 offers a schematic illustration of the problem.

(a)
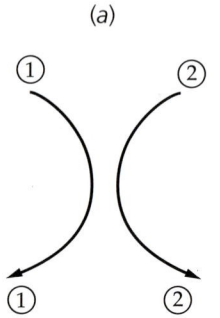

(b)
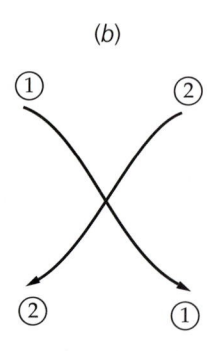

(c)
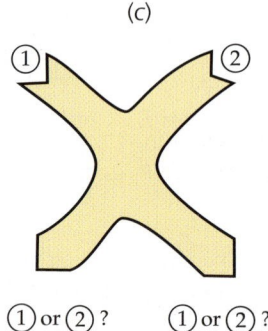

FIGURE 35-19 Two possible
classical electron paths (Figure 35-19*a* and
Figure 35-19*b*). If electrons were classical
particles, they could be distinguished by
the paths followed. However, because of
the quantum-mechanical wave properties
of electrons, the paths are spread out,
as indicated by the shaded region in
Figure 35-19*c*. It is impossible to
distinguish which electron is
which after they separate.

The indistinguishability of identical particles has important consequences. For instance, consider the very simple case of two identical, noninteracting particles in a one-dimensional infinite square well. The time-independent Schrödinger equation for two particles, each of mass m, is

$$-\frac{\hbar^2}{2m}\frac{\partial^2\psi(x_1, x_2)}{\partial x_1^2} - \frac{\hbar^2}{2m}\frac{\partial^2\psi(x_1, x_2)}{\partial x_2^2} + U\psi(x_1, x_2) = E\psi(x_1, x_2) \qquad 35\text{-}35$$

where x_1 and x_2 are the coordinates of the two particles. If the particles interact, the potential energy U contains terms with both x_1 and x_2 that cannot be separated into separate terms containing only x_1 or x_2. For example, the electrostatic repulsion of two electrons in one dimension is represented by the potential energy $ke^2/|x_2 - x_1|$. However, if the particles do not interact (as we are assuming here), we can write $U = U_1(x_1) + U_2(x_2)$. For the infinite square well, we need only solve the Schrödinger equation inside the well where $U = 0$, and require that the wave function be zero at the walls of the well. With $U = 0$, Equation 35-35 looks just like the expression for a particle in a two-dimensional well (Equation 35-30, with no z and with y replaced by x_2).

Solutions of this equation can be written in the form[†]

$$\psi_{n,m} = \psi_n(x_1)\psi_m(x_2) \qquad 35\text{-}36$$

where ψ_n and ψ_m are the single-particle wave functions for a particle in an infinite well and n and m are the quantum numbers of particles 1 and 2, respectively. For example, for $n = 1$ and $m = 2$, the wave function is

$$\psi_{1,2} = A \sin\frac{\pi x_1}{L} \sin\frac{2\pi x_2}{L} \qquad 35\text{-}37$$

The probability of finding particle 1 in dx_1 and particle 2 in dx_2 is $\psi_{n,m}^2(x_1, x_2)\, dx_1\, dx_2$, which is just the product of the separate probabilities $\psi_n^2(x_1)\, dx_1$ and $\psi_m^2(x_2)\, dx_2$. However, even though we have labeled the particles 1 and 2, we cannot distinguish which is in dx_1 and which is in dx_2 if they are identical. The mathematical descriptions of identical particles must be the same if we interchange the labels. The probability density $\psi^2(x_1, x_2)$ must therefore be the same as $\psi^2(x_2, x_1)$:

$$\psi^2(x_2, x_1) = \psi^2(x_1, x_2) \qquad 35\text{-}38$$

Equation 35-38 is satisfied if $\psi(x_2, x_1)$ is either **symmetric** or **antisymmetric** on the exchange of particles—that is, if either

$$\psi(x_2, x_1) = \psi(x_1, x_2), \quad \text{symmetric} \qquad 35\text{-}39$$

or

$$\psi(x_2, x_1) = -\psi(x_1, x_2), \quad \text{antisymmetric} \qquad 35\text{-}40$$

Note that the wave functions given by Equations 35-36 and 35-37 are neither symmetric nor antisymmetric. If we interchange x_1 and x_2 in these wave functions, we get a different wave function, which implies that the particles can be distinguished.

We can find symmetric and antisymmetric wave functions that are solutions of the Schrödinger equation by adding or subtracting $\psi_{n,m}$ and $\psi_{m,n}$. Adding them, we obtain

---

[†] Again, this result can be obtained by solving Equation 35-35, but it also can be understood in terms of our knowledge of probability. The probability of electron 1 being in region dx_1 and electron 2 being in region dx_2 is the product of the individual probabilities.

$$\psi_S = A'[\psi_n(x_1)\psi_m(x_2) + \psi_n(x_2)\psi_m(x_1)], \quad \text{symmetric} \qquad \text{35-41}$$

and subtracting them, we obtain

$$\psi_A = A'[\psi_n(x_1)\psi_m(x_2) - \psi_n(x_2)\psi_m(x_1)], \quad \text{antisymmetric} \qquad \text{35-42}$$

For example, the symmetric and antisymmetric wave functions for the first excited state of two identical particles in an infinite square well would be

$$\psi_S = A'\left(\sin\frac{\pi x_1}{L}\sin\frac{2\pi x_2}{L} + \sin\frac{\pi x_2}{L}\sin\frac{2\pi x_1}{L}\right) \qquad \text{35-43}$$

and

$$\psi_A = A'\left(\sin\frac{\pi x_1}{L}\sin\frac{2\pi x_2}{L} - \sin\frac{\pi x_2}{L}\sin\frac{2\pi x_1}{L}\right) \qquad \text{35-44}$$

There is an important difference between antisymmetric and symmetric wave functions. If $n = m$, the antisymmetric wave function is identically zero for all values of x_1 and x_2, whereas the symmetric wave function is not. Thus, if the wave function describing two identical particles is antisymmetric, the quantum numbers n and m of two particles cannot be the same. This is an example of the **Pauli exclusion principle,** which was first stated by Wolfgang Pauli for electrons in an atom:

No two electrons in an atom can have the same quantum numbers.

PAULI EXCLUSION PRINCIPLE

It is found that electrons, protons, neutrons, and some other particles have antisymmetric wave functions and obey the Pauli exclusion principle. These particles are called **fermions.** Other particles (e.g., α particles, deuterons, photons, and mesons) have symmetric wave functions and do not obey the Pauli exclusion principle. These particles are called **bosons.**

SUMMARY

1. The Schrödinger equation is a differential equation that relates the second spatial derivative of a wave function to its first time derivative. Wave functions that describe physical situations are solutions of this differential equation.

2. Because a wave function must be normalizable, it must be well behaved; that is, it must approach zero as x approaches infinity. For bound systems such as a particle in a box, a simple harmonic oscillator, or an electron in an atom, this requirement leads to energy quantization.

3. The well-behaved wave functions for bound systems describe standing waves.

Topic	**Relevant Equations and Remarks**	
1. **Time-Independent Schrödinger Equation**	$-\dfrac{\hbar^2}{2m}\dfrac{d^2\psi(x)}{dx^2} + U(x)\psi(x) = E\psi(x)$	35-4

Allowable solutions	In addition to satisfying the Schrödinger equation, a wave function $\psi(x)$ must be continuous and (if U is not infinite) must have a continuous first derivative $d\psi/dx$. Because the probability of finding an electron somewhere must be 1, the wave function must obey the normalization condition $$\int_{-\infty}^{\infty} |\psi|^2 \, dx = 1$$ This condition implies the boundary condition that ψ must approach 0 as x approaches $\pm\infty$. Such boundary conditions lead to the quantization of energy.
2. Confined Particles	When the total energy E is greater than the potential energy $U(x)$ in some region (the classically allowed region) and less than $U(x)$ outside that region, the wave function oscillates within the classically allowed region and increases or decreases exponentially outside that region. The wave function approaches zero as x approaches ∞ only for certain values of the total energy E. The energy is thus quantized.
In a finite square well	In a finite well of height U_0, there are only a finite number of allowed energies, and these are slightly less than the corresponding energies in an infinite well.
In the simple harmonic oscillator	In the oscillator potential energy function $U(x) = \frac{1}{2}m\omega_0^2 x^2$, the allowed energies are equally spaced and given by $$E_n = (n + \tfrac{1}{2})\hbar\omega_0 \qquad\qquad \textbf{35-26}$$ The ground-state wave function is given by $$\psi_0(x) = A_0 e^{-ax^2} \qquad\qquad \textbf{35-23}$$ where A_0 is the normalization constant and $a = m\omega_0/(2\hbar)$.
3. Reflection and Barrier Penetration	When the potential changes abruptly over a small distance, a particle may be reflected even though $E > U(x)$. A particle may penetrate a region in which $E < U(x)$. Reflection and penetration of electron waves are similar to those for other kinds of waves.
4. The Schrödinger Equation in Three Dimensions	The wave function for a particle in a three-dimensional box can be written $$\psi(x, y, z) = \psi_1(x)\psi_2(y)\psi_3(z)$$ where ψ_1, ψ_2, and ψ_3 are wave functions for a one-dimensional box.
Degeneracy	When more than one wave function is associated with the same energy level, the energy level is said to be degenerate. Degeneracy arises because of spatial symmetry.
5. The Schrödinger Equation for Two Identical Particles	A wave function that describes two identical particles must be either symmetric or antisymmetric when the coordinates of the particles are exchanged. Fermions (which include electrons, protons, and neutrons) are described by antisymmetric wave functions and obey the Pauli exclusion principle, which states that no two particles can have the same values for their quantum number. Bosons (which include α particles, deuterons, photons, and mesons) have symmetric wave functions and do not obey the Pauli exclusion principle.

PROBLEMS

- Single-concept, single-step, relatively easy
- •• Intermediate-level, may require synthesis of concepts
- ••• Challenging
- **SSM** Solution is in the *Student Solutions Manual*
- **iSOLVE** Problems available on iSOLVE online homework service
- **iSOLVE**✓ These "Checkpoint" online homework service problems ask students additional questions about their confidence level, and how they arrived at their answer.

In a few problems, you are given more data than you actually need; in a few other problems, you are required to supply data from your general knowledge, outside sources, or informed estimates.

Conceptual Problems

1 • True or false: Boundary conditions on the wave function lead to energy quantization.

2 • Sketch (a) the wave function and (b) the probability distribution for the $n = 4$ state for the finite square-well potential.

3 • Sketch (a) the wave function and (b) the probability distribution for the $n = 5$ state for the finite square-well potential.

Estimation and Approximation

4 • **SSM** The Schrödinger equation could be applied equally well to baseballs as to electrons; yet, we would never analyze the motion of a baseball with a wave function. Explain why this is the case by estimating the quantum mechanically predicted lowest energy level of a baseball trapped inside a locker. You can treat the locker as if it were a one-dimensional infinite potential well. What value of the quantum number n would you need for a ball rolling around in a locker, after you toss the ball in, so that the kinetic energy is approximately equal to the quantum mechanically calculated energy?

The Schrödinger Equation

5 •• Show that if $\psi_1(x)$ and $\psi_2(x)$ are each solutions to the time-independent Schrödinger equation (Equation 35-4), then $\psi_3(x) = \psi_1(x) + \psi_2(x)$ is also a solution. This is known as the superposition principle and it applies to the solutions of all linear differential equations.

The Harmonic Oscillator

6 •• The harmonic oscillator problem may be used to describe molecules. For example, the hydrogen molecule H_2 is found to have equally spaced energy levels separated by 8.7×10^{-20} J. What value of spring constant would be needed to get this energy spacing, assuming that the molecule can be modeled as a single hydrogen atom attached to a spring?

7 •• Show that the expectation value $<x> = \int x|\psi|^2\, dx$ is zero for both the ground state and the first excited state of the harmonic oscillator.

8 •• **SSM** Use the procedure of Example 35-1 to verify that the energy of the first excited state of the harmonic oscillator is $E_1 = \frac{3}{2}\hbar\omega_0$. (Note: *Rather than solve for a again, use the result* $a = m\omega_0/(2\hbar)$ *obtained in Example 35-1.*)

9 •• Show that the normalization constant A_0 of Equation 35-23 is $A_0 = (2m\omega_0/h)^{1/4}$.

10 •• Show that for the ground state of the harmonic oscillator $<x^2> = \int x^2|\psi|^2\, dx = \hbar/(2m\omega_0) = 1/(4a)$. Use this result to show that the average potential energy equals half the total energy.

11 •• The quantity $\sqrt{<x^2> - <x>^2}$ is a measure of the average spread in the location of a particle. (a) Consider an electron trapped in a harmonic oscillator potential. Its lowest energy level is found to be 2.1×10^{-4} eV. Calculate $\sqrt{<x^2> - <x>^2}$ for this electron. (See Problems 7 and 10.) (b) Now consider an electron trapped in an infinite square-well potential. If the width of the well is equal to $\sqrt{<x^2> - <x>^2}$, what would be the lowest energy level for this electron?

12 ••• Classically, the average kinetic energy of the harmonic oscillator equals the average potential energy. We may assume that this is also true for the quantum mechanical harmonic oscillator. Use this condition to determine the expectation value of p^2 for the ground state of the harmonic oscillator.

13 ••• We know that for the classical harmonic oscillator, $p_{av} = 0$. It can be shown that for the quantum mechanical harmonic oscillator, $<p> = 0$. Use the results of Problem 7, Problem 10, and Problem 12 to determine the uncertainty product $\Delta x\, \Delta p$ for the ground state of the harmonic oscillator.

Reflection and Transmission of Electron Waves: Barrier Penetration

14 •• **SSM** A particle of mass m with wave number k_1 is traveling to the right along the negative x axis. The potential energy of the particle is equal to zero everywhere on the negative x axis and is equal to U_0 everywhere on the positive x axis, $U_0 > 0$. (a) Show that if the total energy is $E = \alpha U_0$, where $\alpha \geq 1$, wave number k_2 in the region $x > 0$ is given by

$$k_2 = \sqrt{\frac{\alpha - 1}{\alpha}}\, k_1$$

(b) Using a spreadsheet program or graphing calculator, graph the reflection coefficient R and the transmission coefficient T for $1 \leq \alpha \leq 5$.

15 •• **iSOLVE** ✓ Suppose that the potential in Problem 14 drops from zero to $-U_0$ at $x = 0$ so that the particle speeds up instead of slowing down. The wave number for the incident particle is again k_1, and the total energy is $2U_0$. (a) What is the wave number for the particle in the region of positive x? (b) Calculate the reflection coefficient R at $x = 0$. (c) What is the transmission coefficient T? (d) If one million particles with wave number k_1 are incident upon the potential drop, how many particles are expected to continue along in the positive x direction? How does this compare with the classical prediction?

16 •• **iSOLVE** ✓ A particle of energy E approaches a step barrier of height U_0. What should be the ratio E/U_0 so that the reflection coefficient is $1/2$?

17 •• **iSOLVE** Use Equation 35-29 to calculate the order of magnitude of the probability that a proton will tunnel out of a nucleus in one collision with the nuclear barrier if it has energy 6 MeV below the top of the potential barrier and the barrier thickness is 10^{-15} m.

18 •• **SSM** A 10-eV electron is incident on a potential barrier of height 25 eV and width of 1 nm. (a) Use Equation 35-29 to calculate the order of magnitude of the probability that the electron will tunnel through the barrier. (b) Repeat your calculation for a width of 0.1 nm.

19 ••• To understand how a small change in α-particle energy can dramatically change the tunneling probability from a nucleus, consider an α particle emitted by a uranium nucleus ($Z = 92$). (a) Referring to Figure 35-17, calculate the distance of closest approach r_1 that α particles with kinetic energies of 4 MeV and 7 MeV could make to the uranium nucleus. (b) Use the result from Part (a) to calculate the relative transmission coefficient $e^{-2\alpha a}$ for the same α particles. (Note: The actual half-lives of uranium nuclei vary over nine orders of magnitude. Your calculation will show a smaller range than this; however, to find half-life, you must also include the frequency with which the α particle strikes the barrier.)

The Schrödinger Equation in Three Dimensions

20 • A particle is confined to a three-dimensional box that has sides L_1, $L_2 = 2L_1$, and $L_3 = 3L_1$. Give the quantum numbers n_1, n_2, n_3 that correspond to the lowest ten quantum states of this box.

21 • Give the wave functions for the lowest ten quantum states of the particle in Problem 20.

22 • (a) Repeat Problem 20 for the case $L_2 = 2L_1$ and $L_3 = 4L_1$. (b) What quantum numbers correspond to degenerate energy levels?

23 • **SSM** Give the wave functions for the lowest ten quantum states of the particle in Problem 22.

24 • A particle moves in a potential well given by $U(x, y, z) = 0$ for $-L/2 < x < L/2$, $0 < y < L$, and $0 < z < L$; and $U = \infty$ outside these ranges. (a) Write an expression for the ground-state wave function for this particle. (b) How do the allowed energies compare with those for a well having $U = 0$ for $0 < x < L$, rather than for $-L/2 < x < L/2$?

25 •• A particle moves freely in the two-dimensional region defined by $0 \le x \le L$ and $0 \le y \le L$. (a) Find the wave functions satisfying the Schrödinger equation. (b) Find the

corresponding energies. (c) Find the lowest two states that are degenerate. Give the quantum numbers for this case. (d) Find the lowest three states that have the same energy. Give the quantum numbers for the three states having the same energy.

The Schrödinger Equation for Two Identical Particles

26 • Show that Equation 35-37 satisfies Equation 35-35 with $U = 0$, and find the energy of this state.

27 • **iSOLVE** What is the ground-state energy of ten noninteracting bosons in a one-dimensional box of length L?

28 • **SSM** **iSOLVE** What is the ground-state energy of ten noninteracting fermions, such as neutrons, in a one-dimensional box of length L? (Because the quantum number associated with spin can have two values, each spatial state can hold two neutrons.)

Orthogonality of Wave Functions

The integral of two functions over some space interval is somewhat analogous to the dot product of two vectors. If this integral is zero, the functions are said to be orthogonal, which is analogous to two vectors being perpendicular. The following problems illustrate the general principle that any two wave functions corresponding to different energy levels in the same potential are orthogonal. A general hint for all these problems is that the integral of an antisymmetric integrand over symmetric limits is equal to zero.

29 •• Show that the ground-state wave function and the wave function of the first excited state of the harmonic oscillator are orthogonal; that is, show that $\int \psi_0(x) \psi_1(x)\, dx = 0$.

30 •• The wave function for the state $n = 2$ of the harmonic oscillator is $\psi_2(x) = A_2(2ax^2 - \frac{1}{2})e^{-ax^2}$, where A_2 is the normalization constant for this wave function. Show that the wave functions for the states $n = 1$ and $n = 2$ of the harmonic oscillator are orthogonal.

31 •• For the wave functions $\psi_n(x) = \sqrt{2/L}\,\sin(n\pi x/L)$ corresponding to a particle in an infinite square-well potential from 0 to L, show that $\int \psi_n(x) \psi_m(x)\, dx = 0$; that is, show that ψ_n and ψ_m are orthogonal.

General Problems

32 •• Consider a particle in a one-dimensional box of length L that is centered at the origin (see Problem 65 in Chapter 34). (a) What are the values of $\psi_1(0)$ and $\psi_2(0)$? (b) What are the values of $<x>$ for the states $n = 1$ and $n = 2$? (c) Evaluate $<x^2>$ for the states $n = 1$ and $n = 2$.

33 •• **SSM** Eight identical noninteracting fermions (e.g., neutrons) are confined to a two-dimensional square box of side length L. Determine the energies of the three lowest states. (See Problem 26.)

34 •• **iSOLVE** A particle is confined to a two-dimensional box defined by the following boundary conditions: $U(x, y) = 0$, for $-L/2 \le x \le L/2$ and $-3L/2 \le y \le 3L/2$; and $U(x, y) = \infty$ elsewhere. (a) Determine the energies of the lowest three bound states. Are any of these states degenerate? (b) Identify the quantum numbers of the lowest doubly degenerate bound state and determine its energy.

35 ••• The classical probability distribution function for a particle in a one-dimensional box of length L is $P = 1/L$. (See Example 34-5.) (a) Show that the classical expectation value of x^2 for a particle in a one-dimensional box of length L centered at the origin (Problem 32) is $L^2/12$. (b) Find the quantum expectation value of x^2 for the nth state of a particle in the one-dimensional box of Problem 32, and show that it approaches the classical limit $L^2/12$ for $n \gg 1$.

36 •• Show that Equation 35-27 and Equation 35-28 imply that the transmission coefficient for particles of energy E incident on a step barrier $U_0 < E$ is given by

$$T = \frac{4k_1 k_2}{(k_1 + k_2)^2} = \frac{4r}{(1 + r)^2}$$

where $r = k_2/k_1$.

37 •• (a) Show that for the case of a particle of energy E incident on a step barrier $U_0 < E$, the wave numbers k_1 and k_2 are related by

$$\frac{k_2}{k_1} = r = \sqrt{1 - \frac{U_0}{E}}$$

(b) Use this result to show that $R = (1 - r)^2/(1 + r)^2$

38 •• (a) Using a spreadsheet program or graphing calculator and the results of Problem 36 and Problem 37, graph the transmission coefficient T and reflection coefficient R as a function of incident energy E for values of E ranging from $E = U_0$ to $E = 10.0U_0$. (b) What limiting values do your graphs indicate?

39 ••• Determine the normalization constant A_2 in Problem 30.

40 ••• Consider the time-independent, one-dimensional Schrödinger equation when the potential function is symmetric about the origin, that is, when $U(x)$ is even.[†] (a) Show that if $\psi(x)$ is a solution of the Schrödinger equation with energy E, then $\psi(-x)$ is also a solution with the same energy E, and that, therefore, $\psi(x)$ and $\psi(-x)$ can differ by only a multiplicative constant. (b) Write $\psi(x) = C\psi(-x)$, and show that $C = \pm 1$. Note that $C = +1$ means that $\psi(x)$ is an even function of x, and $C = -1$ means that $\psi(x)$ is an odd function of x.

41 ••• SSM In this problem you will derive the ground-state energy of the harmonic oscillator using the precise form of the uncertainty principle, $\Delta x \, \Delta p \geq \hbar/2$, where Δx and Δp are defined to be the standard deviations $(\Delta x)^2 = [(x - x_{av})^2]_{av}$ and $(\Delta p)^2 = [(p - p_{av})^2]_{av}$ (see Equation 17-35a). Proceed as follows:

1. Write the total classical energy in terms of the position x and momentum p using $U(x) = m\omega_0^2 x^2$ and $K = p^2/2m$.

2. Use the result of Equation 17-35 to write $(\Delta x)^2 = [(x - x_{av})^2]_{av} = (x^2)_{av} - x_{av}^2$ and $(\Delta p)^2 = [(p - p_{av})^2]_{av} = (p^2)_{av} - p_{av}^2$.

3. Use the symmetry of the potential energy function to argue that x_{av} and p_{av} must be zero, so that $(\Delta x)^2 = (x^2)_{av}$ and $(\Delta p)^2 = (p^2)_{av}$.

4. Assume that $\Delta p = \hbar/(2\Delta x)$ to eliminate $(p^2)_{av}$ from the average energy $E_{av} = (p^2)_{av}/(2m) + \frac{1}{2}m\omega^2(x^2)_{av}$ and write E_{av} as $E_{av} = \hbar^2/(8mZ) + \frac{1}{2}m\omega^2 Z$, where $Z = (x^2)_{av}$.

5. Set $dE/dZ = 0$ to find the value of Z for which E is a minimum.

6. Show that the minimum energy is given by $(E_{av})_{min} = +\frac{1}{2}\hbar\omega_0$.

42 ••• A particle of mass m near the earth's surface at $z = 0$ can be described by the potential energy

$$U = mgz, \quad z > 0$$

$$U = \infty, \quad z < 0$$

For some positive value of total energy E, indicate the classically allowed region on a sketch of $U(z)$ versus z. Sketch also the classical kinetic energy versus z. The Schrödinger equation for this problem is quite difficult to solve. Using arguments similar to those in Section 35-2 about the curvature of the wave function as given by the Schrödinger equation, sketch your guesses for the shape of the wave function for the ground state and for the first two excited states.

---

† A function $f(x)$ is even if $f(x) = f(-x)$ for all x.

Atoms

AT A DISTANCE OF 6,000 LIGHT YEARS FROM EARTH, THE STAR CLUSTER RCW 38 IS A RELATIVELY CLOSE STAR-FORMING REGION. THIS IMAGE COVERS AN AREA ABOUT 5 LIGHT YEARS ACROSS AND CONTAINS THOUSANDS OF HOT, VERY YOUNG STARS FORMED LESS THAN A MILLION YEARS AGO. X RAYS FROM THE HOT UPPER ATMOSPHERES OF 190 OF THESE STARS WERE DETECTED BY CHANDRA, AN X-RAY OBSERVATORY ORBITING EARTH. THE MECHANISMS GENERATING THESE X RAYS ~~IS~~ *are* NOT KNOWN.

ON EARTH, X-RAY MACHINES PRODUCE X RAYS BY BOMBARDING A TARGET WITH HIGH-ENERGY ELECTRONS. THE ATOMIC NUMBER OF THE ATOMS THAT MAKE UP THE TARGET CAN BE DETERMINED BY ANALYZING THE RESULTING X-RAY SPECTRA.

 **How is the atomic number obtained from the spectral analysis? Example 36-8 shows one way to accomplish this task.**

There are 113 chemical elements that have been discovered, and there are a couple of additional chemical elements that recently have been reported. Each element is characterized by an atom that contains a number of protons Z, an equal number of electrons, and a number of neutrons N. The number of protons Z is called the **atomic number.** The lightest atom, hydrogen (H), has $Z = 1$; the next lightest atom, helium (He), has $Z = 2$; the third lightest, lithium (Li), has $Z = 3$; and so forth. Nearly all the mass of the atom is concentrated in a tiny nucleus, which contains the protons and neutrons. Typically, the nuclear radius is approximately 1 fm to 10 fm (1 fm = 10^{-15} m). The distance between the nucleus and the electrons is approximately 0.1 nm = 100,000 fm. This distance determines the *size* of the atom.

The chemical properties and physical properties of an element are determined by the number and arrangement of the electrons in the atom. Because each proton has a positive charge $+e$, the nucleus has a total positive charge $+Ze$. The electrons are negatively charged ($-e$), so they are attracted to the nucleus and repelled by each other. Since electrons and protons have equal but opposite charges, and there are an equal number of electrons and protons in an atom, atoms are electrically neutral. Atoms that lose or gain one or more electrons are then electrically charged and are called *ions*.

➤ We will begin our study of atoms by discussing the Bohr model, a semi-classical model developed by Niels Bohr in 1913 to explain the spectra emitted by hydrogen atoms. Although this *prequantum mechanics* model has many shortcomings, it provides a useful framework for the discussion of atomic phenomena. For example, we now know that the electron does not circle the nucleus in well-defined orbits, as in the Bohr model, but instead is described by a wave function that satisfies the Schrödinger equation. However, the probability distributions that follow from the full quantum theory do in fact have maxima at the positions of the Bohr orbits. After discussing the Bohr model, we will apply our knowledge of quantum mechanics from Chapter 35 to give a qualitative description of the hydrogen atom. We will then discuss the structure of other atoms and the periodic table of the elements. Finally, we will discuss both optical and X-ray spectra.

36-1 The Nuclear Atom

Atomic Spectra

By the beginning of the twentieth century, a large body of data had been collected on the emission of light by atoms in a gas when the atoms are excited by an electric discharge. Viewed through a spectroscope with a narrow-slit aperture, this light appears as a discrete set of lines of different colors or wavelengths; the spacing and intensities of the lines are characteristic of the element. The wavelengths of these spectral lines could be accurately determined, and much effort went into finding regularities in the spectra. Figure 36-1 shows line spectra for hydrogen and for mercury.

In 1885 a Swiss schoolteacher, Johann Balmer, found that the wavelengths of the lines in the visible spectrum of hydrogen can be represented by the formula

$$\lambda = (364.6 \text{ nm})\frac{m^2}{m^2 - 4}, \quad m = 3, 4, 5, \dots \qquad \text{36-1}$$

Balmer suggested that this might be a special case of a more general expression that would be applicable to the spectra of other elements. Such an expression, found by Johannes R. Rydberg and Walter Ritz and known as the **Rydberg–Ritz formula,** gives the reciprocal wavelength as

(a)

(b)

FIGURE 36-1 (*a*) Line spectrum of hydrogen and (*b*) line spectrum of mercury.

$$\frac{1}{\lambda} = R\left(\frac{1}{n_2^2} - \frac{1}{n_1^2}\right) \qquad\qquad 36\text{-}2$$

where n_1 and n_2 are integers, with $n_1 > n_2$, and R is the Rydberg constant. The Rydberg constant is the same for all spectral series of the same element and varies only slightly in a regular way from element to element. For hydrogen, R has the value

$$R_H = 1.097776 \times 10^7 \text{ m}^{-1}$$

The Rydberg–Ritz formula gives the wavelengths for all the lines in the spectra of hydrogen, as well as alkali elements such as lithium and sodium. The hydrogen Balmer series given by Equation 36-1 is also given by Equation 36-2, with $R = R_H$, $n_2 = 2$, and $n_1 = m$.

Many attempts were made to construct a model of the atom that would yield these formulas for its radiation spectrum. The most popular model, due to J. J. Thomson, considered various arrangements of electrons embedded in some kind of fluid that contained most of the mass of the atom and had enough positive charge to make the atom electrically neutral. Thomson's model, called the "plum pudding" model, is illustrated in Figure 36-2. Since classical electromagnetic theory predicted that a charge oscillating with frequency f would radiate electromagnetic energy of that frequency, Thomson searched for configurations that were stable and that had normal modes of vibration of frequencies equal to those of the spectrum of the atom. A difficulty of this model and all other models was that, according to classical physics, electric forces alone cannot produce stable equilibrium. Thomson was unsuccessful in finding a model that predicted the observed frequencies for any atom.

The Thomson model was essentially ruled out by a set of experiments by H. W. Geiger and E. Marsden, under the supervision of E. Rutherford in approximately 1911, in which alpha particles from radioactive radium were scattered by atoms in a gold foil. Rutherford showed that the number of alpha particles scattered at large angles could not be accounted for by an atom in which the positive charge was distributed throughout the atomic size (known to be about 0.1 nm in diameter) but required that the positive charge and most of the mass of the atom be concentrated in a very small region, now called the nucleus, of diameter of the order of 10^{-6} nm = 1 fm.

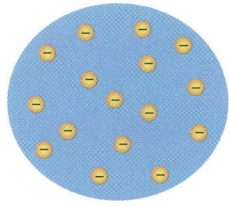

FIGURE 36-2 J. J. Thomson's plum pudding model of the atom. In this model, the negative electrons are embedded in a fluid of positive charge. For a given configuration in such a system, the resonance frequencies of oscillations of the electrons can be calculated. According to classical theory, the atom should radiate light of frequency equal to the frequency of oscillation of the electrons. Thomson could not find any configuration that would give frequencies in agreement with the measured frequencies of the spectrum of any atom.

36-2 The Bohr Model of the Hydrogen Atom

Niels Bohr, working in the Rutherford laboratory in 1912, proposed a model of the hydrogen atom that extended the work of Planck, Einstein, and Rutherford and successfully predicted the observed spectra. According to Bohr's model, the electron of the hydrogen atom moves under the influence of the Coulomb attraction to the positive nucleus according to classical mechanics, which predicts circular or elliptical orbits with the force center at one focus, as in the motion of the planets around the sun. For simplicity, Bohr chose a circular orbit, as shown in Figure 36-3.

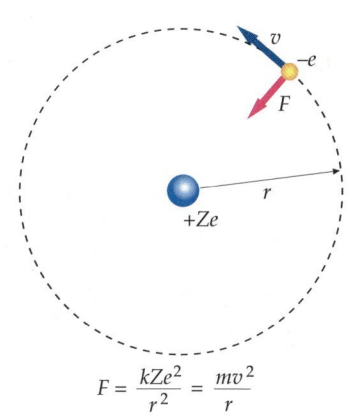

$$F = \frac{kZe^2}{r^2} = \frac{mv^2}{r}$$

FIGURE 36-3 Electron of charge $-e$ traveling in a circular orbit of radius r around the nuclear charge $+Ze$. The attractive electrical force kZe^2/r^2 keeps the electron in its orbit.

Energy for a Circular Orbit

Consider an electron of charge $-e$ moving in a circular orbit of radius r about a positive charge Ze such as the nucleus of a hydrogen atom ($Z = 1$) or of a singly ionized helium atom ($Z = 2$). The total energy of the electron can be related to the radius of the orbit. The potential energy of the electron of charge $-e$ at a distance r from a positive charge Ze is

$$U = \frac{kq_1q_2}{r} = \frac{k(Ze)(-e)}{r} = -\frac{kZe^2}{r} \qquad \text{36-3}$$

where k is the Coulomb constant. The kinetic energy K can be obtained as a function of r by using Newton's second law, $F_{\text{net}} = ma$. Setting the Coulomb attractive force equal to the mass times the centripetal acceleration gives

$$\frac{kZe^2}{r^2} = m\frac{v^2}{r} \qquad \text{36-4}a$$

Multiplying both sides by $r/2$ gives

$$K = \frac{1}{2}mv^2 = \frac{1}{2}\frac{kZe^2}{r} \qquad \text{36-4}b$$

Thus, the kinetic energy and the potential energy vary inversely with r. Note that the magnitude of the potential energy is twice that of the kinetic energy:

$$U = -2K \qquad \text{36-5}$$

This is a general result in $1/r^2$ force fields. It also holds for circular orbits in a gravitational field (see Example 11-6 in Section 11-3). The total energy is the sum of the kinetic energy and the potential energy:

$$E = K + U = \frac{1}{2}\frac{kZe^2}{r} - \frac{kZe^2}{r}$$

or

$$E = -\frac{1}{2}\frac{kZe^2}{r} \qquad \text{36-6}$$

ENERGY IN A CIRCULAR ORBIT FOR A $1/r^2$ FORCE

Although mechanical stability is achieved because the Coulomb attractive force provides the centripetal force necessary for the electron to remain in orbit, classical *electromagnetic* theory says that such an atom would be unstable electrically. The atom would be unstable because the electron must accelerate when moving in a circle and therefore radiate electromagnetic energy of frequency equal to that of its motion. According to the classical theory, such an atom would quickly collapse, with the electron spiraling into the nucleus as it radiates away its energy.

Bohr's Postulates

Bohr circumvented the difficulty of the collapsing atom by *postulating* that only certain orbits, called stationary states, are allowed, and that in these orbits the electron does not radiate. An atom radiates only when the electron makes a transition from one allowed orbit (stationary state) to another.

The electron in the hydrogen atom can move only in certain nonradiating, circular orbits called stationary states.

The second postulate relates the frequency of radiation to the energies of the stationary states. If E_i and E_f are the initial and final energies of the atom, the frequency of the emitted radiation during a transition is given by

$$f = \frac{E_i - E_f}{h} \qquad \text{36-7}$$

BOHR'S SECOND POSTULATE—PHOTON FREQUENCY FROM ENERGY CONSERVATION

where h is Planck's constant. This postulate is equivalent to the assumption of conservation of energy with the emission of a photon of energy hf. Combining Equation 36-6 and Equation 36-7, we obtain for the frequency

$$f = \frac{E_1 - E_2}{h} = \frac{1}{2}\frac{kZe^2}{h}\left(\frac{1}{r_2} - \frac{1}{r_1}\right) \qquad \text{36-8}$$

where r_1 and r_2 are the radii of the initial and final orbits.

To obtain the frequencies implied by the Rydberg–Ritz formula, $f = c/\lambda = cR(1/n_2^2 - 1/n_1^2)$, it is evident that the radii of stable orbits must be proportional to the squares of integers. Bohr searched for a quantum condition for the radii of the stable orbits that would yield this result. After much trial and error, Bohr found that he could obtain it if he postulated that the angular momentum of the electron in a stable orbit equals an integer times $\hbar$ ("bar," Planck's constant divided by 2π). Since the angular momentum of a circular orbit is just mvr, this postulate is

$$mvr = \frac{nh}{2\pi} = n\hbar, \quad n = 1, 2, 3, \ldots \qquad \text{36-9}$$

BOHR'S THIRD POSTULATE—QUANTIZED ANGULAR MOMENTUM

where $\hbar = h/2\pi = 1.055 \times 10^{-34}$ J·s $= 6.582 \times 10^{-16}$ eV·s.

Equation 36-9 relates the speed v to the radius r. Equation 36-4a, from Newton's second law, gives us another equation relating the speed to the radius:

$$\frac{kZe^2}{r^2} = m\frac{v^2}{r}$$

or

$$v^2 = \frac{kZe^2}{mr} \qquad \text{36-10}$$

We can determine r by eliminating v between Equations 36-9 and 36-10. Solving Equation 36-9 for v and squaring gives

$$v^2 = n^2\frac{\hbar^2}{m^2r^2}$$

Equating this expression for v^2 with the expression given by Equation 36-10, we get

$$n^2 \frac{\hbar^2}{m^2 r^2} = \frac{kZe^2}{mr}$$

Solving for r, we obtain

$$r = n^2 \frac{\hbar^2}{mkZe^2} = n^2 \frac{a_0}{Z} \qquad \text{36-11}$$

RADIUS OF THE BOHR ORBITS

where a_0 is called the **first Bohr radius.**

$$a_0 = \frac{\hbar^2}{mke^2} \approx 0.0529 \text{ nm} \qquad \text{36-12}$$

FIRST BOHR RADIUS

Substituting the expressions for r in Equation 36-11 into Equation 36-8 for the frequency gives

$$f = \frac{1}{2} \frac{kZe^2}{h} \left(\frac{1}{r_2} - \frac{1}{r_1} \right) = Z^2 \frac{mk^2e^4}{4\pi\hbar^3} \left(\frac{1}{n_2^2} - \frac{1}{n_1^2} \right) \qquad \text{36-13}$$

If we compare this expression with $Z = 1$ for $f = c/\lambda$ with the empirical Rydberg–Ritz formula (Equation 36-2), we obtain for the Rydberg constant

$$R = \frac{mk^2e^4}{4\pi c\hbar^3} \qquad \text{36-14}$$

Using the values of m, e, and $\hbar$ known in 1913, Bohr calculated R and found his result to agree (within the limits of the uncertainties of the constants) with the value obtained from spectroscopy.

STANDING-WAVE CONDITION IMPLIES QUANTIZATION OF ANGULAR MOMENTUM

EXAMPLE 36-1

For waves in a circle, the standing-wave condition is that there is an integral number of wavelengths in the circumference. That is, $n\lambda = 2\pi r$, where $n = 1, 2, 3$, and so on. Show that this condition for electron waves implies quantization of angular momentum.

1. Write the standing-wave condition:

$$n\lambda = 2\pi r$$

2. Use the de Broglie relation (Equation 34-10) to relate the momentum p to λ:

$$p = \frac{h}{\lambda} = \frac{nh}{2\pi r} = n\frac{\hbar}{r}$$

3. The angular momentum of an electron in a circular orbit is $mvr = pr$, where $p = mv$:

$$L = mvr = pr = \boxed{n\hbar}$$

Energy Levels

The total mechanical energy of the electron in the hydrogen atom is related to the radius of the circular orbit by Equation 36-6. If we substitute the quantized values of r as given by Equation 36-11, we obtain

$$E_n = -\frac{1}{2}\frac{kZe^2}{r} = -\frac{1}{2}\frac{kZ^2e^2}{n^2a_0} = -\frac{1}{2}\frac{mk^2Z^2e^4}{n^2\hbar^2}$$

or

$$E_n = -Z^2\frac{E_0}{n^2} \qquad\qquad 36\text{-}15$$

ENERGY LEVELS IN THE HYDROGEN ATOM

where

$$E_0 = \frac{mk^2e^4}{2\hbar^2} = \frac{1}{2}\frac{ke^2}{a_0} \approx 13.6\text{ eV} \qquad\qquad 36\text{-}16$$

The energies E_n with $Z = 1$ are the quantized allowed energies for the hydrogen atom.

Transitions between these allowed energies result in the emission or absorption of a photon whose frequency is given by $f = (E_i - E_f)/h$, and whose wavelength is

$$\lambda = \frac{c}{f} = \frac{hc}{E_i - E_f} \qquad\qquad 36\text{-}17$$

As we found in Chapter 34, it is convenient to have the value of hc in electron-volt nanometers:

$$hc = 1240\text{ eV·nm} \qquad\qquad 36\text{-}18$$

Since the energies are quantized, the frequencies and wavelengths of the radiation emitted by the hydrogen atom are quantized in agreement with the observed line spectrum.

Figure 36-4 shows the energy-level diagram for hydrogen. The energy of the hydrogen atom in the ground state is $E_1 = -13.6$ eV. As n approaches infinity the energy approaches zero. The process of removing an electron from an atom is called ionization, and the energy required to remove the electron is the **ionization energy.** The ionization energy of the ground-state hydrogen atom, which is also its binding energy, is 13.6 eV. A few transitions from a higher state to a lower state are indicated in Figure 36-4. When Bohr published his model of the hydrogen atom, the Balmer series, corresponding to $n_2 = 2$ and $n_1 = 3, 4, 5$, and so on; and the Paschen series, corresponding to $n_2 = 3$ and $n_1 = 4, 5, 6$, and so on, were known. In 1916, T. Lyman found the series corresponding to $n_2 = 1$, and in 1922 and 1924, F. Brackett and H. A. Pfund, respectively, found the series corresponding to $n_2 = 4$ and $n_2 = 5$. Only the Balmer series lies in the visible portion of the electromagnetic spectrum.

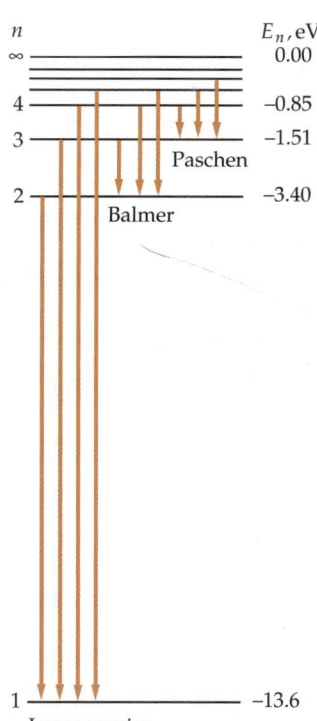

FIGURE 36-4 Energy-level diagram for hydrogen showing the first few transitions in each of the Lyman, Balmer, and Paschen series. The energies of the levels are given by Equation 36-15.

EXAMPLE 36-2

Find (a) the energy and (b) the wavelength of the line with the longest wavelength in the Lyman series.

PICTURE THE PROBLEM From Figure 36-4, we can see that the Lyman series corresponds to transitions ending at the ground-state energy, $E_f = E_1 = -13.6$ eV. Since λ varies inversely with energy, the transition with the longest wavelength is the transition with the lowest energy, which is that from the first excited state $n = 2$ to the ground state $n = 1$.

1. The energy of the photon is the difference in the energies of the initial and final atomic state:

$$E_{photon} = \Delta E_{atom} = E_i - E_f$$

$$= E_2 - E_1 = \frac{-13.6 \text{ eV}}{2^2} - \frac{-13.6 \text{ eV}}{1^2}$$

$$= -3.40 \text{ eV} + 13.6 \text{ eV} = \boxed{10.2 \text{ eV}}$$

2. The wavelength of the photon is:

$$\lambda = \frac{hc}{E_2 - E_1} = \frac{1240 \text{ eV·nm}}{10.2 \text{ eV}} = \boxed{121.6 \text{ nm}}$$

REMARKS This photon is outside the visible spectrum, in the ultraviolet region. Since all the other lines in the Lyman series have even greater energies and shorter wavelengths, the Lyman series is completely in the ultraviolet region.

 EXERCISE Find the shortest wavelength for a line in the Lyman series. (*Answer* 91.2 nm)

Despite its spectacular successes, the Bohr model of the hydrogen atom had many shortcomings. There was no justification for the postulates of stationary states or for the quantization of angular momentum other than the fact that these postulates led to energy levels that agreed with spectroscopic data. Furthermore, attempts to apply the model to more complicated atoms had little success. The quantum-mechanical theory resolves these difficulties. The stationary states of the Bohr model correspond to the standing-wave solutions of the Schrödinger equation analogous to the standing electron waves for a particle in a box discussed in Chapter 34 and Chapter 35. Energy quantization is a direct consequence of the standing-wave solutions of the Schrödinger equation. For hydrogen, these quantized energies agree with those obtained from the Bohr model and with experiment. The quantization of angular momentum that had to be postulated in the Bohr model is predicted by the quantum theory.

36-3 Quantum Theory of Atoms

The Schrödinger Equation in Spherical Coordinates

In quantum theory, the electron is described by its wave function ψ. The probability of finding the electron in some volume dV of space equals the product of the absolute square of the electron wave function $|\psi|^2$ and dV. Boundary conditions on the wave function lead to the quantization of the wavelengths and frequencies and thereby to the quantization of the electron energy.

Consider a single electron of mass m moving in three dimensions in a region in which the potential energy is U. The time-independent Schrödinger equation for such a particle is given by Equation 35-30:

$$-\frac{\hbar^2}{2m}\left(\frac{\partial^2\psi}{\partial x^2}+\frac{\partial^2\psi}{\partial y^2}+\frac{\partial^2\psi}{\partial z^2}\right)+U\psi=E\psi \qquad\qquad 36\text{-}19$$

For a single isolated atom, the potential energy U depends only on the radial distance $r = \sqrt{x^2 + y^2 + z^2}$. The problem is then most conveniently treated using the spherical coordinates r, θ, and ϕ, which are related to the rectangular coordinates x, y, and z by

$$z = r\cos\theta$$

$$x = r\sin\theta\cos\phi$$

$$y = r\sin\theta\sin\phi \qquad\qquad 36\text{-}20$$

These relations are shown in Figure 36-5. The transformation of the bracketed term in Equation 36-19 is straightforward but involves much tedious calculation, which we will omit. The result is

$$\frac{\partial^2\psi}{\partial x^2}+\frac{\partial^2\psi}{\partial y^2}+\frac{\partial^2\psi}{\partial z^2}=\frac{1}{r^2}\frac{\partial}{\partial r}\left(r^2\frac{\partial\psi}{\partial r}\right)+\frac{1}{r^2}\left[\frac{1}{\sin\theta}\frac{\partial}{\partial\theta}\left(\sin\theta\frac{\partial\psi}{\partial\theta}\right)+\frac{1}{\sin^2\theta}\frac{\partial^2\psi}{\partial\phi^2}\right]$$

Substituting into Equation 36-19 gives

$$-\frac{\hbar^2}{2mr^2}\frac{\partial}{\partial r}\left(r^2\frac{\partial\psi}{\partial r}\right)-\frac{\hbar^2}{2mr^2}\left[\frac{1}{\sin\theta}\frac{\partial}{\partial\theta}\left(\sin\theta\frac{\partial\psi}{\partial\theta}\right)+\frac{1}{\sin^2\theta}\frac{\partial^2\psi}{\partial\phi^2}\right]+U(r)\psi=E\psi$$

$$36\text{-}21$$

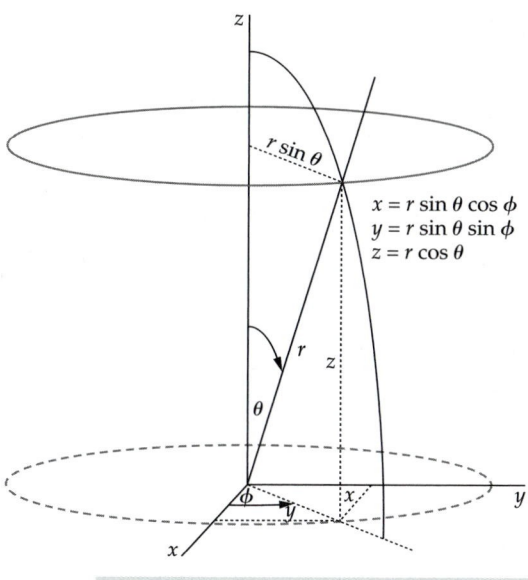

FIGURE 36-5 Geometric relations between spherical coordinates and rectangular coordinates.

Despite the formidable appearance of this equation, it was not difficult for Schrödinger to solve because it is similar to other partial differential equations in classical physics that had been thoroughly studied. We will not solve this equation but merely discuss qualitatively some of the interesting features of the wave functions that satisfy it.

The first step in the solution of a partial differential equation, such as Equation 36-21, is to separate the variables by writing the wave function $\psi(r, \theta, \phi)$ as a product of functions of each single variable:

$$\psi(r, \theta, \phi) = R(r)f(\theta)g(\phi) \qquad\qquad 36\text{-}22$$

where R depends only on the radial coordinate r, f depends only on θ, and g depends only on ϕ. When this form of $\psi(r, \theta, \phi)$ is substituted into Equation 36-21, the partial differential equation can be transformed into three ordinary differential equations, one for $R(r)$, one for $f(\theta)$, and one for $g(\phi)$. The potential energy $U(r)$ appears only in the equation for $R(r)$, which is called the **radial equation.** The particular form of $U(r)$ given in Equation 36-19 therefore has no effect on the solutions of the equations for $f(\theta)$ and $g(\phi)$, and therefore has no effect on the angular dependence of the wave function $\psi(r, \theta, \phi)$. These solutions are applicable to any problem in which the potential energy depends only on r.

Quantum Numbers in Spherical Coordinates

In three dimensions, the requirement that the wave function be continuous and normalizable introduces three quantum numbers, one associated with each spatial dimension. In spherical coordinates the quantum number associated with r is labeled n, that associated with θ is labeled ℓ, and that associated with ϕ is labeled m_ℓ.[†] The quantum numbers n_1, n_2, and n_3 that we found in Chapter 35

[†] For simplicity, m_ℓ is sometimes written as m.

for a particle in a three-dimensional square well in rectangular coordinates x, y, and z were independent of one another, but the quantum numbers associated with wave functions in spherical coordinates are interdependent. The possible values of these quantum numbers are

$$n = 1, 2, 3, \ldots$$

$$\ell = 0, 1, 2, 3, \ldots, n - 1$$

$$m_\ell = -\ell, (-\ell + 1), \ldots, -2, -1, 0, 1, 2, \ldots, (\ell + 1), \ell \qquad 36\text{-}23$$

QUANTUM NUMBERS IN SPHERICAL COORDINATES

That is, n can be any positive integer; ℓ can be 0 or any positive integer up to $n - 1$; and m_ℓ can have $2\ell + 1$ possible values, ranging from $-\ell$ to $+\ell$ in integral steps.

The number n is called the **principal quantum number.** It is associated with the dependence of the wave function on the distance r and therefore with the probability of finding the electron at various distances from the nucleus. The quantum numbers ℓ and m_ℓ are associated with the angular momentum of the electron and with the angular dependence of the electron wave function. The quantum number ℓ is called the **orbital quantum number.** The magnitude L of the orbital angular momentum $\vec{L}$ is related to the orbital quantum number ℓ by

$$L = \sqrt{\ell(\ell + 1)}\hbar \qquad 36\text{-}24$$

The quantum number m_ℓ is called the **magnetic quantum number.** It is related to the component of the angular momentum along some direction in space. All spatial directions are equivalent for an isolated atom, but placing the atom in a magnetic field results in the direction of the magnetic field being separated out from the other directions. The convention is that the z direction is chosen for the magnetic-field direction. Then the z component of the angular momentum of the electron is given by the quantum condition

$$L_z = m_\ell \hbar \qquad 36\text{-}25$$

This quantum condition arises from the boundary condition on the azimuth coordinate ϕ that the probability of finding the electron at some arbitrary angle ϕ_1 must be the same as that of finding the electron at angle $\phi_1 + 2\pi$ because these are the same points in space.

If we measure the angular momentum of the electron in units of $\hbar$, we see that the angular-momentum magnitude is quantized to the value $\sqrt{\ell(\ell + 1)}$ units and that its component along any direction can have only the $2\ell + 1$ values ranging from $-\ell$ to $+\ell$ units. Figure 36-6 shows a vector-model diagram illustrating the possible orientations of the angular-momentum vector for $\ell = 2$. Note that only specific values of θ are allowed; that is, the directions in space are quantized.

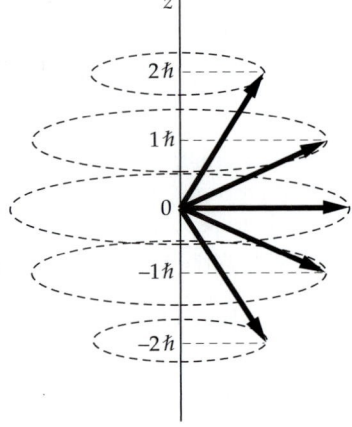

FIGURE 36-6 Vector-model diagram illustrating the possible values of the z component of the angular-momentum vector for the case $\ell = 2$. The magnitude of the angular momentum is $L = \hbar\sqrt{\ell(\ell + 1)} = \hbar\sqrt{2(2 + 1)} = \hbar\sqrt{6}$.

The Directions of the Angular Momentum   **E X A M P L E 3 6 - 3**

If the angular momentum is characterized by the quantum number $\ell = 2$, what are the possible values of L_z, and what is the smallest possible angle between $\vec{L}$ and the z axis?

PICTURE THE PROBLEM The possible orientations of $\vec{L}$ and the z axis are shown in Figure 36-6. The z-axis direction is parallel with that of the external magnetic field in the vicinity of the atom.

1. Write the possible values of L_z:

$$\boxed{L_z = m_\ell \hbar, \text{ where } m_\ell = -2, -1, 0, 1, 2}$$

2. Express the angle θ between $\vec{L}$ and the z axis in terms of L and L_z:

$$\cos\theta = \frac{L_z}{L} = \frac{m\hbar}{\sqrt{\ell(\ell+1)}\hbar} = \frac{m}{\sqrt{\ell(\ell+1)}}$$

3. The smallest angle occurs when $m_\ell = \ell = 2$:

$$\cos\theta_{min} = \frac{2}{\sqrt{2(2+1)}} = \frac{2}{\sqrt{6}} = 0.816$$

$$\theta_{min} = \boxed{35.3°}$$

REMARKS We note the somewhat strange result that the angular-momentum vector cannot lie along the z axis.

EXERCISE An atom in a region with a magnetic field has an angular momentum characterized by the quantum number $\ell = 4$. What are the possible values of m_ℓ? (*Answer* $-4, -3, -2, -1, 0, 1, 2, 3, 4$)

36-4 Quantum Theory of the Hydrogen Atom

We can treat the simplest atom, the hydrogen atom, as a stationary nucleus, a proton, that has a single moving particle, an electron, with kinetic energy $p^2/2m$. The potential energy $U(r)$ due to the electrostatic attraction between the electron and the proton[†] is

$$U(r) = -\frac{kZe^2}{r} \qquad\qquad 36\text{-}26$$

For this potential-energy function, the Schrödinger equation can be solved exactly. In the lowest energy state, which is the ground state, the principal quantum number n has the value 1, ℓ is 0, and m_ℓ is 0.

Energy Levels

The allowed energies of the hydrogen atom that result from the solution of the Schrödinger equation are

$$E_n = -\frac{mk^2e^4}{2\hbar^2 n^2} = -Z^2\frac{E_0}{n^2}, \quad n = 1, 2, 3, \ldots \qquad\qquad 36\text{-}27$$

ENERGY LEVELS FOR HYDROGEN

where

$$E_0 = -\frac{mk^2e^4}{2\hbar^2} \approx 13.6 \text{ eV} \qquad\qquad 36\text{-}28$$

---

† We include the factor Z, which is 1 for hydrogen, so that we can apply our results to other one-electron atoms, such as singly ionized helium He$^+$, for which Z = 2.

These energies are the same as in the Bohr model. Note that the energy is negative, indicating that the electron is bound to the nucleus (thus the term *bound state*), and that the energy depends only on the principal quantum number n. The fact that the energy does not depend on the orbital quantum number ℓ is a peculiarity of the inverse-square force and holds only for an inverse r potential such as Equation 36-26. For more complicated atoms having several electrons, the interaction of the electrons leads to a dependence of the energy on ℓ. In general, the lower the value of ℓ, the lower the energy for such atoms. Since there is usually no preferred direction in space, the energy for any atom does not ordinarily depend on the magnetic quantum number m_ℓ, which is related to the z component of the angular momentum. The energy does depend on m_ℓ if the atom is in a magnetic field.

Figure 36-7 shows an energy-level diagram for hydrogen. This diagram is similar to Figure 36-4, except that the states with the same value of n but with different values of ℓ are shown separately. These states (called *terms*) are referred to by giving the value of n along with a code letter: s for $\ell = 0$, p for $\ell = 1$, d for $\ell = 2$, and f for $\ell = 3$.[†] (Lowercase letters s, p, d, f, and so on, are used to identify the orbital angular momentum of an individual electron; uppercase letters S, P, D, F, and so on, are used to identify the orbital angular momentum for the entire multielectron atom. For hydrogen, either uppercase or lowercase letters will do, but most people use lowercase as we have done.) When an atom makes a transition from one allowed energy state to another, electromagnetic radiation in the form of a photon is emitted or absorbed. Such transitions result in spectral lines that are characteristic of the atom. The transitions obey the **selection rules:**

$$\Delta m_\ell = 0 \text{ or } \pm 1$$

$$\Delta \ell = \pm 1 \qquad\qquad 36\text{-}29$$

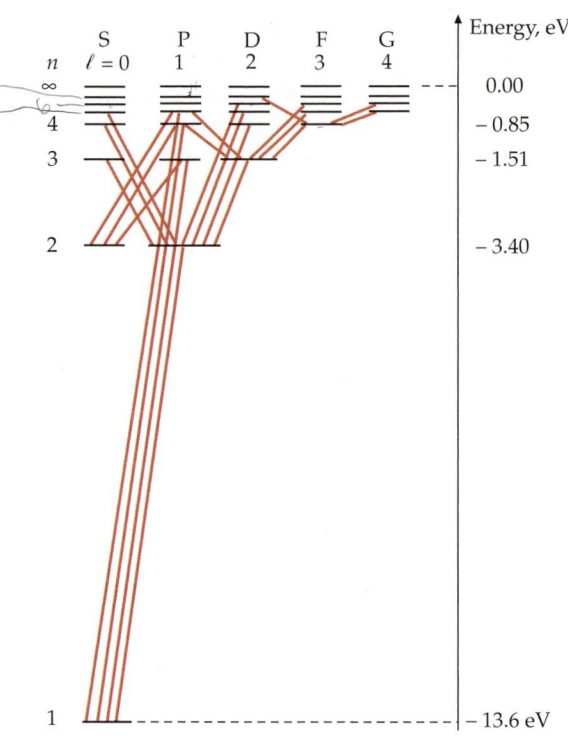

FIGURE 36-7 Energy-level diagram for hydrogen. The diagonal lines show transitions that involve emission or absorption of radiation and obey the selection rule $\Delta\ell = \pm 1$. States with the same value of n but with different values of ℓ have the same energy $-E_0/n^2$, where $E_0 = 13.6$ eV as in the Bohr model.

These selection rules are related to the conservation of angular momentum and to the fact that the photon itself has an intrinsic angular momentum that has a maximum component along any axis of $1\hbar$. The wavelengths of the light emitted by hydrogen (and by other atoms) are related to the energy levels by

$$hf = \frac{hc}{\lambda} = E_i - E_f \qquad\qquad 36\text{-}30$$

where E_i and E_f are the energies of the initial and final states.

Wave Functions and Probability Densities

The solutions of the Schrödinger equation in spherical coordinates are characterized by the quantum numbers n, ℓ, and m_ℓ, and are written $\psi_{n\ell m_\ell}$. For any given value of n, there are n possible values of ℓ ($\ell = 0, 1, \ldots, n - 1$), and for each value of ℓ, there are $2\ell + 1$ possible values of m_ℓ. For hydrogen, the energy depends only on n, so there are generally many different wave functions that correspond to the same energy (except at the lowest energy level, for which $n = 1$ and therefore ℓ and m_ℓ must be 0). These energy levels are therefore degenerate (see Section 35-5). The origins of this degeneracy are the $1/r$ dependence of the potential energy and the fact that, in the absence of any external fields, there is no preferred direction in space.[‡]

[†] These code letters are remnants of spectroscopists descriptions of various spectral lines as *sharp, principal, diffuse,* and *fundamental.* For values greater than 3, the letters follow alphabetically; thus, g is used for $\ell = 4$, and so forth.
[‡] If spin, relativistic effects, the spin of the nucleus, and quantum electrodynamics are considered, the degeneracy is broken.

The Ground State In the lowest energy state, the ground state of hydrogen, the principal quantum number n has the value 1, ℓ is 0, and m_ℓ is 0. The energy is -13.6 eV, and the angular momentum is zero. (In the Bohr model of the atom the angular momentum in the ground state is equal to $\hbar$, not zero.) The wave function for the ground state is

$$\psi_{1,0,0} = C_{1,0,0}e^{-Zr/a_0} \qquad\qquad 36\text{-}31$$

where

$$a_0 = \frac{\hbar^2}{mke^2} = 0.0529 \text{ nm}$$

is the first Bohr radius and $C_{1,0,0}$ is a constant that is determined by normalization. In three dimensions, the normalization condition is

$$\int |\psi|^2\, dV = 1$$

where dV is a volume element and the integration is performed over all space. In spherical coordinates, the volume element (Figure 36-8) is

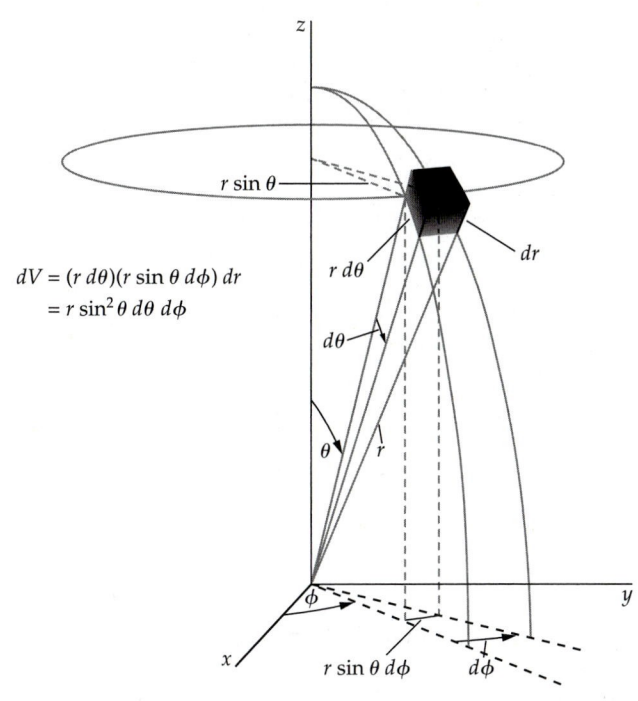

$$dV = (r\, d\theta)(r \sin \theta\, d\phi)\, dr$$
$$= r \sin^2 \theta\, d\theta\, d\phi$$

FIGURE 36-8 Volume element in spherical coordinates.

$$dV = (r \sin \theta\, d\, \phi)(r\, d\, \theta)\, dr = r^2 \sin \theta\, d\theta\, d\phi\, dr$$

We integrate over all space by integrating over ϕ, from $\phi = 0$ to $\phi = 2\pi$, over θ, from $\theta = 0$ to $\theta = \pi$; and over r, from $r = 0$ to $r = \infty$. The normalization condition is thus

$$\int |\psi|^2\, dV = \int_0^\infty \left[\int_0^\pi \left(\int_0^{2\pi} |\psi|^2 r^2 \sin \theta\, d\, \phi \right) d\,\theta \right] dr$$

$$= \int_0^\infty \left[\int_0^\pi \left(\int_0^{2\pi} C_{1,0,0}^2 e^{-2Zr/a_0} r^2 \sin \theta\, d\, \phi \right) d\,\theta \right] dr = 1$$

Since there is no θ or ϕ dependence in $\psi_{1,0,0}$, the triple integral can be factored into the product of three integrals. This gives

$$\int |\psi|^2\, dV = \left(\int_0^{2\pi} d\,\phi \right) \left(\int_0^\pi \sin \theta\, d\,\theta \right) \left(\int_0^\infty C_{1,0,0}^2 e^{-2Zr/a_0} r^2\, dr \right)$$

$$= 2\pi \cdot 2 \cdot C_{1,0,0}^2 \left(\int_0^\infty r^2 e^{-2Zr/a_0}\, dr \right) = 1$$

The remaining integral is of the form $\int_0^\infty x^n e^{-ax}\, dx$, with n a positive integer and with $a > 0$. Using successive integration-by-parts operations[†] yields the result

$$\int_0^\infty x^n e^{-ax}\, dx = \frac{n!}{a^{n+1}}$$

---

† This integral can also be looked up in a table of integrals.

so

$$\int_0^\infty r^2 e^{-2Zr/a_0}\,dr = \frac{a_0^3}{4Z^3}$$

Then

$$4\pi C_{1,0,0}^2\left(\frac{a_0^3}{4Z^3}\right) = 1$$

so

$$C_{1,0,0} = \frac{1}{\sqrt{\pi}}\left(\frac{Z}{a_0}\right)^{3/2} \qquad\qquad 36\text{-}32$$

The normalized ground-state wave function is thus

$$\psi_{1,0,0} = \frac{1}{\sqrt{\pi}}\left(\frac{Z}{a_0}\right)^{3/2} e^{-Zr/a_0} \qquad\qquad 36\text{-}33$$

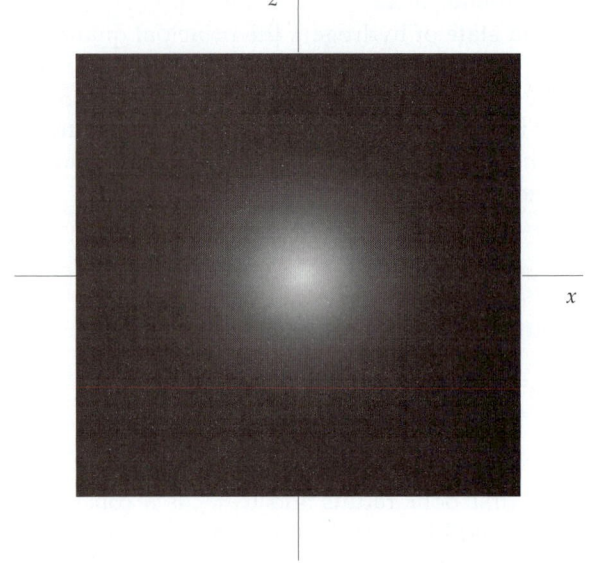

FIGURE 36-9 Computer-generated picture of the probability density $|\psi|^2$ for the ground state of hydrogen. The quantity $-e|\psi|^2$ can be thought of as the electron charge density in the atom. The density is spherically symmetric, is greatest at the origin, and decreases exponentially with r.

The probability of finding the electron in a volume dV is $|\psi|^2\,dV$. The probability density $|\psi|^2$ is illustrated in Figure 36-9. Note that this probability density is spherically symmetric; that is, the probability density depends only on r, and is independent of θ or ϕ. The probability density is maximum at the origin.

We are more often interested in the probability of finding the electron at some radial distance r between r and $r + dr$. This radial probability $P(r)\,dr$ is the probability density $|\psi|^2$ times the volume of the spherical shell of thickness dr, which is $dV = 4\pi r^2\,dr$. The probability of finding the electron in the range from r to $r + dr$ is thus $P(r)\,dr = |\psi|^2 4\pi r^2\,dr$, and the **radial probability density** is

$$P(r) = 4\pi r^2 |\psi|^2 \qquad\qquad 36\text{-}34$$

RADIAL PROBABILITY DENSITY

For the hydrogen atom in the ground state, the radial probability density is

$$P(r) = 4\pi r^2 |\psi|^2 = 4\pi C_{1,0,0}^2 r^2 e^{-2Zr/a_0} = 4\left(\frac{Z}{a_0}\right)^3 r^2 e^{-2Zr/a_0} \qquad\qquad 36\text{-}35$$

Figure 36-10 shows the radial probability density $P(r)$ as a function of r. The maximum value of $P(r)$ occurs at $r = a_0/Z$, which for $Z = 1$ is the first Bohr radius. In contrast to the Bohr model, in which the electron stays in a well-defined orbit at $r = a_0$, we see that it is possible for the electron to be found at any distance from the nucleus. However, the most probable distance is a_0 (assuming $Z = 1$), and the chance of finding the electron at a much different distance is small. It is often useful to think of the electron in an atom as a charged cloud of charge density $\rho = -e|\psi|^2$, but we should remember that when it interacts with matter, an electron is always observed as a single charge.

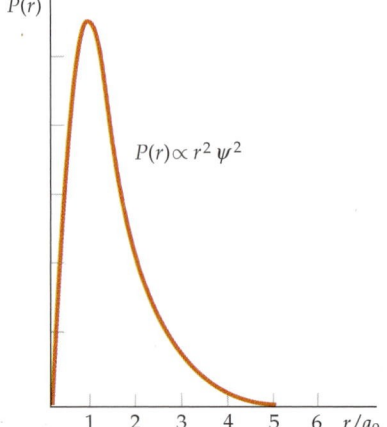

FIGURE 36-10 Radial probability density $P(r)$ versus r/a_0 for the ground state of the hydrogen atom. $P(r)$ is proportional to $r^2\psi^2$. The value of r for which $P(r)$ is maximum is the most probable distance $r = a_0$.

PROBABILITY THAT THE ELECTRON IS IN A THIN SPHERICAL SHELL **EXAMPLE 36-4**

Find the probability of finding the electron in a thin spherical shell of radius r and thickness $\Delta r = 0.06a_0$ at (a) $r = a_0$ and (b) $r = 2a_0$ for the ground state of the hydrogen atom.

PICTURE THE PROBLEM Because the range Δr is so small compared to r, the variation in the radial probability density $P(r)$ in the shell can be neglected. The probability of finding the electron in some small range Δr is then $P(r) \Delta r$.

1. Use Equation 36-35 with $Z = 1$ and $r = a_0$:

$$P(r) \Delta r = \left[4\left(\frac{1}{a_0}\right)^3 r^2 e^{-2r/a_0} \right] \Delta r = \left[4\left(\frac{1}{a_0}\right)^3 a_0^2 e^{-2} \right](0.06a_0) = \boxed{0.0325}$$

2. Use Equation 36-35 with $Z = 1$ and $r = 2a_0$:

$$P(r) \Delta r = \left[4\left(\frac{1}{a_0}\right)^3 r^2 e^{-2r/a_0} \right] \Delta r = \left[4\left(\frac{1}{a_0}\right)^3 4a_0^2 e^{-4} \right](0.06a_0) = \boxed{0.0176}$$

REMARKS There is approximately a 3 percent chance of finding the electron in this range at $r = a_0$, but at $r = 2a_0$ the chance is slightly less than 2 percent.

The First Excited State In the first excited state, $n = 2$ and ℓ can be either 0 or 1. For $\ell = 0$, $m_\ell = 0$, and we again have a spherically symmetric wave function, this time given by

$$\psi_{2,0,0} = C_{2,0,0}\left(2 - \frac{Zr}{a_0} \right) e^{-Zr/(2a_0)}$$ 36-36

For $\ell = 1$, m_ℓ can be $+1$, 0, or -1. The corresponding wave functions are

$$\psi_{2,1,0} = C_{2,1,0} \frac{Zr}{a_0} e^{-Zr/(2a_0)} \cos \theta$$ 36-37

$$\psi_{2,1,\pm1} = C_{2,1,1} \frac{Zr}{a_0} e^{-Zr/(2a_0)} \sin \theta\, e^{\pm i\phi}$$ 36-38

where $C_{2,0,0}$, $C_{2,1,0}$, and $C_{2,1,1}$ are normalization constants. The probability densities are given by

$$\psi_{2,0,0}^2 = C_{2,0,0}^2 \left(2 - \frac{Zr}{a_0} \right)^2 e^{-Zr/a_0}$$ 36-39

$$\psi_{2,1,0}^2 = C_{2,1,0}^2 \left(\frac{Zr}{a_0} \right)^2 e^{-Zr/a_0} \cos^2 \theta$$ 36-40

$$|\psi_{2,1,\pm1}|^2 = C_{2,1,1}^2 \left(\frac{Zr}{a_0} \right)^2 e^{-Zr/a_0} \sin^2 \theta$$ 36-41

The wave functions and probability densities for $\ell \neq 0$ are not spherically symmetric, but instead depend on the angle θ. The probability densities do not depend on ϕ. Figure 36-11 shows the probability density $|\psi|^2$ for $n = 2$, $\ell = 0$, and $m_\ell = 0$ (Figure 36-11a); for $n = 2$, $\ell = 1$, and $m_\ell = 0$ (Figure 36-11b); and for $n = 2$, $\ell = 1$, and $m_\ell = \pm1$ (Figure 36-11c). An important feature of these plots is that the electron cloud is spherically symmetric for $\ell = 0$ and is not spherically symmetric for $\ell \neq 0$. These angular distributions of the electron charge density depend only on the values of ℓ and m_ℓ and not on the radial part of the wave function.

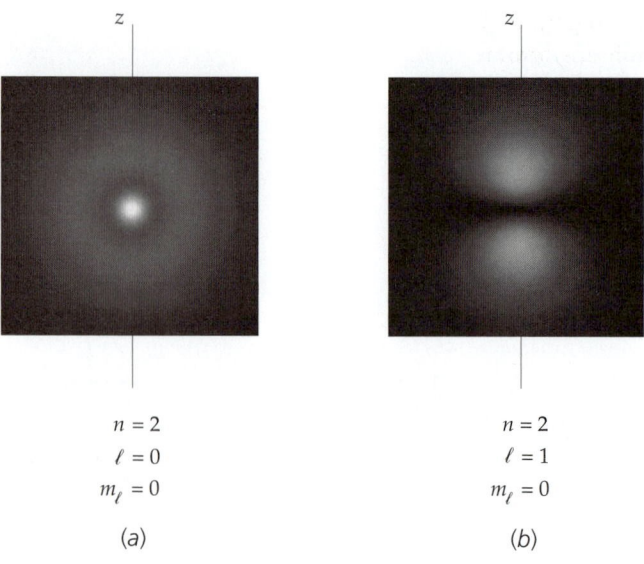

$n = 2$
$\ell = 0$
$m_\ell = 0$

(a)

$n = 2$
$\ell = 1$
$m_\ell = 0$

(b)

$n = 2$
$\ell = 1$
$m_\ell = \pm 1$

(c)

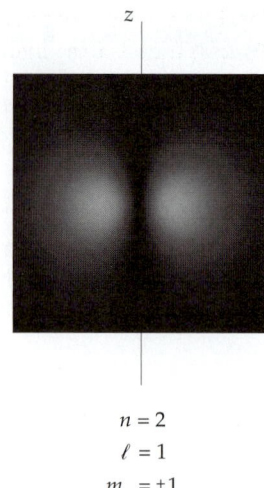

FIGURE 36-11
Computer-generated picture of the probability densities $|\psi|^2$ for the electron in the $n = 2$ states of hydrogen. All three images represent figures of revolution about the z axis. (a) For $\ell = 0$, $|\psi|^2$ is spherically symmetric. (b) For $\ell = 1$ and $m_\ell = 0$, $|\psi|^2$ is proportional to $\cos^2 \theta$. (c) For $\ell = 1$ and $m_\ell = +1$ or -1, $|\psi|^2$ is proportional to $\sin^2 \theta$.

Similar charge distributions for the valence electrons of more complicated atoms play an important role in the chemistry of molecular bonding.

Figure 36-12 shows the probability of finding the electron at a distance r as a function of r for $n = 2$, when $\ell = 1$ and when $\ell = 0$. We can see from the figure that the probability distribution depends on ℓ as well as on n.

For $n = 1$, we found that the most likely distance between the electron and the nucleus is a_0, which is the first Bohr radius, whereas for $n = 2$ and $\ell = 1$, the most likely distance between the electron and the nucleus is $4a_0$. These are the orbital radii for the first and second Bohr orbits (Equation 36-11). For $n = 3$ (and $\ell = 2$),[†] the most likely distance between the electron and nucleus is $9a_0$, which is the radius of the third Bohr orbit.

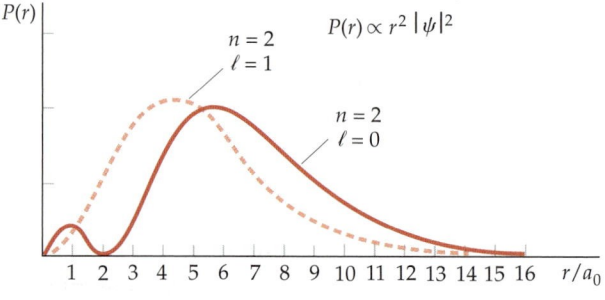

FIGURE 36-12 Radial probability density $P(r)$ versus r/a_0 for the $n = 2$ states of hydrogen. For $\ell = 1$, $P(r)$ is maximum at the Bohr value $r = 2^2 a_0$. For $\ell = 0$, there is a maximum near this value and a much smaller maximum near the origin.

36-5 The Spin–Orbit Effect and Fine Structure

The orbital magnetic moment of an atomic electron can be derived semiclassically, even though it is quantum mechanical in origin.[‡] Consider a particle of mass m and charge q moving with speed v in a circle of radius r. The magnitude of the angular momentum of the particle is $L = mvr$, and the magnitude of the magnetic moment is the product of the current and the area of the circle $\mu = IA = I\pi r^2$. If T is the time for the charge to complete one revolution, the current (charge passing a point per unit time) is q/T. Since the period T is the distance $2\pi r$ divided by the velocity v, the current is $I = q/T = qv/(2\pi r)$. The magnetic moment is then

$$\mu = IA = \frac{qv}{2\pi r} \pi r^2 = \frac{1}{2} qvr = \frac{q}{2m} L$$

where we have substituted L/m for vr. If the charge q is positive, the angular momentum and magnetic moment are in the same direction. We can therefore write

$$\vec{\mu} = \frac{q}{2m} \vec{L} \qquad\qquad 36\text{-}42$$

[†] The correspondence with the Bohr model is closest for the maximum value of ℓ, which is $n - 1$.
[‡] This topic was first presented in Section 27-5.

Equation 36-42 is the general classical relation between magnetic moment and angular momentum. It also holds in the quantum theory of the atom for orbital angular momentum, but not for the intrinsic spin angular momentum of the electron. For electron spin, the magnetic moment is twice that predicted by Equation 36-42.[†] The extra factor of 2 is a result from quantum theory that has no analog in classical mechanics.

The quantum of angular momentum is $\hbar$, so we express the magnetic moment in terms of $\vec{L}/\hbar$:

$$\vec{\mu} = \frac{q\hbar}{2m} \frac{\vec{L}}{\hbar}$$

For an electron, $m = m_e$ and $q = -e$, so the magnetic moment of the electron due to its orbital motion is

$$\vec{\mu}_\ell = -\frac{e\hbar}{2m_e} \frac{\vec{L}}{\hbar} = -\mu_B \frac{\vec{L}}{\hbar}$$

where $\mu_B = e\hbar/(2m_e) = 5.79 \times 10^{-5}$ eV/T is the quantum unit of magnetic moment called a Bohr magneton. The magnetic moment of an electron due to its intrinsic spin angular momentum $\vec{S}$ is

$$\vec{\mu}_S = -2 \times \frac{e\hbar}{2m_e} \frac{\vec{S}}{\hbar} = -2\mu_B \frac{\vec{S}}{\hbar}$$

In general, an electron in an atom has both orbital angular momentum characterized by the quantum number ℓ and spin angular momentum characterized by the quantum number s. Analogous classical systems that have two kinds of angular momentum are the earth, which is spinning about its axis of rotation in addition to revolving about the sun, and a precessing gyroscope that has angular momentum of precession in addition to its spin. The total angular momentum $\vec{J}$ is the sum of the orbital angular momentum $\vec{L}$ and the spin angular momentum $\vec{S}$, where

$$\vec{J} = \vec{L} + \vec{S} \qquad\qquad 36\text{-}43$$

Classically $\vec{J}$ is an important quantity because the resultant torque on a system equals the rate of change of the total angular momentum, and in the case of only central forces, the total angular momentum is conserved. For a classical system, the direction of the total angular momentum $\vec{J}$ is without restrictions and the magnitude of $\vec{J}$ can take on any value between $J_{max} = L + S$ and $J_{min} = |L - S|$. However, in quantum mechanics, the directions of both $\vec{L}$ and $\vec{S}$ are more restricted and the magnitudes L and S are both quantized. Furthermore, like $\vec{L}$ and $\vec{S}$, the direction of the total angular momentum $\vec{J}$ is restricted and the magnitude of $\vec{J}$ is quantized. For an electron with orbital angular momentum characterized by the quantum number ℓ and spin $s = \frac{1}{2}$, the total angular-momentum magnitude J is equal to $\sqrt{j(j + 1)}\hbar$, where the quantum number j is given by

$$j = +\tfrac{1}{2}, \quad \ell = 0$$

---

† This result, and the phenomenon of electron spin itself, was predicted in 1927 by Paul Dirac, who combined special relativity and quantum mechanics into a relativistic wave equation called the Dirac equation. Precise measurements indicate that the magnetic moment of the electron due to its spin is 2.00232 times that predicted by Equation 36-42. The fact that the intrinsic magnetic moment of the electron is approximately twice what we would expect makes it clear that the simple model of the electron as a spinning ball is not to be taken literally.

We have already discussed the lightest element, hydrogen, which has just one electron. In the ground (lowest energy) state, the electron has $n = 1$ and $\ell = 0$, with $m_\ell = 0$ and $m_s = +\frac{1}{2}$ or $-\frac{1}{2}$. We call this a 1s electron. The 1 signifies that $n = 1$, and the s signifies that $\ell = 0$.

As electrons are added to make the heavier atoms, the electrons go into those states that will give the lowest total energy consistent with the Pauli exclusion principle.

Helium (Z = 2)

The next element after hydrogen is helium ($Z = 2$), which has two electrons. In the ground state, both electrons are in the K shell with $n = 1$, $\ell = 0$, and $m_\ell = 0$; one electron has $m_s = +\frac{1}{2}$ and the other has $m_s = -\frac{1}{2}$. This configuration is lower in energy than any other two-electron configuration. The resultant spin of the two electrons is zero. Since the orbital angular momentum is also zero, the total angular momentum is zero. The electron configuration for helium is written 1s². The 1 signifies that $n = 1$, the s signifies that $\ell = 0$, and the superscript 2 signifies that there are two electrons in this state. Since ℓ can be only 0 for $n = 1$, these two electrons fill the K ($n = 1$) shell. The energy required to remove the most loosely bound electron from an atom in the ground state is called the **ionization energy.** This energy is the binding energy of the last electron placed in the atom. For helium, the ionization energy is 24.6 eV, which is relatively large. Helium is therefore basically inert.

ELECTRON INTERACTION ENERGY IN HELIUM   **E X A M P L E   3 6 - 6**

(a) Use the measured ionization energy to calculate the energy of interaction of the two electrons in the ground state of the helium atom. (b) Use your result to estimate the average separation of the two electrons.

PICTURE THE PROBLEM The energy of one electron in the ground state of helium is E_1 (which is negative) given by Equation 36-27, with $n = 1$ and $Z = 2$. If the electrons did not interact, the energy of the second electron would also be E_1, the same as that of the first electron. Thus, for an atom with noninteracting electrons, the ionization energy would be $|E_1|$ and the ground-state energy would be $E_{non} = 2E_1$. This is represented by the lowest level in Figure 36-16. Because of the interaction energy, the ground-state energy is greater than $2E_1$. This is represented by the higher level labeled E_g in the figure. When we add $E_{ion} = 24.6$ eV to ionize He, we obtain ionized helium, written He⁺, which has just one electron and therefore energy E_1.

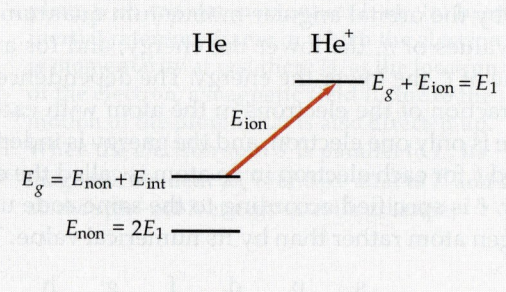

FIGURE 36-16

(a) 1. The energy of interaction plus the energy of two noninteracting electrons equals the ground-state energy of helium:

$$E_{int} + E_{non} = E_g$$

2. Solve for E_{int} and substitute $E_{non} = 2E_1$:

$$E_{int} = E_g - E_{non} = E_g - 2E_1$$

3. Use Equation 36-27 to calculate the energy E_1 of one electron in the ground state:

$$E_n = -Z^2 \frac{E_0}{n^2}$$

so

$$E_1 = -(2)^2 \frac{13.6 \text{ eV}}{1^2} = -54.4 \text{ eV}$$

4. Substitute this value for E_1:

$$E_{\text{int}} = E_{\text{g}} - 2E_1 = E_{\text{g}} - 2(-54.4 \text{ eV})$$

$$= E_{\text{g}} + 108.8 \text{ eV}$$

5. The ground-state energy of He, E_{g}, plus the ionization energy equals the ground-state energy of He$^+$, which is E_1:

$$E_{\text{g}} + E_{\text{ion}} = E_1 = -54.4 \text{ eV}$$

6. Substitute $E_{\text{ion}} = 24.6$ eV to calculate E_{g}:

$$E_{\text{g}} = E_1 - E_{\text{ion}} = -54.4 \text{ eV} - 24.6 \text{ eV}$$

$$= -79 \text{ eV}$$

7. Substitute this result for E_{g} to obtain E_{int}:

$$E_{\text{int}} = E_{\text{g}} + 108.8 \text{ eV} = -79 \text{ eV} + 108.8 \text{ eV}$$

$$= \boxed{29.8 \text{ eV}}$$

(b) 1. The energy of interaction of two electrons separated by distance r_{s} apart is the potential energy:

$$U = +\frac{ke^2}{r_{\text{s}}}$$

2. Set U equal to 29.8 eV, and solve for r. It is convenient to express r in terms of a_0, the radius of the first Bohr orbit in hydrogen, and to use Equation 36-16:

$$r_{\text{s}} = \frac{ke^2}{U} = \frac{ke^2}{a_0}\frac{a_0}{U} = 2\frac{ke^2}{2a_0}\frac{a_0}{U} = 2\frac{E_0}{U}a_0$$

$$= 2\frac{13.6 \text{ eV}}{29.8 \text{ eV}}a_0 = \boxed{0.913a_0}$$

PLAUSIBILITY CHECK This separation is approximately the size of the diameter d_1 of the first Bohr orbit for an electron in helium, which is $d_1 = 2r_1 = 2a_0/Z = a_0$.

Lithium (Z = 3)

The next element, lithium, has three electrons. Since the K shell ($n = 1$) is completely filled with two electrons, the third electron must go into a higher energy shell. The next lowest energy shell after $n = 1$ is the $n = 2$ or L shell. The outer electron is much farther from the nucleus than are the two inner $n = 1$ electrons. It is most likely to be found at a radius near that of the second Bohr orbit, which is four times the radius of the first Bohr orbit.

The nuclear charge is partially screened from the outer electron by the two inner electrons. Recall that the electric field outside a spherically symmetric charge density is the same as if all the charge were at the center of the sphere. If the outer electron were completely outside the charge cloud of the two inner electrons, the electric field the outer electron would see would be that of a single charge $+e$ at the center due to the nuclear charge of $+3e$ and the charge $-2e$ of the inner electron cloud. However, the outer electron does not have a well-defined orbit; instead, it is itself a charge cloud that penetrates the charge cloud of the inner electrons to some extent. Because of this penetration, the effective nuclear charge $Z'e$ is somewhat greater than $+1e$. The energy of the outer electron at a distance r from a point charge $+Z'e$ is given by Equation 36-6, with the nuclear charge $+Z$ replaced by $+Z'$.

$$E = -\frac{1}{2}\frac{kZ'e^2}{r} \qquad\qquad\qquad 36\text{-}46$$

The greater the penetration of the inner electron cloud, the greater the effective nuclear charge $Z'e$ and the lower the energy. Because the penetration is greater for ℓ values closer to zero (see Figure 36-12), the energy of the outer electron in lithium is lower for the s state ($\ell = 0$) than for the p state ($\ell = 1$). The electron configuration of lithium in the ground state is therefore 1s^{2}2s. The ionization energy of lithium is only 5.39 eV. Because its outer electron is so loosely bound to the atom, lithium is very active chemically. It behaves like a one-electron atom, similar to hydrogen.

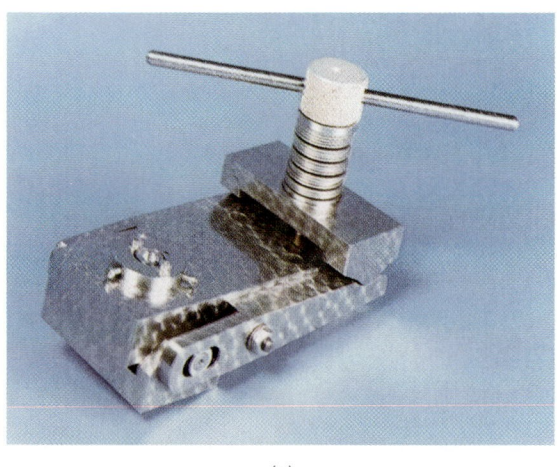

(a)

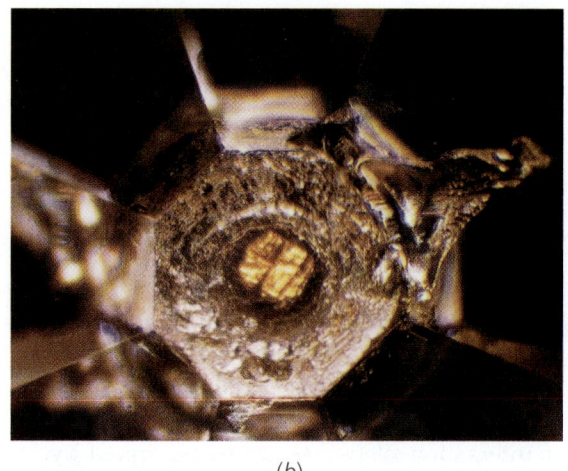

(b)

(*a*) A diamond anvil cell, in which the facets of two diamonds (approximately 1 mm² each) are used to compress a sample substance, subjecting it to very high pressure. (*b*) Samarium monosulfide (SmS) is normally a black, dull-looking semiconductor. When it is subjected to pressure above 7000 atm, an electron from the 4f state is dislocated into the 5d state. The resulting compound glitters like gold and behaves like a metal.

EFFECTIVE NUCLEAR CHARGE FOR AN OUTER ELECTRON **E X A M P L E   3 6 - 7**

Suppose the electron cloud of the outer electron in the lithium atom in the ground state were completely outside the electron clouds of the two inner electrons, the nuclear charge would be shielded by the two inner electrons and the effective nuclear charge would be $Z'e = 1e$. Then the energy of the outer electron would be $-(13.6 \text{ eV})/2^2 = -3.4 \text{ eV}$. However, the ionization energy of lithium is 5.39 eV, not 3.4 eV. Use this fact to calculate the effective nuclear charge Z' seen by the outer electron in lithium.

PICTURE THE PROBLEM Because the outer electron is in the $n = 2$ shell, we will take $r = 4a_0$ for its average distance from the nucleus. We can then calculate Z' from Equation 36-46. Since r is given in terms of a_0, it will be convenient to use the fact that $E_0 = ke^2/(2a_0) = 13.6 \text{ eV}$ (Equation 36-16).

1. Equation 36-46 relates the energy of the outer electron to its average distance r and the effective nuclear charge Z':

$$E = \frac{1}{2}\frac{kZ'e^2}{r}$$

2. Substitute the given values $r = 4a_0$ and $E = -5.39 \text{ eV}$:

$$-5.39 \text{ eV} = -\frac{1}{2}\frac{kZ'e^2}{4a_0}$$

3. Use $ke^2/(2a_0) = E_0 = 13.6 \text{ eV}$ and solve for Z':

$$-5.39 \text{ eV} = -\frac{Z'}{4}\frac{ke^2}{2a_0} = -\frac{Z'}{4}(13.6 \text{ eV})$$

so

$$Z' = 4\frac{5.39 \text{ eV}}{13.6 \text{ eV}} = \boxed{1.59}$$

REMARKS This calculation is interesting but not very rigorous. We essentially used the radius ($r = 4a_0$) for the circular orbit from the semiclassical Bohr model and the measured ionization energy to calculate the effective inner charge seen by the outer electron. We know, of course, that this outer electron does not move in a circular orbit of constant radius, but is better represented by a stationary charged cloud of charge density $|\psi|^2$ that penetrates the charged clouds of the inner electrons.

Beryllium (Z = 4)

The energy of the beryllium atom is a minimum if both outer electrons are in the 2s state. There can be two electrons with $n = 2$, $\ell = 0$, and $m_\ell = 0$ because of the two possible values for the spin quantum number m_s. The configuration of beryllium is thus $1s^2 2s^2$.

Hydrogen

Boron to Neon (Z = 5 to Z = 10)

Since the 2s subshell is filled, the fifth electron must go into the next available (lowest energy) subshell, which is the 2p subshell, with $n = 2$ and $\ell = 1$. Since there are three possible values of m ($+1$, 0, and -1) and two values of m_s for each value of m_ℓ, there can be six electrons in this subshell. The electron configuration for boron is $1s^2 2s^2 2p$. The electron configurations for the elements carbon ($Z = 6$) to neon ($Z = 10$) differ from that for boron only in the number of electrons in the 2p subshell. The ionization energy increases with Z for these elements, reaching the value of 21.6 eV for the last element in the group, neon. Neon has the maximum number of electrons allowed in the $n = 2$ shell. The electron configuration for neon is $1s^2 2s^2 2p^6$. Because of its very high ionization energy, neon, like helium, basically is chemically inert. The element just before neon, fluorine, has a hole in the 2p subshell; that is, it has room for one more electron. It readily combines with elements such as lithium that have one outer electron. Lithium, for example, will donate its single outer electron to the fluorine atom to make an F^- ion and a Li^+ ion. These ions then bond together to form a molecule of lithium fluoride.

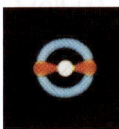

Carbon

Silicon

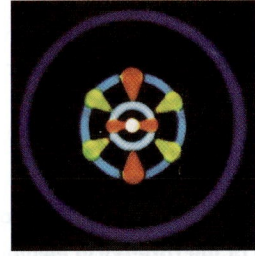

Iron

Sodium to Argon (Z = 11 to Z = 18)

The eleventh electron must go into the $n = 3$ shell. Since this electron is very far from the nucleus and from the inner electrons, it is weakly bound in the sodium ($Z = 11$) atom. The ionization energy of sodium is only 5.14 eV. Sodium therefore combines readily with atoms such as fluorine. With $n = 3$, the value of ℓ can be 0, 1, or 2. Because of the lowering of the energy due to penetration of the electron shield formed by the other ten electrons (similar to that discussed for lithium) the 3s state is lower than the 3p or 3d states. This energy difference between subshells of the same n value becomes greater as the number of electrons increases. The electron configuration of sodium is $1s^2 2s^2 2p^6 3s^1$. As we move to elements with higher values of Z, the 3s subshell and then the 3p subshell fill. These two subshells can accommodate $2 + 6 = 8$ electrons. The configuration of argon ($Z = 18$) is $1s^2 2s^2 2p^6 3s^2 3p^6$. One might expect the nineteenth electron to go into the third subshell (the d subshell with $\ell = 2$), but the penetration effect is now so strong that the energy of the next electron is lower in the 4s subshell than in the 3d subshell. There is thus another large energy difference between the eighteenth and nineteenth electrons, and so argon, with its full 3p subshell, is basically stable and inert.

Silver

Elements With Z > 18

The nineteenth electron in potassium ($Z = 19$) and the twentieth electron in calcium ($Z = 20$) go into the 4s subshell rather than the 3d subshell. The electron configurations of the next ten elements, scandium ($Z = 21$) through zinc ($Z = 30$),

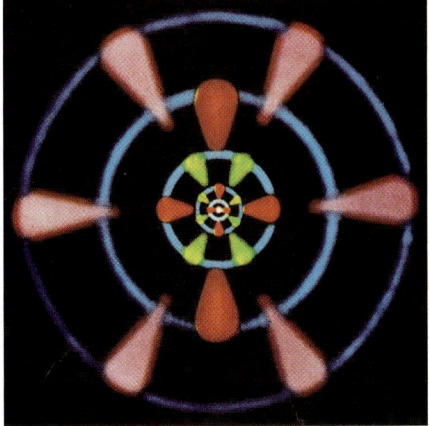

Europium

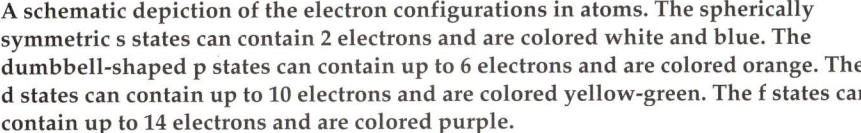

A schematic depiction of the electron configurations in atoms. The spherically symmetric s states can contain 2 electrons and are colored white and blue. The dumbbell-shaped p states can contain up to 6 electrons and are colored orange. The d states can contain up to 10 electrons and are colored yellow-green. The f states can contain up to 14 electrons and are colored purple.

TABLE 36-1 (continued)

Electron Configurations of the Atoms in Their Ground States

For some of the rare-earth elements (Z = 57 to 71) and the heavy elements (Z > 89) the configurations are not firmly established.

		K (1)	L (2)		M (3)			N (4)				O (5)				P (6)			Q (7)
		s	s	p	s	p	d	s	p	d	f	s	p	d	f	s	p	d	s
Z	Element	(0)	(0)	(1)	(0)	(1)	(2)	(0)	(1)	(2)	(3)	(0)	(1)	(2)	(3)	(0)	(1)	(2)	(1)
58	Ce cerium	2	2	6	2	6	10	2	6	10	1	2	6	1	.	2			
59	Pr praseodymium	2	2	6	2	6	10	2	6	10	3	2	6	.	.	2			
60	Nd neodymium	2	2	6	2	6	10	2	6	10	4	2	6	.	.	2			
61	Pm promethium	2	2	6	2	6	10	2	6	10	5	2	6	.	.	2			
62	Sm samarium	2	2	6	2	6	10	2	6	10	6	2	6	.	.	2			
63	Eu europium	2	2	6	2	6	10	2	6	10	7	2	6	.	.	2			
64	Gd gadolinium	2	2	6	2	6	10	2	6	10	7	2	6	1	.	2			
65	Tb terbium	2	2	6	2	6	10	2	6	10	9	2	6	.	.	2			
66	Dy dysprosium	2	2	6	2	6	10	2	6	10	10	2	6	.	.	2			
67	Ho holmium	2	2	6	2	6	10	2	6	10	11	2	6	.	.	2			
68	Er erbium	2	2	6	2	6	10	2	6	10	12	2	6	.	.	2			
69	Tm thulium	2	2	6	2	6	10	2	6	10	13	2	6	.	.	2			
70	Yb ytterbium	2	2	6	2	6	10	2	6	10	14	2	6	.	.	2			
71	Lu lutetium	2	2	6	2	6	10	2	6	10	14	2	6	1	.	2			
72	Hf hafnium	2	2	6	2	6	10	2	6	10	14	2	6	2	.	2			
73	Ta tantalum	2	2	6	2	6	10	2	6	10	14	2	6	3	.	2			
74	W tungsten (wolfram)	2	2	6	2	6	10	2	6	10	14	2	6	4	.	2			
75	Re rhenium	2	2	6	2	6	10	2	6	10	14	2	6	5	.	2			
76	Os osmium	2	2	6	2	6	10	2	6	10	14	2	6	6	.	2			
77	Ir iridium	2	2	6	2	6	10	2	6	10	14	2	6	7	.	2			
78	Pt platinum	2	2	6	2	6	10	2	6	10	14	2	6	9	.	1			
79	Au gold	2	2	6	2	6	10	2	6	10	14	2	6	10	.	1			
80	Hg mercury	2	2	6	2	6	10	2	6	10	14	2	6	10	.	2			
81	Tl thallium	2	2	6	2	6	10	2	6	10	14	2	6	10	.	2	1		
82	Pb lead	2	2	6	2	6	10	2	6	10	14	2	6	10	.	2	2		
83	Bi bismuth	2	2	6	2	6	10	2	6	10	14	2	6	10	.	2	3		
84	Po polonium	2	2	6	2	6	10	2	6	10	14	2	6	10	.	2	4		
85	At astatine	2	2	6	2	6	10	2	6	10	14	2	6	10	.	2	5		
86	Rn radon	2	2	6	2	6	10	2	6	10	14	2	6	10	.	2	6		
87	Fr francium	2	2	6	2	6	10	2	6	10	14	2	6	10	.	2	6	.	1
88	Ra radium	2	2	6	2	6	10	2	6	10	14	2	6	10	.	2	6	.	2
89	Ac actinium	2	2	6	2	6	10	2	6	10	14	2	6	10	.	2	6	1	2
90	Th thorium	2	2	6	2	6	10	2	6	10	14	2	6	10	.	2	6	2	2
91	Pa protactinium	2	2	6	2	6	10	2	6	10	14	2	6	10	1	2	6	2	2
92	U uranium	2	2	6	2	6	10	2	6	10	14	2	6	10	3	2	6	1	2
93	Np neptunium	2	2	6	2	6	10	2	6	10	14	2	6	10	4	2	6	1	2
94	Pu plutonium	2	2	6	2	6	10	2	6	10	14	2	6	10	6	2	6	.	2
95	Am americium	2	2	6	2	6	10	2	6	10	14	2	6	10	7	2	6	.	2

TABLE 36-1 (continued)

Electron Configurations of the Atoms in Their Ground States

For some of the rare-earth elements (Z = 57 to 71) and the heavy elements (Z > 89) the configurations are not firmly established.

		Shell (n):	K (1)	L (2)		M (3)			N (4)				O (5)				P (6)			Q (7)
			s	s	p	s	p	d	s	p	d	f	s	p	d	f	s	p	d	s
Z	Element	Subshell (ℓ):	(0)	(0)	(1)	(0)	(1)	(2)	(0)	(1)	(2)	(3)	(0)	(1)	(2)	(3)	(0)	(1)	(2)	(1)
96	Cm curium		2	2	6	2	6	10	2	6	10	14	2	6	10	7	2	6	1	2
97	Bk berkelium		2	2	6	2	6	10	2	6	10	14	2	6	10	8	2	6	1	2
98	Cf californium		2	2	6	2	6	10	2	6	10	14	2	6	10	10	2	6	.	2
99	Es einsteinium		2	2	6	2	6	10	2	6	10	14	2	6	10	11	2	6	.	2
100	Fm fermium		2	2	6	2	6	10	2	6	10	14	2	6	10	12	2	6	.	2
101	Md mendelevium		2	2	6	2	6	10	2	6	10	14	2	6	10	13	2	6	.	2
102	No nobelium		2	2	6	2	6	10	2	6	10	14	2	6	10	14	2	6	.	2
103	Lr lawrencium		2	2	6	2	6	10	2	6	10	14	2	6	10	14	2	6	1	2
104	Rf rutherfordium		2	2	6	2	6	10	2	6	10	14	2	6	10	14	2	6	2	2
105	Db dubnium		2	2	6	2	6	10	2	6	10	14	2	6	10	14	2	6	3	2
106	Sg seaborgium		2	2	6	2	6	10	2	6	10	14	2	6	10	14	2	6	4	2
107	Bh bohrium		2	2	6	2	6	10	2	6	10	14	2	6	10	14	2	6	5	2
108	Hs hassium		2	2	6	2	6	10	2	6	10	14	2	6	10	14	2	6	6	2
109	Mt meitnerium		2	2	6	2	6	10	2	6	10	14	2	6	10	14	2	6	7	2

36-7 Optical Spectra and X-Ray Spectra

When an atom is in an excited state (i.e., when it is in an energy state above the ground state), it makes transitions to lower energy states, and in doing so emits electromagnetic radiation. The wavelength of the electromagnetic radiation emitted is related to the initial and final states by the Bohr formula (Equation 36-17), $\lambda = hc/(E_i - E_f)$, where E_i and E_f are the initial and final energies and h is Planck's constant. The atom can be excited to a higher energy state by bombarding the atom with a beam of electrons, as in a spectral tube with a high voltage across it. Since the excited energy states of an atom form a discrete (rather than continuous) set, only certain wavelengths are emitted. These wavelengths of the emitted radiation constitute the emission spectrum of the atom.

Optical Spectra

To understand atomic spectra we need to understand the excited states of the atom. The situation for an atom with many electrons is, in general, much more complicated than that of hydrogen with just one electron. An excited state of the atom may involve a change in the state of any one of the electrons, or even two or more electrons. Fortunately, in most cases, an excited state of an atom involves the excitation of just one of the electrons in the atom. The energies of excitation of the outer, valence electrons of an atom are of the order of a few electron volts. Transitions involving these electrons result in photons in or near the visible or **optical spectrum.** (Recall that the energies of visible photons range from approximately 1.5 eV to 3 eV.) The excitation energies can often be calculated from a simple model in which the atom is pictured as a single electron plus a stable core consisting of the nucleus plus the other inner electrons. This model works particularly well for the alkali metals: Li, Na, K, Rb, and Cs. These elements are in the first column of the periodic table. The optical spectra of these elements are similar to the optical spectra of hydrogen.

Figure 36-18 shows an energy-level diagram for the optical transitions in sodium, whose electrons form a neon core plus one outer electron. Since the total spin angular momentum of the core adds up to zero, the spin of each state of sodium is $\frac{1}{2}$. Because of the spin–orbit effect, the states with $J = L - \frac{1}{2}$ have a slightly different energy than those with $J = L + \frac{1}{2}$ (except for states with $L = 0$). Each state (except for the $L = 0$ states) is therefore split into two states, called a doublet. The doublet splitting is very small and not evident on the energy scale of this diagram. The usual spectroscopic notation is that these states are labeled with a superscript given by $2S + 1$, followed by a letter denoting the orbital angular momentum, followed by a subscript denoting the total angular momentum J. For states with a total spin angular momentum $S = \frac{1}{2}$ the superscript is 2, indicating the state is a doublet. Thus $^2P_{3/2}$, read as "doublet P three halves," denotes a state in which $L = 1$ and $J = \frac{3}{2}$. (The $L = 0$, or S, states are customarily labeled as if they were doublets even though they are not.) In the first excited state, the outer electron is excited from the 3s level to the 3p level, which is approximately 2.1 eV above the ground state. The energy difference between the $P_{3/2}$ and $P_{1/2}$

A neon sign outside a Chinatown restaurant in Paris. Neon atoms in the tube are excited by an electron current passing through the tube. The excited neon atoms emit light in the visible range as they decay toward their ground states. The colors of neon signs result from the characteristic red-orange spectrum of neon plus the color of the glass tube itself.

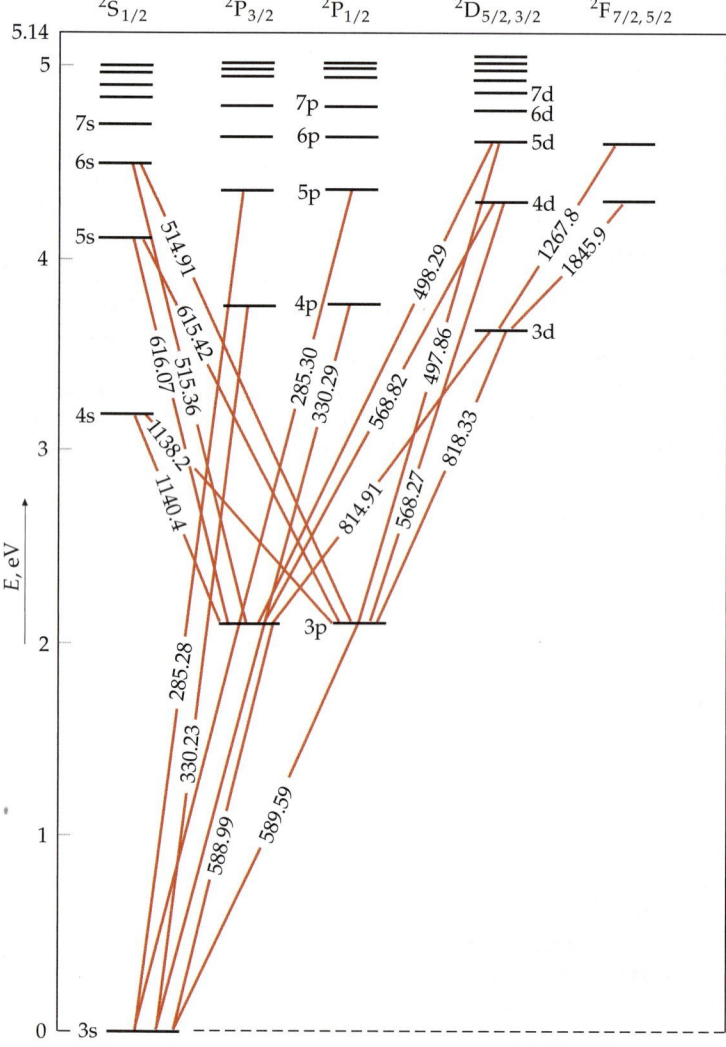

FIGURE 36-18 Energy-level diagram for sodium. The diagonal lines show observed optical transitions, with wavelengths given in nanometers. The energy of the ground state has been chosen as the zero point for the scale on the left.

states due to the spin–orbit effect is about 0.002 eV. Transitions from these states to the ground state give the familiar sodium yellow doublet:

$$3p(^2P_{1/2}) \rightarrow 3s(^2S_{1/2}), \quad \lambda = 589.6 \text{ nm}$$

$$3p(^2P_{3/2}) \rightarrow 3s(^2S_{1/2}), \quad \lambda = 589.0 \text{ nm}$$

The energy levels and spectra of other alkali metal atoms are similar to those for sodium. The optical spectrum for atoms such as helium, beryllium, and magnesium that have two outer electrons is considerably more complex because of the interaction of the two outer electrons.

X-Ray Spectra

X rays are usually produced in the laboratory by bombarding a target element with a high-energy beam of electrons in an X-ray tube. The result (Figure 36-19) consists of a continuous spectrum that depends only on the energy of the bombarding electrons and a line spectrum that is characteristic of the target element. The characteristic spectrum results from excitation of the inner core electrons in the target element.

The energy needed to excite an inner core electron—for example, an electron in the $n = 1$ state (K shell)—is much greater than the energy required to excite an outer, valence electron. An inner electron cannot be excited to any of the filled states (e.g., the $n = 2$ states in sodium) because of the exclusion principle. The energy required to excite an inner core electron to an unoccupied state is typically of the order of several kilo-electron volts. If an electron is knocked out of the $n = 1$ state (K shell), there is a vacancy left in this shell. This vacancy can be filled if an electron in the L shell (or in a higher shell) makes a transition into the K shell. The photons emitted by electrons making such transitions also have energies of the order of kilo-electron volts and produce the sharp peaks in the X-ray spectrum, as shown in Figure 36-19. The K_α line arises from transitions from the $n = 2$ (L) shell to the $n = 1$ (K) shell. The K_β line arises from transitions from the $n = 3$ shell to the $n = 1$ shell. These and other lines arising from transitions ending at the $n = 1$ shell make up the K series of the characteristic X-ray spectrum of the target element. Similarly, a second series, the L series, is produced by transitions from higher energy states to a vacated place in the $n = 2$ (L) shell. The letters K, L, M, and so on, designate the final shell of the electron making the transition and the series α, β, and so on, designates the number of shells above the final shell for the initial state of the electron.

In 1913, the English physicist Henry Moseley measured the wavelengths of the characteristic K_α X-ray spectra for approximately forty elements. Using this data, Moseley showed that a plot of $\lambda^{-1/2}$ versus the order in which the elements appeared in the periodic table resulted in a straight line (with a few gaps and a few outliers). From his data, Moseley was able to accurately determine the atomic number Z for each known element, and to predict the existence of some elements that were later discovered. The equation of the straight line of his plot is given by

$$\frac{1}{\sqrt{\lambda_{K_\alpha}}} = a(Z - 1)$$

The work of Bohr and Moseley can be combined to obtain an equation relating the wavelength of the emitted photon and the atomic number. According to the Bohr model of a single-electron atom (see Equation 36-13), the wavelength of the emitted photon when the electron makes the transition from $n = 2$ to $n = 1$ is given by

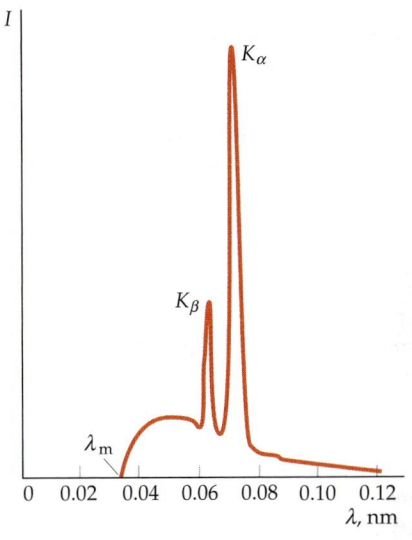

FIGURE 36-19 X-ray spectrum of molybdenum. The sharp peaks labeled K_α and K_β are characteristic of the element. The cutoff wavelength λ_m is independent of the target element and is related to the voltage V of the X-ray tube by $\lambda_m = hc/eV$.

$$\frac{1}{\lambda} = Z^2 \frac{E_0}{hc}\left(1 - \frac{1}{2^2}\right)$$

where $E_0 = 13.6$ eV is the binding energy of the ground-state hydrogen atom. Taking the square root of both sides gives

$$\frac{1}{\sqrt{\lambda_{K_\alpha}}} = \left[\frac{E_0}{hc}\left(1 - \frac{1}{2^2}\right)\right]^{1/2} Z$$

Moseley's equation and this equation are in agreement if $Z - 1$ is substituted for Z in Moseley's equation and if $a = 3E_0/(4hc)$. This raises the question, why a factor of $Z - 1$ instead of a factor of Z? Part of the explanation is that the formula from the Bohr theory ignores the shielding of the nuclear charge. In a multi-electron atom, electrons in the $n = 2$ states are electrically shielded from the nuclear charge by the two electrons in the $n = 1$ state, so the $n = 2$ state electrons are attracted by an effective nuclear charge of about $(Z - 2)e$. However, when there is only one electron in the K shell, the $n = 2$ electrons are attracted by an effective nuclear charge of about $(Z - 1)e$. When an electron from state n drops into the vacated state in the $n = 1$ shell, a photon of energy $E_2 - E_1$ is emitted. The wavelength of this photon is

$$\lambda_{K_\alpha} = \frac{hc}{(Z - 1)^2 E_0\left(1 - \frac{1}{2^2}\right)} \qquad \text{36-47}$$

which is obtained from the previous equation with $Z - 1$ substituted for Z.

IDENTIFYING THE ELEMENT FROM THE K_α X-RAY LINE   **EXAMPLE 36-8**

The wavelength of the K_α X-ray line for a certain element is $\lambda = 0.0721$ nm. What is the element?

PICTURE THE PROBLEM The K_α line corresponds to a transition from $n = 2$ to $n = 1$. The wavelength is related to the atomic number Z by Equation 36-47.

1. Solve Equation 36-47 for $(Z - 1)^2$:

$$\lambda_{K_\alpha} = \frac{hc}{(Z - 1)^2 E_0\left(1 - \frac{1}{2^2}\right)}$$

so

$$(Z - 1)^2 = \frac{4hc}{3\lambda_{K_\alpha} E_0}$$

2. Substitute the given data and solve for Z:

$$(Z - 1)^2 = \frac{4(1240 \text{ eV·nm})}{3(0.0721 \text{ nm})(13.6 \text{ eV})} = 1686$$

so

$$Z = 1 + \sqrt{1686} = 42.06$$

3. Since Z is an integer, we round to the nearest integer:

$$Z = 42$$

> The element is molybdenum.

1. The Bohr model is important historically because it was the first model to succeed at explaining the discrete optical spectrum of atoms in terms of the quantization of energy. It has been superceded by the quantum-mechanical model.

2. The quantum theory of atoms results from the application of the Schrödinger equation to a bound system consisting of nucleus of charge $+Ze$ and Z electrons of charge $-e$.

3. For the simplest atom, hydrogen, consisting of one proton and one electron, the time independent Schrödinger equation can be solved exactly to obtain the wave functions ψ, which depend on the quantum numbers n, ℓ, m_ℓ, and m_s.

4. The electron configuration of atoms is governed by the Pauli exclusion principle, which states that no two electrons in an atom can have the same set of values for the quantum numbers n, ℓ, m_ℓ, and m_s. Using the exclusion principle and the restrictions on the quantum numbers, we can understand much of the structure of the periodic table.

Topic	Relevant Equations and Remarks	
1. The Bohr Model of the Hydrogen Atom		
Postulates for the hydrogen atom		
Nonradiating orbits	The electron moves in a circular nonradiating orbit around the proton.	
Photon frequency from energy conservation	$f = \dfrac{E_i - E_f}{h}$	36-7
Quantized angular momentum	$mvr = n\hbar, \qquad n = 1, 2, 3, \ldots$	36-9
First Bohr radius	$a_0 = \dfrac{\hbar^2}{mke^2} \approx 0.0529 \text{ nm}$	36-12
Radius of the Bohr orbits	$r = n^2 \dfrac{a_0}{Z}$	36-11
Energy levels in the hydrogen atom	$E_n = -\dfrac{mk^2e^4}{2\hbar^2} \dfrac{Z^2}{n^2} = -Z^2 \dfrac{E_0}{n^2}$	36-15
	where	
	$E_0 = \dfrac{mk^2e^4}{2\hbar^2} = \dfrac{1}{2}\dfrac{ke^2}{a_0} \approx 13.6 \text{ eV}$	36-16
Wavelengths emitted by the hydrogen atom	$\lambda = \dfrac{c}{f} = \dfrac{hc}{E_i - E_f} = \dfrac{1240 \text{ eV·nm}}{E_i - E_f}$	36-17, 36-18
2. Quantum Theory of Atoms	The electron is described by a wave function ψ that is a solution of the Schrödinger equation. Energy quantization arises from standing-wave conditions. ψ is described by the quantum numbers n, ℓ, and m_ℓ, and the spin quantum number $m_s = \pm\frac{1}{2}$.	
Schrödinger equation in spherical coordinates	$-\dfrac{\hbar^2}{2mr^2}\dfrac{\partial}{\partial r}\left(r^2 \dfrac{\partial\psi}{\partial r}\right) - \dfrac{\hbar^2}{2mr^2}\left[\dfrac{1}{\sin\theta}\dfrac{\partial}{\partial\theta}\left(\sin\theta\dfrac{\partial\psi}{\partial\theta}\right) + \dfrac{1}{\sin^2\theta}\dfrac{\partial^2\psi}{\partial\phi^2}\right] + U(r)\psi = E\psi$	36-21

10 •• SSM Why is the energy of the 3s state considerably lower than the energy of the 3p state for sodium, whereas in hydrogen these states have essentially the same energy?

11 •• Discuss the evidence from the periodic table of the need for a fourth quantum number. How would the properties of helium differ if there were only three quantum numbers, n, ℓ, and m_ℓ?

12 •• Separate the following six elements—potassium, calcium, titanium, chromium, manganese, and copper—into two groups of three each, so that those in a group have similar properties.

13 • What element has the electron configuration (a) $1s^2 2s^2 2p^6 3s^2 3p^3$ and (b) $1s^2 2s^2 2p^6 3s^2 3p^6 3d^5 4s^1$?

14 • SSM For the principal quantum number $n = 3$, what are the possible values of the quantum numbers ℓ and m_ℓ?

15 • An electron in the L shell means that (a) $\ell = 0$, (b) $\ell = 1$, (c) $n = 1$, (d) $n = 2$, or (e) $m_\ell = 2$.

16 •• The Bohr theory and the Schrödinger theory of the hydrogen atom give the same results for the energy levels. Discuss the advantages and disadvantages of each model.

17 •• The Sommerfeld–Hosser displacement theorem states that the optical spectrum of any neutral atom is very similar to the spectrum of the singly charged positive ion of the element immediately following it in the periodic table. Discuss why this is true.

18 •• SSM The Ritz combination principle states that for any atom, one can find different spectral lines λ_1, λ_2, λ_3, and λ_4, so that $1/\lambda_1 + 1/\lambda_2 = 1/\lambda_3 + 1/\lambda_4$. Show why this is true using an energy-level diagram.

19 • Using the triplet of numbers (n, ℓ, m_ℓ) to represent an electron with principal quantum number n, orbital quantum number ℓ, and magnetic quantum number m_ℓ, which of the following transitions is allowed? (a) $(5, 2, 2) \rightarrow (3, 1, 2)$; (b) $(2, 0, 0) \rightarrow (3, 0, 1)$; (c) $(4, 3, -2) \rightarrow (3, 2, 0)$; (d) $(1, 0, 0) \rightarrow (2, 1, -1)$; or (e) $(2, 1, 0) \rightarrow (3, 0, 0)$.

Estimation and Approximation

20 •• SSM In laser cooling and trapping, atoms in a beam traveling in one direction are slowed by interaction with an intense laser beam in the opposite direction. The photons scatter off the atoms via resonance absorption, a process by which the incident photon is absorbed by the atom, and a short time later a photon of equal energy is emitted in a random direction. The net result of a single such scattering event is a transfer of momentum to the atom in a direction opposite to the motion of the atom, followed by a second transfer of momentum to the atom in a random direction. Thus, during photon absorption the atom loses speed, but during photon emission the change in speed of the atom is, on average, zero (because the directions of the emitted photons are random). An analogy often made to this process is that of slowing down a bowling ball by bouncing ping-pong balls off of it. (a) Given a typical photon energy used in these experiments of about 1 eV and a momentum typical for an atom with a

thermal speed appropriate to a temperature of about 500 K (a typical temperature for an atomic beam), estimate the number of photon-atom collisions that are required to bring an atom to rest. (The average kinetic energy of an atom is equal to $\frac{3}{2}kT$, where k is the Boltzmann constant. Use this to estimate the speed of the atoms.) (b) Compare this with the number of ping-pong ball–bowling ball collisions that are required to bring the bowling ball to rest. (Assume the speeds of the incident ping-pong balls are all equal to the initial speed of the bowling ball.) (c) ^{85}Rb is a type of atom often used in cooling experiments. The wavelength of the light resonant with the cooling transition is $\lambda = 780.24$ nm. Estimate the number of photons needed to slow down an ^{85}Rb atom from a typical thermal velocity of 300 m/s to a stop.

21 •• (a) We can define a thermal de Broglie wavelength λ_T for an atom in a gas at temperature T as being the de Broglie wavelength for an atom moving at the rms speed appropriate to that temperature. (The average kinetic energy of an atom is equal to $\frac{3}{2}kT$, where k is the Boltzman constant. Use this to calculate the rms speed of the atoms.) Show that

$$\lambda_T = \sqrt{\frac{h^2}{3mkT}},$$ where m is the mass of the atom. (b) Cooled neutral atoms can form a *Bose condensate* (a new state of matter) when their thermal de Broglie wavelength becomes larger than the average interatomic spacing. From this criterion, estimate the temperature needed to create a Bose condensate in a gas of ^{85}Rb atoms whose number density is 10^{12} atoms/cm^3.

The Bohr Model of the Hydrogen Atom

22 • Use the known values of the constants in Equation 36-12 to show that a_0 is approximately 0.0529 nm.

23 • iSOLVE ✔ The longest wavelength in the Lyman series was calculated in Example 36-2. Find the wavelengths for the transitions (a) $n_1 = 3$ to $n_2 = 1$ and (b) $n_1 = 4$ to $n_2 = 1$.

24 • Find the photon energy for the three longest wavelengths in the Balmer series and calculate the wavelengths.

25 •• (a) Find the photon energy and wavelength for the series limit (shortest wavelength) in the Paschen series ($n_2 = 3$). (b) Calculate the wavelength for the three longest wavelengths in this series and indicate their positions on a horizontal linear scale.

26 •• SSM Repeat Problem 25 for the Brackett series, $n_2 = 4$.

27 •• The hydrogen spectrum is found by collimating the light from a hydrogen discharge tube and shining it on a grating to disperse the light into its various colors. The grating spacing is $d = 3.377$ μm. A bright red line ($m = 1$) is seen at an angle of 11.233° from the center of the spectroscope (see Figure 36-1). (a) What is the wavelength of this spectral line? (b) Assuming this line is from a transition from level $n_1 = 3$ to level $n_2 = 2$ (i.e., the longest wavelength Balmer series transition), what do you calculate for the value of the Rydberg constant?

28 ••• In this problem, you will estimate the radius and the energy of the lowest stationary state of the hydrogen atom using the uncertainty principle. The total energy of the electron with momentum p and mass m a distance r from the proton in the hydrogen atom is given by

$$E = p^2/2m - ke^2/r$$

where k is the Coulomb constant. Assume that the minimum value of $p^2 \approx (\Delta p)^2 = \hbar^2/r^2$, where Δp is the uncertainty in p and we have taken $\Delta r \sim r$ for the order of magnitude of the uncertainty in position; the energy is then

$$E = \hbar^2/2mr^2 - ke^2/r$$

Find the radius r for which this energy is a minimum, and calculate the minimum value of E in electron volts.

29 ••• **SSM** In the center-of-mass reference frame of a hydrogen atom, the electron and nucleus have equal and opposite momenta of magnitude p. (a) Show that the total kinetic energy of the electron and nucleus can be written $K = p^2/(2\mu)$ where $\mu = m_e M/(M + m_e)$ is called the reduced mass, m_e is the mass of the electron, and M is the mass of the nucleus. (b) For the equations for the Bohr model of the atom, the motion of the nucleus can be taken into account by replacing the mass of the electron with the reduced mass. Use Equation 36-14 to calculate the Rydberg constant for a hydrogen atom with a nucleus of mass $M = m_p$. Find the approximate value of the Rydberg constant by letting M go to infinity in the reduced mass formula. To how many figures does this approximate value agree with the actual value? (c) Find the percentage correction for the ground-state energy of the hydrogen atom by using the reduced mass in Equation 36-16. *Remark: In general, the reduced mass for a two-body problem with masses m_1 and m_2 is given by*

$$\mu = \frac{m_1 m_2}{m_1 + m_2}$$

30 •• **SSM** The Pickering series of the spectrum of He⁺ (singly-ionized helium) consists of spectral lines due to transitions to the $n = 4$ state of He⁺. Experimentally, every other line of the Pickering series is very close to a spectral line in the Balmer series for hydrogen transitions to $n = 2$. (a) Show that this is true. (b) Calculate the wavelength of a transition from the $n = 6$ level to the $n = 4$ level of He⁺, and show that it corresponds to one of the Balmer lines.

Quantum Numbers in Spherical Coordinates

31 • For $\ell = 1$, find (a) the magnitude of the angular momentum L and (b) the possible values of m_ℓ. (c) Draw a vector diagram to scale showing the possible orientations of $\vec{L}$ with the z axis.

32 • Repeat Problem 31 for $\ell = 3$.

33 • If $n = 3$, (a) what are the possible values of ℓ? (b) For each value of ℓ in Part (a), list the possible values of m_ℓ. (c) Using the fact that there are two quantum states for each value of ℓ and m_ℓ because of electron spin, find the total number of electron states with $n = 3$.

34 • Find the total number of electron states with (a) $n = 4$ and (b) $n = 2$. (See Problem 33.)

35 •• **SSM** **ISOLVE** Find the minimum value of the angle θ between $\vec{L}$ and the z axis for (a) $\ell = 1$, (b) $\ell = 4$, and (c) $\ell = 50$.

36 •• What are the possible values of n and m_ℓ if (a) $\ell = 3$, (b) $\ell = 4$, and (c) $\ell = 0$.

37 •• For an $\ell = 2$ state, find (a) the square magnitude of the angular momentum $\vec{L}^2$, (b) the maximum value of L_z^2, and (c) the smallest value of $L_x^2 + L_y^2$.

Quantum Theory of the Hydrogen Atom

38 • For the ground state of the hydrogen atom, find the values of (a) ψ, (b) ψ^2, and (c) the radial probability density $P(r)$ at $r = a_0$. Give your answers in terms of a_0.

39 • **SSM** **ISOLVE** ✓ (a) If spin is not included, how many different wave functions are there corresponding to the first excited energy level $n = 2$ for hydrogen? (b) List these functions by giving the quantum numbers for each state.

40 •• For the ground state of the hydrogen atom, calculate the probability of finding the electron in the range $\Delta r = 0.03a_0$ at (a) $r = a_0$ and (b) $r = 2a_0$.

41 •• The value of the constant $C_{2,0,0}$ in Equation 36-36 is given by

$$C_{2,0,0} = \frac{1}{4\sqrt{2\pi}} \left(\frac{Z}{a_0}\right)^{3/2}$$

Find the values of (a) ψ, (b) ψ^2, and (c) the radial probability density $P(r)$ at $r = a_0$ for the state $n = 2$, $\ell = 0$, and $m_\ell = 0$ in hydrogen. Give your answers in terms of a_0.

42 ••• Show that the radial probability density for the $n = 2$, $\ell = 1$, and $m_\ell = 0$ state of a one-electron atom can be written as $P(r) = A\cos^2\theta \, r^4 e^{-Zr/a_0}$, where A is a constant.

43 ••• Calculate the probability of finding the electron in the range $\Delta r = 0.02a_0$ for (a) $r = a_0$ and (b) $r = 2a_0$ for the state $n = 2$, $\ell = 0$, and $m_\ell = 0$ in hydrogen. (See Problem 41 for the value of $C_{2,0,0}$.)

44 •• **SSM** Show that the ground-state hydrogen wave function (Equation 36-33) is a solution to Schödinger's equation (Equation 36-21) and the potential energy function equation (Equation 36-26).

45 •• Show by unit cancellation that the expression for the hydrogen ground-state energy given by Equation 36-28 has the dimensions of energy.

46 •• By dimensional analysis, show that the expression for the first Bohr radius given by Equation 36-12 has the dimensions of length.

47 •• The radial probability distribution function for a one-electron atom in its ground state can be written $P(r) = Cr^2 e^{-2Zr/a_0}$, where C is a constant. Show that $P(r)$ has its maximum value at $r = a_0/Z$.

48 ••• Show that the number of states in the hydrogen atom for a given n is $2n^2$.

49 ••• **ISOLVE** ✓ Calculate the probability that the electron in the ground state of a hydrogen atom is in the region $0 < r < a_0$.

The Spin–Orbit Effect and Fine Structure

50 • **SSM** **iSOLVE** The potential energy of a magnetic moment in an external magnetic field is given by $U = -\boldsymbol{\mu} \cdot \boldsymbol{B}$. (a) Calculate the difference in energy between the two possible orientations of an electron in a magnetic field $\boldsymbol{B} = 1.50 \text{ T}\hat{k}$. (b) If these electrons are bombarded with photons of energy equal to this energy difference, "spin flip" transitions can be induced. Find the wavelength of the photons needed for such transitions. This phenomenon is called *electron spin resonance*.

51 • The total angular momentum of a hydrogen atom in a certain excited state has the quantum number $j = \frac{1}{2}$. What can you say about the orbital angular-momentum quantum number ℓ?

52 • A hydrogen atom is in the state $n = 3$, $\ell = 2$. What are the possible values of j?

53 • Using a scaled vector diagram, show how the orbital angular momentum $\vec{L}$ combines with the spin orbital angular momentum $\vec{S}$ to produce the two possible values of total angular momentum $\vec{J}$ for the $\ell = 3$ state of the hydrogen atom.

The Periodic Table

54 • The total number of quantum states of hydrogen with quantum number $n = 4$ is (a) 4, (b) 16, (c) 32, (d) 36, or (e) 48.

55 • How many of oxygen's eight electrons are found in the p state? (a) 0, (b) 2, (c) 4, (d) 6, or (e) 8.

56 • **SSM** Write the electron configuration of (a) carbon and (b) oxygen.

57 • Give the possible values of the z component of the orbital angular momentum of (a) a d electron, and (b) an f electron.

Optical Spectra and X-Ray Spectra

58 • The optical spectra of atoms with two electrons in the same outer shell are similar, but they are quite different from the spectra of atoms with just one outer electron because of the interaction of the two electrons. Separate the following elements into two groups so that those in each group have similar spectra: lithium, beryllium, sodium, magnesium, potassium, calcium, chromium, nickel, cesium, barium.

59 • Write down the possible electron configurations for the first excited state of (a) hydrogen, (b) sodium, and (c) helium.

60 • **iSOLVE** Indicate which of the following elements should have optical spectra similar to hydrogen and which should be similar to helium: Li, Ca, Ti, Rb, Hg, Ag, Cd, Ba, Fr, and Ra.

61 • **SSM** (a) Calculate the next two longest wavelengths in the K series (after the K_α line) of molybdenum. (b) What is the wavelength of the shortest wavelength in this series?

62 • The wavelength of the K_α line for a certain element is 0.3368 nm. What is the element?

63 • Calculate the wavelength of the K_α line in (a) magnesium ($Z = 12$) and (b) copper ($Z = 29$).

General Problems

64 • What is the energy of the shortest wavelength photon emitted by the hydrogen atom?

65 • The wavelength of a spectral line of hydrogen is 97.254 nm. Identify the transition that results in this line under the assumption that the transition is to the ground state.

66 • The wavelength of a spectral line of hydrogen is 1093.8 nm. Identify the transition that results in this line.

67 • Spectral lines of the following wavelengths are emitted by singly ionized helium: 164 nm, 230.6 nm, and 541 nm. Identify the transitions that result in these spectral lines.

68 •• **SSM** We are often interested in finding the quantity ke^2/r in electron volts when r is given in nanometers. Show that $ke^2 = 1.44 \text{ eV·nm}$.

69 •• The wavelengths of the photons emitted by potassium corresponding to transitions from the $4P_{3/2}$ and $4P_{1/2}$ states to the ground state are 766.41 nm and 769.90 nm. (a) Calculate the energies of these photons in electron volts. (b) The difference in the energies of these photons equals the difference in energy ΔE between the $4P_{3/2}$ and $4P_{1/2}$ states in potassium. Calculate ΔE. (c) Estimate the magnetic field that the 4p electron in potassium experiences.

70 •• To observe the characteristic K lines of the X-ray spectrum, one of the $n = 1$ electrons must be ejected from the atom. This is generally accomplished by bombarding the target material with electrons of sufficient energy to eject this tightly bound electron. What is the minimum energy required to observe the K lines of (a) tungsten, (b) molybdenum, and (c) copper?

71 •• **SSM** The combination of physical constants $\alpha = e^2k/\hbar c$, where k is the Coulomb constant, is known as the *fine-structure constant*. It appears in numerous relations in atomic physics. (a) Show that α is dimensionless. (b) Show that in the Bohr model of hydrogen $v_n = c\alpha/n$, where v_n is the speed of the electron in the stationary state of quantum number n.

72 •• The *positron* is a particle that is identical to the electron except that it carries a positive charge of e. Positronium is the bound state of an electron and positron. (a) Calculate the energies of the five lowest energy states of positronium using the reduced mass, as given in Problem 29. (b) Do transitions between any of the levels found in Part (a) fall in the visible range of wavelengths? If so, which transitions are these?

73 • In 1947, Lamb and Retherford showed that there was a very small energy difference between the $2S_{1/2}$ and the $2P_{1/2}$ states of the hydrogen atom. They measured this difference essentially by causing transitions between the two states using very long wavelength electromagnetic radiation. The energy difference (the Lamb shift) is 4.372×10^{-6} eV and is explained by quantum electrodynamics as being due to fluctuations in the energy level of the vacuum. (a) What is the frequency of a photon whose energy is equal to the Lamb shift energy? (*b*) What is the wavelength of this photon? In what spectral region does it belong?

74 • SSM A Rydberg atom is one in which an outer shell electron is placed into a *very* high excited state (n ≈ 40 or higher). Such atoms are useful for experiments that probe the transition from quantum-mechanical behavior to classical. Furthermore, these excited states have extremely long lifetimes (i.e., the electron will stay in this high excited state for a very long time). A hydrogen atom is in the $n = 45$ state. (*a*) What is the ionization energy of the atom when it is in this state? (*b*) What is the energy level separation (in eV) between this state and the $n = 44$ state? (*c*) What is the wavelength of a photon resonant with this transition? (*d*) How large is the atom in the $n = 45$ state?

75 •• The deuteron, the nucleus of deuterium (heavy hydrogen), was first recognized from the spectrum of hydrogen. The deuteron has a mass twice that of the proton. (*a*) Calculate the Rydberg constant for hydrogen and for deuterium using the reduced mass as given in Problem 29. (*b*) Using the result

obtained in Part (*a*), determine the wavelength difference between the longest wavelength Balmer lines of hydrogen and deuterium.

76 •• The muonium atom is a hydrogen atom with the electron replaced by a μ^- particle. The μ^- is identical to an electron but has a mass 207 times as great as the electron. (*a*) Calculate the energies of the five lowest energy levels of muonium using the reduced mass as given in Problem 29. (*b*) Do transitions between any of the levels found in Part (*a*) fall in the visible range of wavelengths, (i.e., between $\lambda = 700$ nm and 400 nm)? If so, which transitions are these?

77 •• The triton, a nucleus consisting of a proton and two neutrons, is unstable with a fairly long half-life of approximately 12 years. Tritium is the bound state of an electron and a triton. (*a*) Calculate the Rydberg constant of tritium using the reduced mass as given in Problem 29. (*b*) Using the result obtained in Part (*a*) and the result obtained in Part (*b*) of Problem 76, determine the wavelength difference between the longest wavelength Balmer lines of tritium and deuterium and between tritium and hydrogen.

Molecules

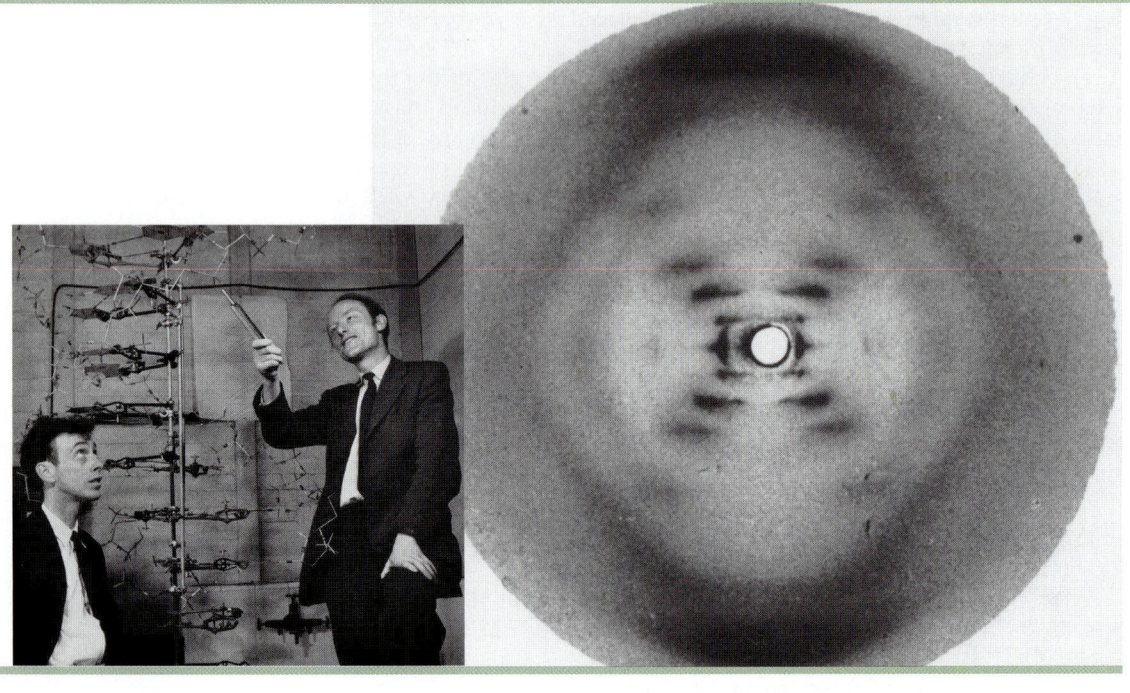

THE DISCOVERERS OF THE STRUCTURE OF DNA JAMES WATSON AT LEFT AND FRANCIS CRICK ARE SHOWN WITH THEIR MODEL OF PART OF A DNA MOLECULE IN 1953. CRICK AND WATSON MET AT THE CAVENDISH LABORATORY, CAMBRIDGE, IN 1951. THEIR WORK ON THE STRUCTURE OF DNA WAS PERFORMED WITH A KNOWLEDGE OF CHARGAFF'S RATIOS OF THE BASES IN DNA AND SOME ACCESS TO THE X-RAY CRYSTALLOGRAPHY OF MAURICE WILKINS AND ROSALIND FRANKLIN AT KING'S COLLEGE LONDON. COMBINING ALL OF THIS WORK LED TO THE DEDUCTION THAT DNA EXISTS AS A DOUBLE HELIX, THUS TO ITS STRUCTURE. CRICK, WATSON, AND WILKINS SHARED THE 1962 NOBEL PRIZE FOR PHYSIOLOGY OR MEDICINE; FRANKLIN DIED FROM CANCER IN 1958.

X-RAY DIFFRACTION PATTERN OF THE B FORM OF DNA ROSALIND FRANKLIN'S COLLEAGUE MAURICE WILKINS, WITHOUT OBTAINING HER PERMISSION, MADE AVAILABLE TO WATSON AND CRICK HER THEN UNPUBLISHED X-RAY DIFFRACTION PATTERN OF THE B FORM OF DNA, WHICH WAS CRUCIAL EVIDENCE FOR THE HELICAL STRUCTURE. IN HIS ACCOUNT OF THIS DISCOVERY, WATSON WROTE: "THE INSTANT I SAW THE PICTURE, MY MOUTH FELL OPEN AND PULSE BEGAN TO RACE. . . . THE BLACK CROSS OF REFLECTIONS WHICH DOMINATED THE PICTURE COULD ARISE ONLY FROM A HELICAL STRUCTURE. . . . MERE INSPECTION OF THE X-RAY PICTURE GAVE SEVERAL OF THE VITAL HELICAL PERAMETERS" (FROM STENT, GUNTHER, *THE DOUBLE HELIX*, NEW YORK: NORTON, 1980).

Most atoms bond together to form molecules or solids. Molecules may exist as separate entities, as in gaseous O_2 or N_2, or they may bond together to form liquids or solids. A molecule is the smallest constituent of a substance that retains its chemical properties.

➤ **In this chapter, we use our understanding of quantum mechanics to discuss molecular bonding and the energy levels and spectra of diatomic molecules. Much of our discussion will be qualitative because, as in atomic physics, the quantum-mechanical calculations are very difficult.**

37-1 Molecular Bonding

There are two extreme views that we can take of a molecule. Consider, for example, H_2. We can think of H_2 either as two H atoms joined together or as a quantum-mechanical system of two protons and two electrons. The latter picture is more fruitful in this case because neither of the electrons in the H_2 molecule can be identified as belonging to either proton. Instead, the wave function for each electron is spread out in space throughout the whole molecule. For more complicated molecules, however, an intermediate picture is useful. For example, the nitrogen molecule N_2 consists of 14 protons and 14 electrons, but only two of the electrons take part in the bonding. We therefore can consider this molecule as two N^+ ions and two electrons that belong to the molecule as a whole. The molecular wave functions for these bonding electrons are called **molecular**

orbitals. In many cases, these molecular wave functions can be constructed from combinations of the atomic wave functions with which we are familiar.

The two principal types of bonds responsible for the formation of molecules are the ionic bond and the covalent bond. Other types of bonds that are important in the bonding of liquids and solids are van der Waals bonds, metallic bonds, and hydrogen bonds. In many cases, bonding is a mixture of these mechanisms.

The Ionic Bond

The simplest type of bond is the **ionic bond,** which is found in salts such as sodium chloride (NaCl). The sodium atom has one 3s electron outside a stable core. The ionization energy of sodium is the energy needed to remove this electron from an isolated sodium atom. This energy is just 5.14 eV (see Figure 36-18). The removal of this electron leaves an isolated positive ion with a spherically symmetric, closed-shell electron core. Chlorine, on the other hand, is one electron short of having a closed shell. The energy released by an isolated atom's acquisition of one electron is called its **electron affinity,** which in the case of chlorine is 3.62 eV. The acquisition of one electron by chlorine results in a negative ion with a spherically symmetric, closed-shell electron core. Thus, the formation of a Na^+ ion and a Cl^- ion by the donation of one electron of sodium to chlorine requires only 5.14 eV − 3.62 eV = 1.52 eV at infinite separation. The electrostatic potential energy U_e of the two ions when they are a distance r apart is $-ke^2/r$. When the separation of the ions is less than approximately 0.95 nm, the negative potential energy of attraction is of greater magnitude than the 1.52 eV of energy needed to create the ions. Thus, at separation distances less than 0.95 nm, it is energetically favorable (i.e., the total energy of the system is reduced) for the sodium atom to donate an electron to the chlorine atom to form NaCl.

Since the electrostatic attraction increases as the ions get closer together, it might seem that equilibrium could not exist. However, when the separation of the ions is very small, there is a strong repulsion that is quantum mechanical in nature and is related to the exclusion principle. This **exclusion-principle repulsion** is responsible for the repulsion of the atoms in all molecules (except H_2)[†] for all bonding mechanisms. We can understand it qualitatively as follows. When the ions are very far apart, the wave function for a core electron in one of the ions does not overlap that of any electron in the other ion. We can distinguish the electrons by the ion to which they belong. This means that electrons in the two ions can have the same quantum numbers because they occupy different regions of space. However, as the distance between the ions decreases, the wave functions of the core electrons begin to overlap; that is, the electrons in the two ions begin to occupy the same region of space. Because of the exclusion principle, some of these electrons must go into higher energy quantum states.[‡] But energy is required to shift the electrons into higher energy quantum states. This increase in energy when the ions are pushed closer together is equivalent to a repulsion of the ions. It is not a sudden process. The energy states of the electrons change gradually as the ions are brought together. A sketch of the potential energy $U(r)$ of the Na^+ and Cl^- ions versus separation distance r is shown in Figure 37-1. The energy is lowest at an equilibrium separation r_0 of approximately 0.236 nm. At smaller separations, the energy rises steeply

FIGURE 37-1 Potential energy for Na^+ and Cl^- ions as a function of separation distance r. The energy at infinite separation is chosen to be 1.52 eV, corresponding to the energy $-\Delta E$ needed to form the ions from neutral atoms. The minimum energy is at the equilibrium separation $r_0 = 0.236$ nm for the ions in the molecule.

[†] In H_2, the repulsion is simply that of the two positively charged protons.

[‡] Recall from our discussion in Chapter 35 that the exclusion principle is related to the fact that the wave function for two identical electrons is antisymmetric on the exchange of the electrons and that an antisymmetric wave function for two electrons with the same quantum numbers is zero if the space coordinates of the electrons are the same.

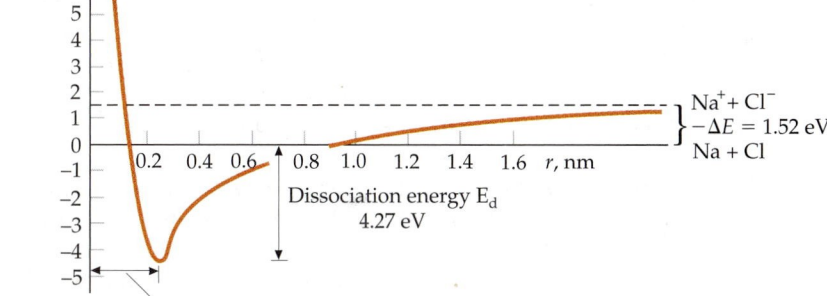

as a result of the exclusion principle. The energy required to separate the ions and form neutral sodium and chlorine atoms is called the **dissociation energy** E_d, which is approximately 4.27 eV for NaCl.

The equilibrium separation distance of 0.236 nm is for gaseous diatomic NaCl, which can be obtained by evaporating solid NaCl. Normally, NaCl exists in a cubic crystal structure, with the Na⁺ and Cl⁻ ions at the alternate corners of a cube. The separation of the ions in a crystal is somewhat larger, approximately 0.28 nm. Because of the presence of neighboring ions of opposite charge, the electrostatic energy per ion pair is lower when the ions are in a crystal.

THE ENERGY OF A SODIUM-FLUORIDE MOLECULE　　　　**E X A M P L E　3 7 - 1**

The electron affinity of fluorine is 3.40 eV, and the equilibrium separation of sodium fluoride (NaF) is 0.193 nm. (a) How much energy is needed to form Na⁺ and F⁻ ions from neutral sodium and fluorine atoms? (b) What is the electrostatic potential energy of the Na⁺ and F⁻ ions at their equilibrium separation? (c) The dissociation energy of NaF is 5.38 eV. What is the energy due to repulsion of the ions at the equilibrium separation?

PICTURE THE PROBLEM (a) The energy ΔE needed to form Na⁺ and F⁻ ions from the neutral sodium and fluorine atoms is the difference between the ionization energy of sodium (5.14 eV) and the electron affinity of fluorine. (b) The electrostatic potential energy with $U = 0$ at infinity is $U_e = -ke^2/r$. (c) If we choose the potential energy at infinity to be ΔE, the total potential energy is $U_{tot} = U_e + \Delta E + U_{rep}$, where U_{rep} is the energy of repulsion, which is found by setting the dissociation energy equal to $-U_{tot}$.

(a) Calculate the energy needed to form Na⁺ and F⁻ ions from the neutral sodium and fluorine atoms (see Picture the Problem):

$$\Delta E = 5.14 \text{ eV} - 3.40 \text{ eV} = \boxed{1.74 \text{ eV}}$$

(b) 1. Calculate the electrostatic potential energy at the equilibrium separation of $r = 0.193$ nm:

$$U_e = -\frac{ke^2}{r}$$

$$= -\frac{(8.99 \times 10^9 \text{ N·m}^2/\text{C}^2)(1.60 \times 10^{-19} \text{ C})^2}{1.93 \times 10^{-10} \text{ m}}$$

$$= -1.19 \times 10^{-18} \text{ J}$$

2. Convert from joules to electron volts:

$$U_e = -1.19 \times 10^{-18} \text{ J}\left(\frac{1 \text{ eV}}{1.60 \times 10^{-19} \text{ J}}\right) = \boxed{-7.45 \text{ eV}}$$

(c) The dissociation energy equals the negative of the total potential energy:

$$E_d = -U_{tot} = -(U_e + \Delta E + U_{rep})$$

so

$$U_{rep} = -(E_d + \Delta E + U_e)$$

$$= -(5.38 \text{ eV} + 1.74 \text{ eV} - 7.45 \text{ eV})$$

$$= \boxed{0.33 \text{ eV}}$$

The Covalent Bond

A completely different mechanism, the **covalent bond,** is responsible for the bonding of identical or similar atoms to form such molecules as gaseous hydrogen (H_2), nitrogen (N_2), and carbon monoxide (CO). If we calculate the energy needed to form H⁺ and H⁻ ions by the transfer of an electron from one atom to the other and then add this energy to the electrostatic potential energy, we find that there is no separation distance for which the total energy is negative. The bond thus cannot be ionic. Instead, the attraction of two hydrogen atoms is an entirely quantum-mechanical effect. The decrease in energy when two hydrogen atoms approach each other is due to the sharing of the two electrons by both atoms. It is intimately connected with the symmetry properties of the wave functions of electrons.

We can gain some insight into covalent bonding by considering a simple, one-dimensional quantum-mechanics problem of two finite square wells. We first consider a single electron that is equally likely to be in either well. Because the wells are identical, the probability distribution, which is proportional to $|\psi^2|$, must be symmetric about the midpoint between the wells. Then ψ must be either symmetric or antisymmetric with respect to the two wells. The two possibilities for the ground state are shown in Figure 37-2a for the case in which the wells are far apart and in Figure 37-2b for the case in which the wells are close together. An important feature of Figure 37-2b is that in the region between the wells the symmetric wave function is large and the antisymmetric wave function is small.

Now consider adding a second electron to the two wells. We saw in Chapter 35 that the wave functions for particles that obey the exclusion principle are antisymmetric on exchange of the particles. Thus the total wave function for the two electrons must be antisymmetric on exchange of the electrons. Note that exchanging the electrons in the wells here is the same as exchanging the wells. The total wave function for two electrons can be written as a product of a space part and a spin part. So, an antisymmetric wave function can be the product of a symmetric space part and an antisymmetric spin part or of a symmetric spin part and an antisymmetric space part.

To understand the symmetry of the total wave function, we must therefore understand the symmetry of the spin part of the wave function. The spin of a single electron can have two possible values for its quantum number m_S: $m_S = +\frac{1}{2}$, which we call spin up, or $m_S = -\frac{1}{2}$, which we call spin down. We will use arrows to designate the spin wave function for a single electron: $\uparrow_1$ or $\uparrow_2$ for electron 1 or electron 2 with spin up and $\downarrow_1$ or $\downarrow_2$ for electron 1 or electron 2 with spin down. The total spin quantum number for two electrons can be $S = 1$, with $m_S = +1, 0$, or -1; or $S = 0$, with $m_S = 0$. We use ϕ_{S, m_S} to denote the spin wave function for two electrons. The spin state $\phi_{1, +1}$, corresponding to $S = 1$ and $m_S = +1$, can be written

$$\phi_{1,+1} = \uparrow_1\uparrow_2, \quad S = 1, m_S = +1 \qquad 37\text{-}1$$

Similarly, the spin state for $S = 1, m_S = -1$ is

$$\phi_{1,-1} = \downarrow_1\downarrow_2, \quad S = 1, m_S = -1 \qquad 37\text{-}2$$

Note that both of these states are symmetric upon exchange of the electrons. The spin state corresponding to $S = 1$ and $m_S = 0$ is not quite so obvious. It turns out to be proportional to

$$\phi_{1,0} = \uparrow_1\downarrow_2 + \uparrow_2\downarrow_1, \quad S = 1, m_S = 0 \qquad 37\text{-}3$$

This spin state is also symmetric upon exchange of the electrons. The spin state for two electrons with antiparallel spins ($S = 0$) is

$$\phi_{0,0} = \uparrow_1\downarrow_2 - \uparrow_2\downarrow_1, \quad S = 0, m_S = 0 \qquad 37\text{-}4$$

This spin state is antisymmetric upon exchange of electrons.

We thus have the important result that the *spin* part of the wave function is symmetric for parallel spins ($S = 1$) and antisymmetric for antiparallel spins

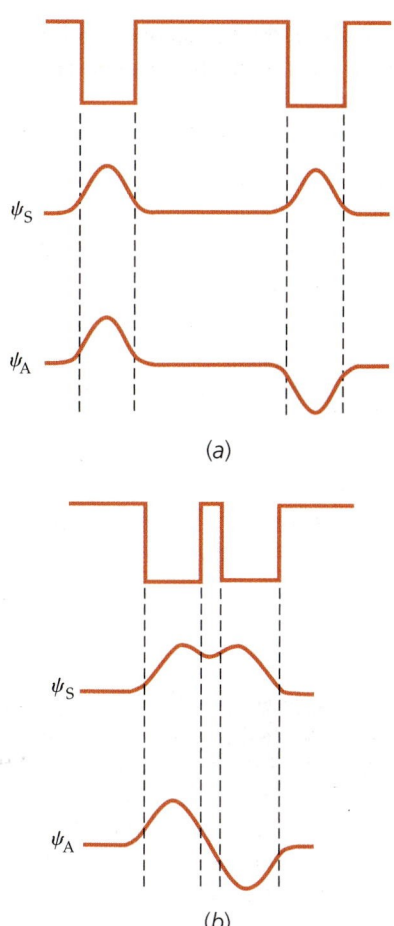

(a)

(b)

FIGURE 37-2 (*a*) Two square wells far apart. The electron wave function can be either symmetric (ψ_S) or antisymmetric (ψ_A) in space. The probability distributions and energies are the same for the two wave functions when the wells are far apart. (*b*) Two square wells that are close together. Between the wells, the antisymmetric space wave function is approximately zero, whereas the symmetric space wave function is quite large.

$(S = 0)$. Because the total wave function is the product of the space function and spin function, we have the following important result:

> For the total wave function of two electrons to be antisymmetric, the space part of the wave function must be antisymmetric for parallel spins $(S = 1)$ and symmetric for antiparallel spins $(S = 0)$.

SPIN ALIGNMENT AND WAVE-FUNCTION SYMMETRY

We can now consider the problem of two hydrogen atoms. Figure 37-3a shows a spatially symmetric wave function ψ_S and a spatially antisymmetric wave function ψ_A for two hydrogen atoms that are far apart, and Figure 37-3b shows the same two wave functions for two hydrogen atoms that are close together. The squares of these two wave functions are shown in Figure 37-3c. Note that the probability distribution $|\psi|^2$ in the region between the protons is large for the symmetric wave function and small for the antisymmetric wave function. Thus, when the space part of the wave function is symmetric $(S = 0)$, the electrons are often found in the region between the protons. The negatively charged electron cloud representing these electrons is concentrated in the space between the protons, as shown in the upper part of Figure 37-3c, and the protons are bound together by this negatively charged cloud. Conversely, when the space part of the wave function is antisymmetric $(S = 1)$, the electrons spend little time between the protons, and the atoms do not bind together to form a molecule. In this case, the electron cloud is not concentrated in the space between the protons, as shown in the lower part of Figure 37-3c.

The total electrostatic potential energy for the H_2 molecule consists of the positive energy of repulsion of the two electrons and the negative potential energy of attraction of each electron for each proton. Figure 37-4 shows the electrostatic potential energy for two hydrogen atoms versus separation for the case in which the space part of the electron wave function is symmetric (U_S) and for the case in which it is antisymmetric (U_A). We can see that the potential energy for the symmetric state is the lower of the two and that the shape of this potential energy curve is similar to that for ionic bonding. The equilibrium separation for H_2 is $r_0 = 0.074$ nm, and the binding energy is 4.52 eV. For the antisymmetric state, the potential energy is never negative and there is no bonding.

We can now see why three hydrogen atoms do not bond to form H_3. If a third hydrogen atom is brought near an H_2 molecule, the third electron cannot be in a 1s state and have its spin antiparallel to the spin of both of the other electrons. If this electron is in an antisymmetric space state with respect to exchange with one

FIGURE 37-3 One-dimensional symmetric and antisymmetric wave functions for two hydrogen atoms (a) far apart and (b) close together. (c) Electron probability distributions ($|\psi|^2$) for the wave functions in Figure 37-3b. For the symmetric wave function, the electron charge density is large between the protons. This negative charge density holds the protons together in the hydrogen molecule H_2. For the antisymmetric wave function, the electron charge density is not large between the protons.

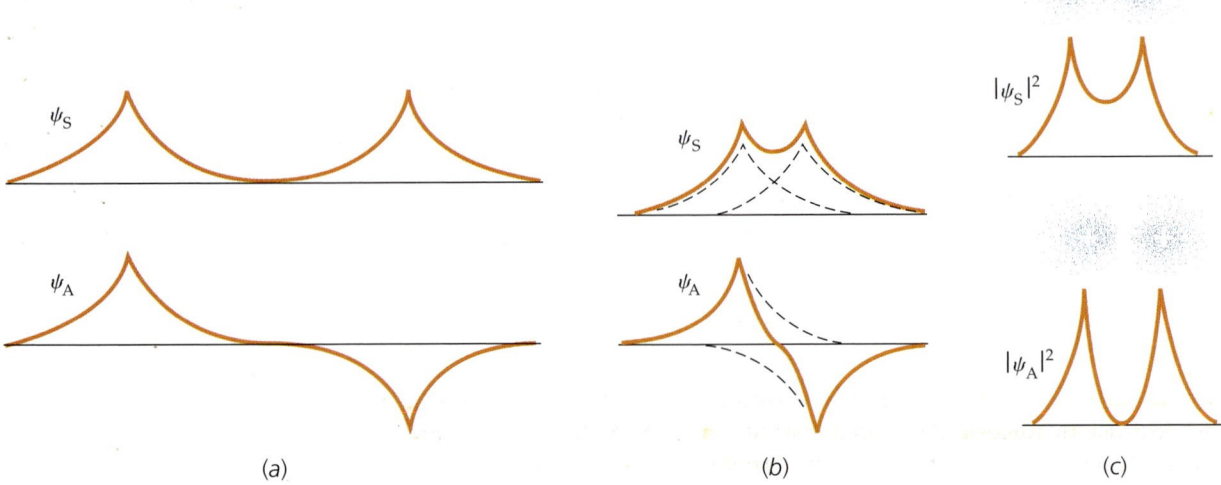

(a) (b) (c)

of the electrons, the repulsion of this atom is greater than the attraction of the other. As the three atoms are pushed together, the third electron is, in effect, forced into a higher quantum-energy state by the exclusion principle. The bond between two hydrogen atoms is called a **saturated bond** because there is no room for another electron. The two shared electrons essentially fill the 1s states of both atoms.

We can also see why two helium atoms do not normally bond together to form the He_2 molecule. There are no valence electrons that can be shared. The electrons in the closed shells are forced into higher energy states when the two atoms are brought together. At low temperatures or high pressures, helium atoms do bond together due to van der Waals forces, which we will discuss next. This bonding is so weak that at atmospheric pressure helium boils at 4 K, and it does not form a solid at any temperature unless the pressure is greater than about 20 atm.

When two identical atoms bond, as in O_2 or N_2, the bonding is purely covalent. However, the bonding of two dissimilar atoms is often a mixture of covalent and ionic bonding. Even in NaCl, the electron donated by sodium to chlorine has some probability of being at the sodium atom because its wave function in the vicinity of the sodium atom, while small, is not zero. Thus, this electron is partially shared in a covalent bond, although this bonding is only a small part of the total bond, which is mainly ionic.

A measure of the degree to which a bond is ionic or covalent can be obtained from the electric dipole moment of the molecule. For example, if the bonding in NaCl were purely ionic, the center of positive charge would be at the Na^+ ion and the center of negative charge would be at the Cl^- ion. The electric dipole moment would have the magnitude

$$p_{ionic} = er_0 \qquad\qquad 37\text{-}5$$

where $r_0 = 2.36 \times 10^{-10}$ m is the equilibrium separation of the ions. Thus, the dipole moment of NaCl would be (from Figure 37-1)

$$p_{ionic} = er_0$$
$$= (1.60 \times 10^{-19}\,\text{C})(2.36 \times 10^{-10}\,\text{m}) = 3.78 \times 10^{-29}\,\text{C·m}$$

The actual measured electric dipole moment of NaCl is

$$p_{measured} = 3.00 \times 10^{-29}\,\text{C·m}$$

We can define the ratio of $p_{measured}$ to p_{ionic} as the fractional amount of ionic bonding. For NaCl, this ratio is $3.00/3.78 = 0.79$. Thus, the bonding in NaCl is about 79 percent ionic.

EXERCISE The equilibrium separation of HCl is 0.128 nm and its measured electric dipole moment is 3.60×10^{-30} C·m. What is the percentage of ionic bonding in HCl? (*Answer* 18 percent)

Other Bonding Types

The van der Waals Bond Any two separated molecules will be attracted to one another by electrostatic forces called van der Waals forces. So will any two atoms that do not form ionic or covalent bonds. The **van der Waals bonds** due to these

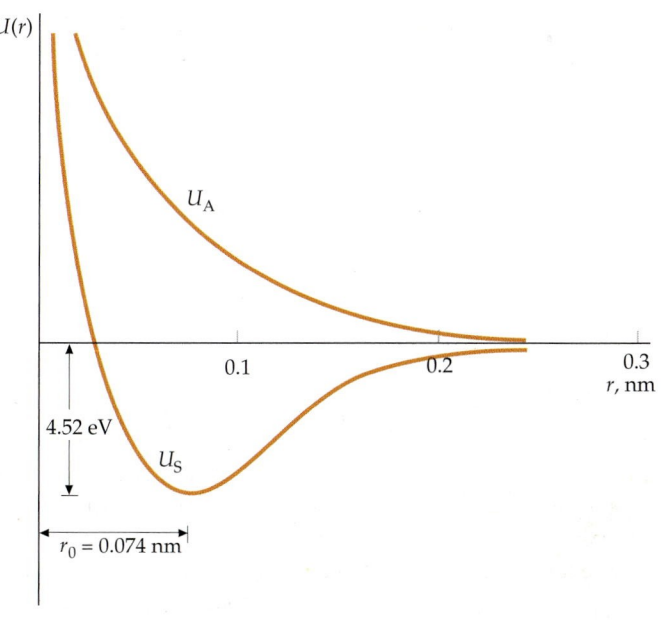

FIGURE 37-4 Potential energy versus separation for two hydrogen atoms. The curve labeled U_S is for a wave function with a symmetric space part, and the curve labeled U_A is for a wave function with an antisymmetric space part.

forces are much weaker than the bonds already discussed. At high enough temperatures, these forces are not strong enough to overcome the ordinary thermal agitation of the atoms or molecules, but at sufficiently low temperatures, thermal agitation becomes negligible, and the van der Waals forces will cause virtually all substances to condense into a liquid and then a solid form.[†] The van der Waals forces arise from the interaction of the instantaneous electric dipole moments of the molecules.

Figure 37-5 shows how two polar molecules—molecules with *permanent* electric dipole moments, such as H_2O—can bond. The electric field due to the dipole moment of one molecule orients the other molecule so that the two dipole moments attract. Nonpolar molecules also attract other nonpolar molecules via the van der Waals forces. Although nonpolar molecules have zero electric dipole moments on the average, they have instantaneous dipole moments that are generally not zero because of fluctuations in the positions of the charges. When two nonpolar molecules are near each other, the fluctuations in the instantaneous dipole moments tend to become correlated so as to produce attraction. This is illustrated in Figure 37-6.

The Hydrogen Bond Another bonding mechanism of great importance is the hydrogen bond, which is formed by the sharing of a proton (the nucleus of the hydrogen atom) between two atoms, frequently two oxygen atoms. This sharing of a proton is similar to the sharing of electrons responsible for the covalent bond already discussed. It is facilitated by the small mass of the proton and by the absence of inner core electrons in hydrogen. The hydrogen bond often holds groups of molecules together and is responsible for the cross-linking that allows giant biological molecules and polymers to hold their fixed shapes. The well-known helical structure of DNA is due to hydrogen-bond linkages across turns of the helix (Figure 37-7).

The Metallic Bond In a metal, two atoms do not bond together by exchanging or sharing an electron to form a molecule. Instead, each valence electron is shared by many atoms. The bonding is thus distributed throughout the entire metal. A metal can be thought of as a lattice of positive ions held together by a *gas* of essentially free electrons that roam throughout the solid. In the quantum-mechanical picture, these free electrons form a cloud of negative charge density between the positively charged lattice ions that holds the ions together. In this

---

[†] Helium is the only element that does not solidify at any temperature at atmospheric pressure.

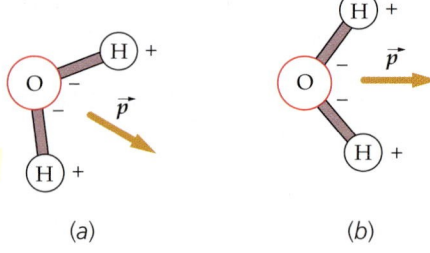

(a)　　　　　　(b)

FIGURE 37-5 Bonding of H_2O molecules because of the attraction of the electric dipoles. The dipole moment of each molecule is indicated by $\vec{p}$. The field of one dipole orients the other dipole so the moments tend to be parallel. When the dipole moments are approximately parallel, the center of negative charge of one molecule is close to the center of positive charge of the other molecule, and the molecules attract.

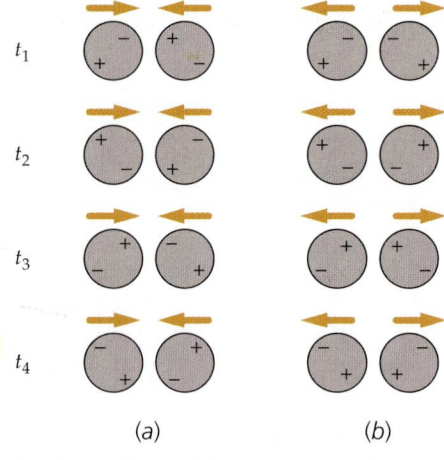

(a)　　　　　　(b)

FIGURE 37-6 van der Waals attraction of molecules with zero average dipole moments. (*a*) Possible orientations of instantaneous dipole moments at different times leading to attraction. (*b*) Possible orientations leading to repulsion. The electric field of the instantaneous dipole moment of one molecule tends to polarize the other molecule; thus the orientations leading to attraction (Figure 37-6*a*) are much more likely than those leading to repulsion (Figure 37-6*b*).

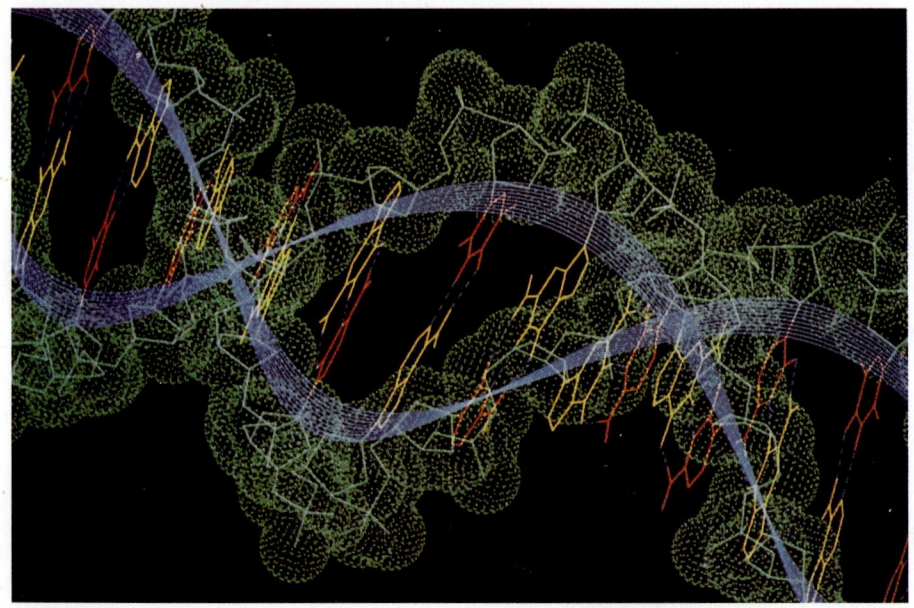

FIGURE 37-7 The DNA molecule.

respect, the metallic bond is somewhat similar to the covalent bond. However, with the metallic bond, there are far more than just two atoms involved, and the negative charge is distributed uniformly throughout the volume of the metal. The number of free electrons varies from metal to metal but is of the order of one per atom.

*37-2 Polyatomic Molecules

Molecules with more than two atoms range from such relatively simple molecules as water, which has a molecular mass number of 18, to such giants as proteins and DNA, which can have molecular masses of hundreds of thousands up to many millions. As with diatomic molecules, the structure of polyatomic molecules can be understood by applying basic quantum mechanics to the bonding of individual atoms. The bonding mechanisms for most polyatomic molecules are the covalent bond and the hydrogen bond. We will discuss only some of the simplest polyatomic molecules—H_2O, NH_3, and CH_4—to illustrate both the simplicity and complexity of the application of quantum mechanics to molecular bonding.

The basic requirement for the sharing of electrons in a covalent bond is that the wave functions of the valence electrons in the individual atoms must overlap as much as possible. As our first example, we will consider the water molecule. The ground-state configuration of the oxygen atom is $1s^2 2s^2 2p^4$. The 1s and 2s electrons are in closed-shell states and do not contribute to the bonding. The 2p shell has room for six electrons, two in each of the three space states (orbitals) corresponding to $\ell = 1$. In an isolated atom, we describe these space states by the hydrogen-like wave functions corresponding to $\ell = 1$ and $m = +1, 0,$ and -1. Since the energy is the same for these three space states, we could equally well use any linear combination of these wave functions. When an atom participates in molecular bonding, certain combinations of these atomic wave functions are important. These combinations are called the p_x, p_y, and p_z **atomic orbitals.** The angular dependence of these orbitals is

$$p_x \propto \sin \theta \cos \phi \qquad\qquad 37\text{-}6$$

$$p_y \propto \sin \theta \sin \phi \qquad\qquad 37\text{-}7$$

$$p_z \propto \cos \phi \qquad\qquad 37\text{-}8$$

The electron charge distribution for these orbitals is maximum along the x, y, or z axis, respectively, as shown in Figure 37-8.

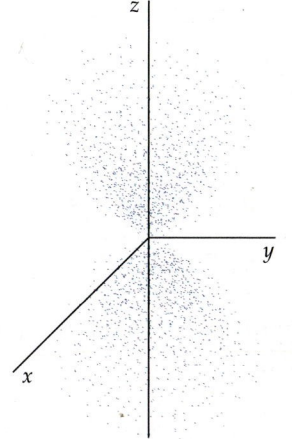

FIGURE 37-8 Computer-generated dot plot illustrating the spatial dependence of the electron charge distribution in the p_x, p_y, and p_z atomic orbitals.

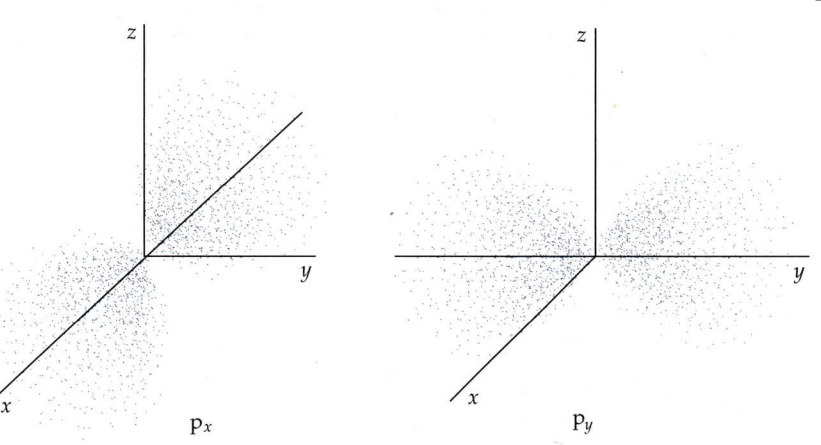

For the oxygen in an H_2O molecule, maximum overlap of the electron wave functions occurs when two of the four 2p electrons are paired with their spins antiparallel in one of the atomic orbitals (for this example, assume the p_z orbital), one of the other electrons is in a second orbital (the p_x orbital), and the other electron is in the third orbital (the p_y orbital). Each of the unpaired electrons (in the p_x and p_y orbitals, in this illustration) forms a bond with the electron of a hydrogen atom, as shown in Figure 37-9. Because of the repulsion of the two hydrogen atoms, the angle between the O–H bonds is actually greater than 90°. The effect of this repulsion can be calculated, and the result is in agreement with the measured angle of 104.5°.

Similar reasoning leads to an understanding of the bonding in NH_3. In the ground state, nitrogen has three electrons in the 2p state. When these three electrons are in the p_x, p_y, and p_z atomic orbitals, they bond to the electrons of hydrogen atoms. Again, because of the repulsion of the hydrogen atoms, the angles between the bonds are somewhat larger than 90°.

The bonding of carbon atoms is somewhat more complicated. Carbon forms a wide variety of different types of molecular bonds, leading to a great diversity in the kinds of organic molecules. The ground-state configuration of carbon is $1s^2 2s^2 2p^2$. From our previous discussion, we might expect carbon to be divalent—that is, bonding only through its two 2p electrons—with the two bonds forming at approximately 90°. However, one of the most important features of the chemistry of carbon is that tetravalent carbon compounds, such as CH_4, are overwhelmingly favored.

The observed valence of 4 for carbon comes about in an interesting way. One of the first excited states of carbon occurs when a 2s electron is excited to a 2p state, giving a configuration of $1s^2 2s^1 2p^3$. In this excited state, we can have four unpaired electrons, one each in the 2s, $2p_x$, $2p_y$, and $2p_z$ atomic orbitals. We might expect there to be three similar bonds corresponding to the three p orbitals and one different bond corresponding to the s orbital. However, when carbon forms tetravalent bonds, these four atomic orbitals become mixed and form four new *equivalent* molecular orbitals called **hybrid orbitals.** This mixing of atomic orbitals, called **hybridization,** is among the most important features involved in the physics of complex molecular bonds. Figure 37-10 shows the tetrahedral structure of the methane molecule (CH_4), and Figure 37-11 shows the structure of the ethane molecule (CH_3–CH_3), which is similar to two joined methane molecules in which one of the C–H bonds is replaced with a C–C bond.

Carbon orbitals can also hybridize, with the s, p_x, and p_y orbitals combining to form three hybrid orbitals in the *xy* plane with 120° bonds and the p_z orbital remaining unmixed. An example of this configuration is graphite, in which the bonds in the *xy* plane provide the strongly layered structure characteristic of the material.

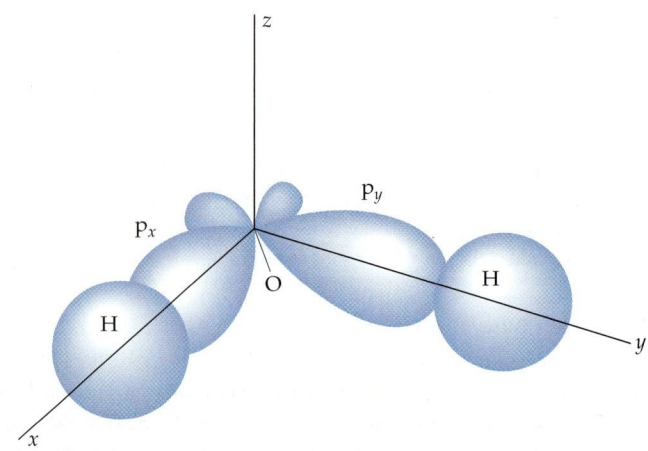

FIGURE 37-9 Electron charge distribution in the H_2O molecule.

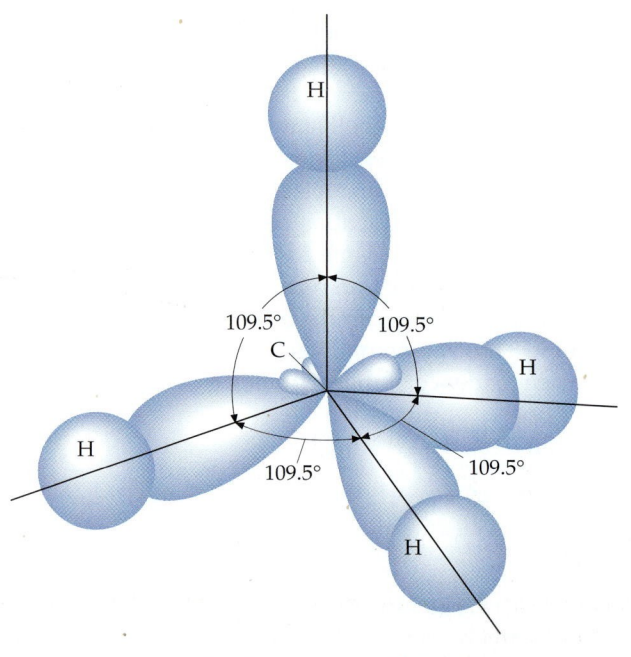

FIGURE 37-10 Electron charge distribution in the CH_4 (methane) molecule.

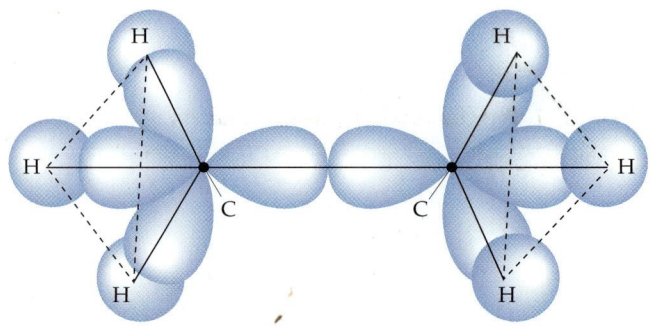

FIGURE 37-11 Electron charge distribution in the CH_3–CH_3 (ethane) molecule.

37-3 Energy Levels and Spectra of Diatomic Molecules

As is the case with an atom, a molecule often emits electromagnetic radiation when it makes a transition from an excited energy state to a state of lower energy. Conversely, a molecule can absorb radiation and make a transition from a lower energy state to a higher energy state. The study of molecular emission and absorption spectra thus provides us with information about the energy states of molecules. For simplicity, we will consider only diatomic molecules here.

The energy of a molecule can be conveniently separated into three parts: electronic, due to the excitation of the electrons of the molecule; vibrational, due to the oscillations of the atoms of the molecule; and rotational, due to the rotation of the molecule about its center of mass. The magnitudes of these energies are sufficiently different that they can be treated separately. The energies due to the electronic excitations of a molecule are of the order of magnitude of 1 eV, the same as for the excitation of an atom. The energies of vibration and rotation are much smaller than this.

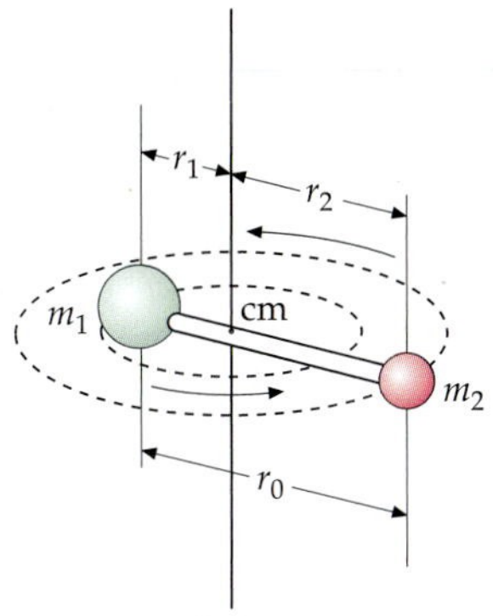

FIGURE 37-12 Diatomic molecule rotating about an axis through its center of mass.

Rotational Energy Levels

Figure 37-12 shows a simple schematic model of a diatomic molecule consisting of a mass m_1 and a mass m_2 separated by a distance r and rotating about its center of mass. Classically, the kinetic energy of rotation (see Equation 9-11) is

$$E = \tfrac{1}{2} I\omega^2 \qquad\qquad 37\text{-}9$$

where I is the moment of inertia and ω is the angular frequency of rotation. If we write this in terms of the angular momentum $L = I\omega$, we have

$$E = \frac{(I\omega)^2}{2I} = \frac{L^2}{2I} \qquad\qquad 37\text{-}10$$

The solution of the Schrödinger equation for rotation leads to quantization of the angular momentum with values given by

$$L^2 = \ell(\ell + 1)\hbar^2, \quad \ell = 0, 1, 2, \ldots \qquad\qquad 37\text{-}11$$

where ℓ is the **rotational quantum number**. This is the same quantum condition on angular momentum that holds for the orbital angular momentum of an electron in an atom. Note, however, that L in Equation 37-10 refers to the angular momentum of the entire molecule rotating about its center of mass. The energy levels of a rotating molecule are therefore given by

$$E_\ell = \frac{\ell(\ell + 1)\hbar^2}{2I} = \ell(\ell + 1)E_{0r}, \quad \ell = 0, 1, 2, \ldots \qquad\qquad 37\text{-}12$$

ROTATIONAL ENERGY LEVELS

where E_{0r} is the characteristic rotational energy of a particular molecule, which is inversely proportional to its moment of inertia:

$$E_{0r} = \frac{\hbar^2}{2I} \qquad\qquad 37\text{-}13$$

CHARACTERISTIC ROTATIONAL ENERGY

A measurement of the rotational energy of a molecule from its rotational spectrum can be used to determine the moment of inertia of the molecule, which can then be used to find the separation of the atoms in the molecule. The moment of inertia about an axis through the center of mass of a diatomic molecule (see Figure 37-12) is

$$I = m_1 r_1^2 + m_2 r_2^2$$

Using $m_1 r_1 = m_2 r_2$, where r_1 is the distance of atom 1 from the center of mass, r_2 is the distance of atom 2 from the center of mass, and $r_0 = r_1 + r_2$, we can write the moment of inertia (see Problem 31) as

$$I = \mu r_0^2 \qquad\qquad 37\text{-}14$$

where μ, called the **reduced mass,** is

$$\mu = \frac{m_1 m_2}{m_1 + m_2} \qquad\qquad 37\text{-}15$$

DEFINITION—REDUCED MASS

If the masses are equal ($m_1 = m_2 = m$), as in H_2 and O_2, the reduced mass is $\mu = \frac{1}{2}m$ and

$$I = \frac{1}{2} m r_0^2 \qquad\qquad 37\text{-}16$$

A unit of mass convenient for discussing atomic and molecular masses is the **unified mass unit,** u, which is defined as one-twelfth the mass of the neutral carbon-12 (^{12}C) atom. The mass of one ^{12}C atom is thus 12 u. The mass of an atom in unified mass units is therefore numerically equal to the molar mass of the atom in grams. The unified mass unit is related to the gram and kilogram by

$$1\,u = \frac{1\,g}{N_A} = \frac{10^{-3}\,kg}{6.0221 \times 10^{23}} = 1.6606 \times 10^{-27}\,kg \qquad\qquad 37\text{-}17$$

where N_A is Avogadro's number.

THE REDUCED MASS OF A DIATOMIC MOLECULE **E X A M P L E   3 7 - 2**

Find the reduced mass of the HCl molecule.

PICTURE THE PROBLEM We find the masses of the hydrogen and chlorine atoms in the periodic table[†] in Appendix C and use the definition in Equation 37-15.

1. The reduced mass μ is related to the individual masses m_H and m_{Cl}:

$$\mu = \frac{m_H m_{Cl}}{m_H + m_{Cl}}$$

2. Find the masses in the periodic table:

$$m_H = 1.01\,u, \quad m_{Cl} = 35.5\,u$$

3. Substitute to calculate the reduced mass:

$$\mu = \frac{m_H m_{Cl}}{m_H + m_{Cl}} = \frac{(1.01\,u)(35.5\,u)}{1.01\,u + 35.5\,u}$$

$$= \boxed{0.982\,u}$$

---

† The masses in these tables are weighted according to the natural isotopic distribution. Thus, the mass of carbon is given as 12.011 rather than 12.000 because natural carbon consists of about 98.9 percent ^{12}C and 1.1 percent ^{13}C. Similarly, natural chlorine consists of about 76 percent ^{35}Cl and 24 percent ^{37}Cl.

REMARKS Note that the reduced mass is less than the mass of either atom in the molecule, and that it is approximately equal to the mass of the hydrogen atom. When one atom of a diatomic molecule is much more massive than the other, the center of mass of the molecule is approximately at the center of the more massive atom, and the reduced mass is approximately equal to the mass of the lighter atom.

| *ROTATIONAL KINETIC ENERGY OF A MOLECULE* | **EXAMPLE 37-3** |

Estimate the characteristic rotational energy of an O_2 molecule, assuming that the separation of the atoms is 0.1 nm.

1. The characteristic rotational energy is inversely proportional to the moment of inertia:

$$E_{0r} = \frac{\hbar^2}{2I}$$

2. Calculate the moment of inertia:

$$I = \mu r_0^2 = \tfrac{1}{2} m r_0^2$$

$$r_0^2 = \frac{\hbar^2}{m E_{0r}}$$

3. Substitute this expression for I into the expression for E_{0r}:

$$E_{0r} = \frac{\hbar^2}{m r_0^2}$$

4. Use $m = 16$ u for the mass of oxygen and the given values of the constants to calculate E_{0r}:

$$E_{0r} = \frac{\hbar^2}{m r_0^2}$$

$$= \frac{(1.055 \times 10^{-34} \text{ J·s})^2}{(16 \text{ u})(10^{-10} \text{ m})^2} \times \left(\frac{1 \text{ u}}{1.66 \times 10^{-27} \text{ kg}}\right)$$

$$= 4.19 \times 10^{-23} \text{ J} = \boxed{2.62 \times 10^{-4} \text{ eV}}$$

We can see from Example 37-3 that the rotational energy levels are several orders of magnitude smaller than energy levels due to electron excitation, which have energies of the order of 1 eV or higher. Transitions within a given set of rotational energy levels yield photons in the microwave region of the electromagnetic spectrum. The rotational energies are also small compared with the typical thermal energy kT at normal temperatures. For $T = 300$ K, for example, kT is about 2.6×10^{-2} eV, which is approximately 100 times the characteristic rotational energy as calculated in Example 37-3, and approximately 1 percent of the typical electronic energy. Thus, at ordinary temperatures, a molecule can be easily excited to the lower rotational energy levels by collisions with other molecules. But such collisions cannot excite the molecule to its electronic energy levels above the ground state.

Vibrational Energy Levels

The quantization of energy in a simple harmonic oscillator was one of the first problems solved by Schrödinger in his paper proposing his wave equation. Solving the Schrödinger equation for a simple harmonic oscillator gives

$$E_\nu = (\nu + \tfrac{1}{2})hf, \quad \nu = 0, 1, 2, \ldots \tag{37-18}$$

VIBRATIONAL ENERGY LEVELS

where f is the frequency of the oscillator and ν (lowercase Greek nu) is the **vibrational quantum number.**[†] An interesting feature of this result is that the energy levels are equally spaced with intervals equal to hf. The frequency of vibration

† We use ν here rather than n so as not to confuse the vibrational quantum number with the principal quantum number n for electronic energy levels.

of a diatomic molecule can be related to the force exerted by one atom on the other. Consider two objects of mass m_1 and m_2 connected by a spring of force constant k_F. The frequency of oscillation of this system (see Problem 36) can be shown to be

$$f = \frac{1}{2\pi}\sqrt{\frac{k_F}{\mu}}$$ 37-19

where μ is the reduced mass given by Equation 37-15. The effective force constant k_F of a diatomic molecule can thus be determined from a measurement of the frequency of oscillation of the molecule.

A selection rule on transitions between vibrational states (of the same electronic state) requires that the vibrational quantum number ν can change only by ± 1, so the energy of a photon emitted by such a transition is hf and the frequency is f, the same as the frequency of vibration. There is a similar selection rule that ℓ must change by ± 1 for transitions between rotational states.

A typical measured frequency of a transition between vibrational states is 5×10^{13} Hz, which gives

$$E \approx hf = (4.14 \times 10^{-15}\ eV \cdot s)(5 \times 10^{13}\ s^{-1}) = 0.2\ eV$$

for the order of magnitude of vibrational energies. This typical vibrational energy is approximately 1000 times greater than the typical rotational energy E_{0r} of the O_2 molecule we found in Example 37-3 and about 8 times greater than the typical thermal energy $kT = 0.026$ eV at $T = 300$ K. Thus, the vibrational levels are almost never excited by molecular collisions at ordinary temperatures.

DETERMINING THE FORCE CONSTANT **EXAMPLE 37-4**

The frequency of vibration of the CO molecule is 6.42×10^{13} Hz. What is the effective force constant for this molecule?

PICTURE THE PROBLEM We use Equation 37-19 to relate k_F to the frequency and the reduced mass, and calculate μ from its definition.

1. The effective force constant is related to the frequency and reduced mass by Equation 37-19:

$$f = \frac{1}{2\pi}\sqrt{\frac{k_F}{\mu}}$$

$$k_F = (2\pi f)^2\mu$$

2. Calculate the reduced mass using 12 u for the mass of the carbon atom and 16 u for the mass of the oxygen atom:

$$\mu = \frac{m_1 m_2}{m_1 + m_2} = \frac{(12\ u)(16\ u)}{12\ u + 16\ u} = 6.86\ u$$

3. Substitute this value of μ into the equation for k_F in Step 1 and convert to SI units:

$$k_F = (2\pi f)^2\mu$$
$$= 4\pi^2 (6.42 \times 10^{13}\ Hz)^2 (6.86\ u)$$
$$= 1.12 \times 10^{30}\ u/s^2 \times \left(\frac{1.66 \times 10^{-27}\ kg}{1\ u}\right)$$
$$= \boxed{1.85 \times 10^3\ N/m}$$

MASTER the CONCEPT WEB

Emission Spectra

Figure 37-13 shows schematically some electronic, vibrational, and rotational energy levels of a diatomic molecule. The vibrational levels are labeled with the quantum number ν and the rotational levels are labeled with ℓ. The lower vibrational levels are evenly spaced, with $\Delta E = hf$. For higher vibrational levels, the approximation that the vibration is simple harmonic is not valid and the levels are not quite evenly spaced. Note that the potential energy curves representing the force between the two atoms in the molecule do not have exactly the same shape for the electronic ground and excited states. This implies that the fundamental frequency of vibration f is different for different electronic states. For transitions between vibrational states of different electronic states, the selection rule $\Delta\nu = \pm 1$ does not hold. Such transitions result in the emission of photons of wavelengths in or near the visible spectrum, so the emission spectrum of a molecule for electronic transitions is also sometimes called the optical spectrum.

The spacing of the rotational levels increases with increasing values of ℓ. Since the energies of rotation are so much smaller than those of vibrational excitation or electronic excitation of a molecule, molecular rotation shows up in optical spectra as a fine splitting of the spectral lines. When the fine structure is not resolved, the spectrum appears as bands, as shown in Figure 37-14a. Close inspection of these bands reveals that they have a fine structure due to the rotational energy levels, as shown in the enlargement in Figure 37-14c.

FIGURE 37-13 Electronic, vibrational, and rotational energy levels of a diatomic molecule. The rotational levels are shown in an enlargement of the $\nu = 0$ and $\nu = 1$ vibrational levels of the electronic ground state.

(a)

(b)

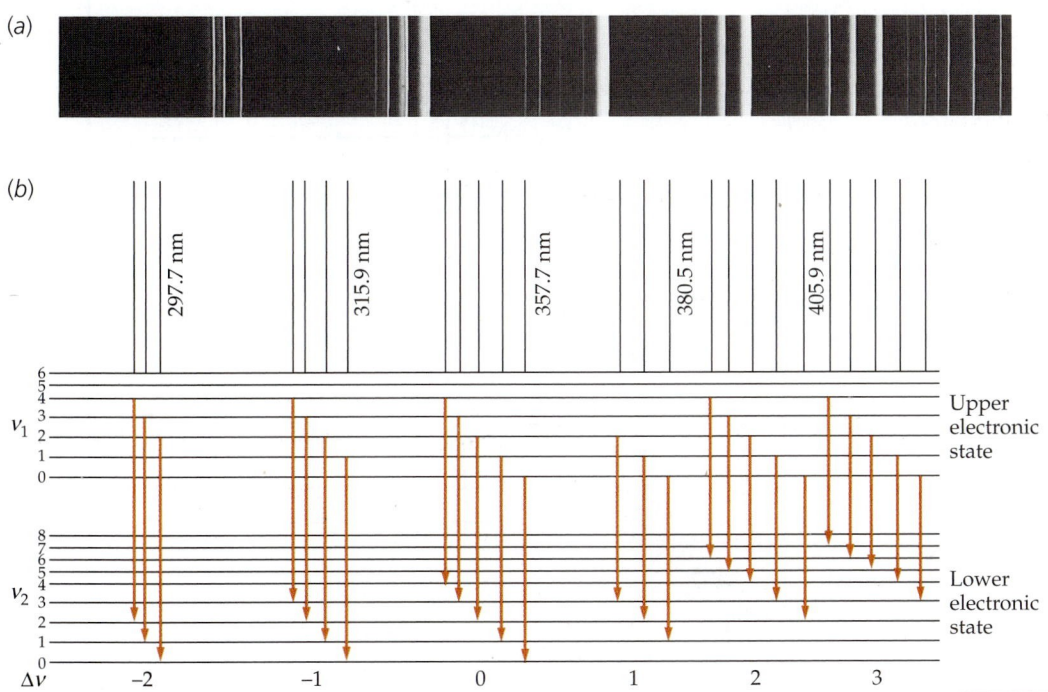

(c)

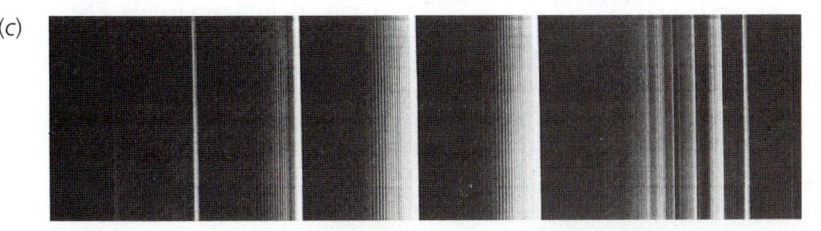

FIGURE 37-14 (a) Part of the emission spectrum of N_2. The spectral lines are due to transitions between the vibrational levels of two electronic states, as indicated in the energy level diagram (b). (c) An enlargement of part of Figure 37-14a shows that the apparent lines are in fact bands with structure caused by rotational levels.

Absorption Spectra

Much molecular spectroscopy is done using infrared absorption techniques in which only the vibrational and rotational energy levels of the ground-state electronic level are excited. For ordinary temperatures, the vibrational energies are sufficiently large in comparison with the thermal energy kT that most of the molecules are in the lowest vibrational state $\nu = 0$, for which the energy is $E_0 = \frac{1}{2}hf$. The transition from $\nu = 0$ to $\nu = 1$ is the predominant transition in absorption. The rotational energies, however, are sufficiently less than the thermal energy kT that the molecules are distributed among several rotational energy states. If the molecule is originally in a vibrational state characterized by $\nu = 0$ and a rotational state characterized by the quantum number ℓ, the molecule's initial energy is

$$E_\ell = \tfrac{1}{2}hf + \ell(\ell + 1)E_{0r} \qquad 37\text{-}20$$

where E_{0r} is given by Equation 37-13. From this state, two transitions are permitted by the selection rules. For a transition to the next higher vibrational state $\nu = 1$ and a rotational state characterized by $\ell + 1$, the final energy is

$$E_{\ell+1} = \tfrac{3}{2}hf + (\ell + 1)(\ell + 2)E_{0r} \qquad 37\text{-}21$$

For a transition to the next higher vibrational state and to a rotational state characterized by $\ell - 1$, the final energy is

$$E_{\ell-1} = \tfrac{3}{2}hf + (\ell - 1)\ell E_{0r} \qquad 37\text{-}22$$

The energy differences are

$$\Delta E_{\ell \to \ell+1} = E_{\ell+1} - E_\ell = hf + 2(\ell + 1)E_{0r} \qquad 37\text{-}23$$

where $\ell = 0, 1, 2$, and so on, and

$$\Delta E_{\ell \to \ell-1} = E_{\ell-1} - E_\ell = hf - 2\ell E_{0r} \qquad 37\text{-}24$$

where $\ell = 1, 2, 3$, and so on. (In Equation 37-24, ℓ begins at $\ell = 1$ because from $\ell = 0$ only the transition $\ell \to \ell + 1$ is possible.) Figure 37-15 illustrates these transitions. The frequencies of these transitions are given by

$$f_{\ell \to \ell+1} = \frac{\Delta E_{\ell \to \ell+1}}{h} = f + \frac{2(\ell + 1)E_{0r}}{h}, \quad \ell = 0, 1, 2, \dots \quad 37\text{-}25$$

and

$$f_{\ell \to \ell-1} = \frac{\Delta E_{\ell \to \ell-1}}{h} = f - \frac{2\ell E_{0r}}{h}, \quad \ell = 1, 2, 3, \dots \quad 37\text{-}26$$

The frequencies for the transitions $\ell \to \ell + 1$ are thus $f + 2(E_{0r}/h)$, $f + 4(E_{0r}/h)$, $f + 6(E_{0r}/h)$, and so forth; those corresponding to the transition $\ell \to \ell - 1$ are $f - 2(E_{0r}/h)$, $f - 4(E_{0r}/h)$, $f - 6(E_{0r}/h)$, and so forth. We thus expect the absorption spectrum to contain frequencies equally spaced by $2E_{0r}/h$ except for a gap of $4E_{0r}/h$ at the vibrational frequency f, as shown in Figure 37-16. A measurement of the position of the gap gives f and a measurement of the spacing of the absorption peaks gives E_{0r}, which is inversely proportional to the moment of inertia of the molecule.

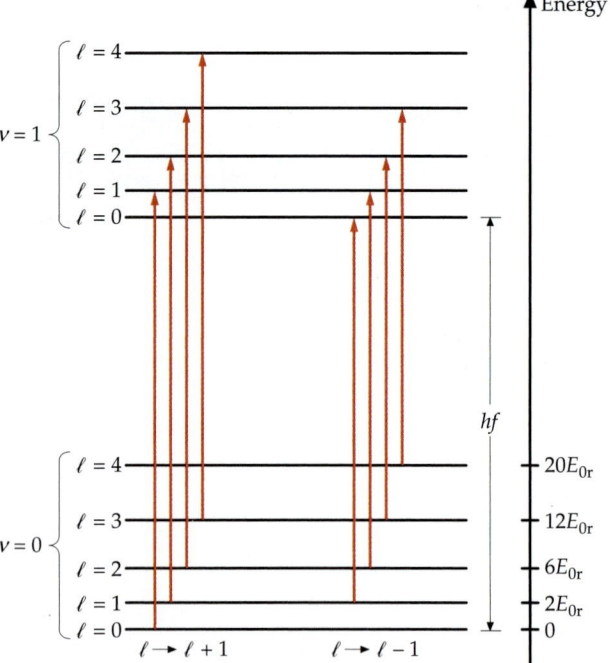

FIGURE 37-15 Absorptive transitions between the lowest vibrational states $\nu = 0$ and $\nu = 1$ in a diatomic molecule. These transitions obey the selection rule $\Delta\ell \pm 1$ and fall into two bands. The energies of the $\ell \to \ell + 1$ band are $hf + 2E_{0r}$, $hf + 4E_{0r}$, $hf + 6E_{0r}$, and so forth; whereas the energies of the $\ell \to \ell - 1$ band are $hf - 2E_{0r}$, $hf - 4E_{0r}$, $hf - 6E_{0r}$, and so forth.

Figure 37-17 shows the absorption spectrum of HCl. The double-peak structure results from the fact that chlorine occurs naturally in two isotopes, ^{35}Cl and ^{37}Cl, which gives HCl with two different moments of inertia. If all the rotational levels were equally populated initially, we would expect the intensities of each absorption line to be equal. However, the population of a rotational level is proportional to the degeneracy of the level, that is, to the number of states with the same value of ℓ, which is $2\ell + 1$, and to the Boltzmann factor $e^{-E/kT}$, where E is the energy of the state. For low values of ℓ, the population increases slightly because of the degeneracy factor, whereas for higher values of ℓ, the population decreases because of the Boltzmann factor. The intensities of the absorption lines therefore increase with ℓ for low values of ℓ and then decrease with ℓ for high values of ℓ, as can be seen from the figure.

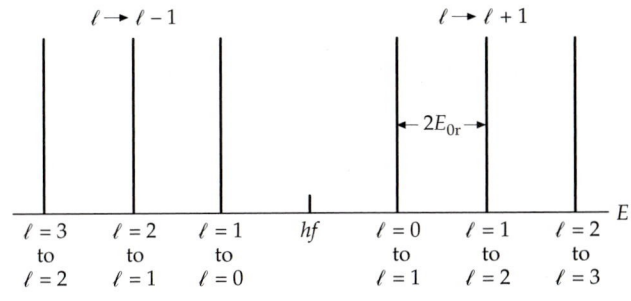

FIGURE 37-16 Expected absorption spectrum of a diatomic molecule. The right branch corresponds to transitions $\ell \to \ell + 1$ and the left branch corresponds to the transitions $\ell \to \ell - 1$. The lines are equally spaced by $2E_{0r}$. The energy midway between the branches is hf, where f is the frequency of vibration of the molecule.

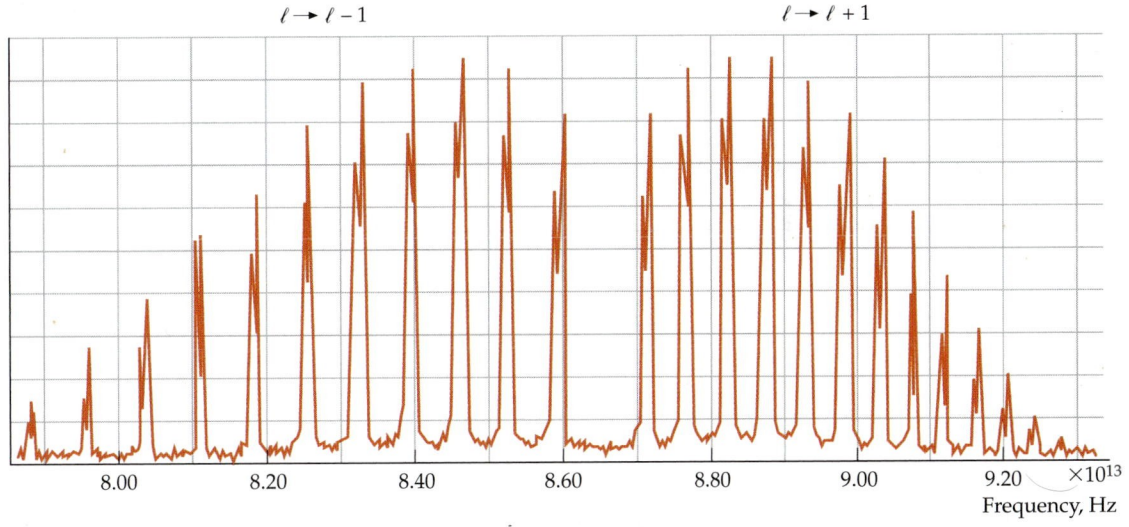

FIGURE 37-17 Absorption spectrum of the diatomic molecule HCl. The double-peak structure results from the two isotopes of chlorine, ^{35}Cl (abundance 75.5 percent) and ^{37}Cl (abundance 24.5 percent). The intensities of the peaks vary because the population of the initial state depends on ℓ.

SUMMARY

1. Atoms are usually found in nature bonded to form molecules or in the lattices of crystalline solids.

2. Ionic bonds and covalent bonds are the principal mechanisms responsible for forming molecules. van der Waals bonds and metallic bonds are important in the formation of liquids and solids. Hydrogen bonds enable large biological molecules to maintain their shape.

3. Like atoms, molecules emit electromagnetic radiation when making a transition from a higher energy state to a lower energy state. The internal energy of a molecule can be separated into three parts: electronic, vibrational, and rotational energy.

4. The molecules in liquids are characterized by a temporary short-range order. The molecules or ions in solids have a more lasting order. Amorphous solids maintain a short-range order similar to the short-range order of a liquid. Crystalline solids display a long-range order determined by their minimum potential energy state.

Topic	Relevant Equations and Remarks
1. Molecular Bonding	
Ionic	Ionic bonds result when an electron is transferred from one atom to another, resulting in a positive ion and a negative ion that bond together.
Covalent	The covalent bond is a quantum-mechanical effect that arises from the sharing of one or more electrons by atoms.
van der Waals	The van der Waals bonds are weak bonds that result from the interaction of the instantaneous electric dipole moments of molecules.
Hydrogen	The hydrogen bond results from the sharing of a proton of the hydrogen atom by other atoms.
Metallic	In the metallic bond, the positive lattice ions of the metal are held together by a cloud of negative charge comprised of free electrons.
Mixed	A diatomic molecule formed from two identical atoms, such as O_2, must bond by covalent bonding. The bonding of two nonidentical atoms is often a mixture of covalent and ionic bonding. The percentage of ionic bonding can be found from the ratio of the measured electric dipole moment to the ionic electric dipole moment defined by $$p_{\text{ionic}} = er_0 \qquad \text{37-5}$$ where r_0 is the equilibrium separation of the ions.
2. *Polyatomic Molecules	The shapes of such polyatomic molecules as H_2O and NH_3 can be understood from the spatial distribution of the atomic-orbital or molecular-orbital wave functions. The tetravalent nature of the carbon atom is a result of the hybridization of the 2s and 2p atomic orbitals.
3. Diatomic Molecules	
Moment of inertia	$$I = \mu r_0^2 \qquad \text{37-14}$$ where r_0 is the equilibrium separation, and μ is the reduced mass.
Reduced mass	$$\mu = \frac{m_1 m_2}{m_1 + m_2} \qquad \text{37-15}$$
Rotational energy levels	$$E_\ell = \frac{\ell(\ell + 1)\hbar^2}{2I} = \ell(\ell + 1)E_{0r}, \quad \ell = 0, 1, 2, \ldots \qquad \text{37-12}$$ where $$E_{0r} = \frac{\hbar^2}{2I} \qquad \text{37-13}$$
Vibrational energy levels	$$E_\nu = (\nu + \tfrac{1}{2})hf, \quad \nu = 0, 1, 2, \ldots \qquad \text{37-18}$$
Effective force constant k_F	$$f = \frac{1}{2\pi}\sqrt{\frac{k_F}{\mu}} \qquad \text{37-19}$$
4. Molecular Spectra	The optical spectra of molecules have a band structure due to transitions between rotational levels. Information about the structure and bonding of a molecule can be found from its rotational and vibrational absorption spectrum involving transitions from one vibrational–rotational level to another. These transitions obey the selection rules $$\Delta\nu = \pm 1, \quad \Delta\ell = \pm 1$$

PROBLEMS

- Single-concept, single-step, relatively easy
- •• Intermediate-level, may require synthesis of concepts
- ••• Challenging
- SSM Solution is in the *Student Solutions Manual*
- iSOLVE Problems available on iSOLVE online homework service
- iSOLVE ✓ These "Checkpoint" online homework service problems ask students additional questions about their confidence level, and how they arrived at their answer.

In a few problems, you are given more data than you actually need; in a few other problems, you are required to supply data from your general knowledge, outside sources, or informed estimates.

Conceptual Problems

1 • SSM Would you expect the NaCl molecule to be polar or nonpolar?

2 • Would you expect the N_2 molecule to be polar or nonpolar?

3 • Does neon occur naturally as Ne or Ne_2? Why?

4 • What type of bonding mechanism would you expect for (a) the HF molecule, (b) the KBr molecule, (c) the N_2 molecule? (d) Ag atoms in a solid?

5 •• SSM The elements on the far right column of the periodic table are sometimes called noble gases because they virtually never react with other atoms to form molecules. However, this behavior is sometimes modified if the resulting molecule is formed in an electronic excited state. An example is ArF. When it is formed in the excited state, it is written ArF* and is called an excimer (for excited dimer). Refer to Figure 37-13 and discuss how this diagram would look for ArF in which the ArF ground state is unstable but the ArF* excited state is stable. *Remark: Excimers are used in certain kinds of lasers.*

6 • Find other elements with the same subshell electron configuration in the two outermost orbitals as carbon. Would you expect the same type of hybridization for these elements as for carbon?

7 • How does the value of the effective force constant calculated for the CO molecule in Example 37-4 compare with the value of the force constant of the suspension springs on a typical automobile, which is about 1.5 kN/m?

8 • Explain why the moment of inertia of a diatomic molecule increases slightly with increasing angular momentum.

9 • Why would you expect the separation distance between the two protons to be larger in the H_2^+ ion than in the H_2 molecule?

10 • Why does an atom usually absorb radiation only from the ground state, whereas a diatomic molecule can absorb radiation from many different rotational states?

11 •• The vibrational energy levels of diatomic molecules are described by a single vibrational frequency f that is the frequency of vibration between the two atoms of the molecule. What would you expect to see in the case of polyatomic molecules? Consider in particular the water molecule H_2O (Figure 37-9).

Estimation and Approximation

12 •• The Anharmonic Oscillator: The potential energy between the atoms in a diatomic molecule has a minimum as shown in Figure 37-13. Near this minimum the graph for the energy as a function of distance between the atoms may be approximated as a parabola, leading to the harmonic oscillator model for the vibrating molecule. An improved approximation is called the anharmonic oscillator and leads to a modification of the energy formula (Equation 37-18). The improved formula for energy is

$$E_\nu = (\nu + \tfrac{1}{2})hf - (\nu + \tfrac{1}{2})^2 hf\alpha$$

For the O_2 molecule, the constants have the values $f = 4.74 \times 10^{13}$ s^{-1} and $\alpha = 7.6 \times 10^{-3}$. Use this formula to estimate the value of the quantum number ν for which the improved formula corrects the original formula by 10 percent.

13 •• To understand why quantum mechanics is not needed to describe many macroscopic systems, estimate the quantum number ℓ and spacing between adjacent energy levels for a baseball ($m \sim 300$ g, $r \sim 3$ cm) spinning about its own axis at 20 rev/min. *Hint: Pick ℓ so the quantum energy formula (Equation 37-12) gives the correct energy for the given system. Then find the energy increase for the next highest energy level.*

14 •• SSM Repeat Problem 13, finding the quantum number ν and spacing between adjacent energy levels for a 5-kg mass attached to 1500-N/m spring vibrating with an amplitude of 2 cm. *Hint: Pick ν so that the quantum energy formula (Equation 37-18) gives the correct energy for the given system. Then find the energy increase for the next highest energy level.*

Molecular Bonding

15 • iSOLVE ✓ Calculate the separation of Na$^+$ and Cl$^-$ ions, for which the potential energy is -1.52 eV.

16 • The dissociation energy of Cl_2 is 2.48 eV. Consider the formation of NaCl according to the reaction $2Na + Cl_2 \rightleftharpoons 2NaCl$. Does this reaction absorb energy or release energy? How much energy per molecule is absorbed or released?

17 • The dissociation energy is sometimes expressed in kilocalories per mole (kcal/mol). (a) Find the relation between the units eV/molecule and kcal/mol. (b) Find the dissociation energy of molecular NaCl in kcal/mol.

18 • SSM ISOLVE The equilibrium separation of the HF molecule is 0.0917 nm, and its measured electric dipole moment is 6.40×10^{-30} C·m. What percentage of the bonding is ionic?

19 •• The dissociation energy of RbF is 5.12 eV, and the equilibrium separation is 0.227 nm. The electron affinity of fluorine is 3.40 eV, and the ionization energy of rubidium is 4.18 eV. Determine the core-repulsion energy of RbF.

20 •• ISOLVE The equilibrium separation of the K^+ and Cl^- ions in KCl is about 0.267 nm. (a) Calculate the potential energy of attraction of the ions, assuming them to be point charges at this separation. (b) The ionization energy of potassium is 4.34 eV, and the electron affinity of chlorine is 3.62 eV. Find the dissociation energy neglecting any energy of repulsion. (See Figure 37-1.) The measured dissociation energy is 4.49 eV. What is the energy due to repulsion of the ions at the equilibrium separation?

21 •• Indicate the mean value of r for two vibration levels in the potential energy curve for a diatomic molecule. Show that because of the asymmetry in the curve, r_{av} increases with increasing vibration energy, and therefore solids expand when heated.

22 •• ISOLVE Calculate the potential energy of attraction between the Na^+ and Cl^- ions at the equilibrium separation $r_0 = 0.236$ nm. Compare this result with the dissociation energy given in Figure 37-1. What is the energy due to repulsion of the ions at the equilibrium separation?

23 •• ISOLVE ✓ The equilibrium separation of the K^+ and F^- ions in KF is about 0.217 nm. (a) Calculate the potential energy of attraction of the ions, assuming them to be point charges at this separation. (b) The ionization energy of potassium is 4.34 eV, and the electron affinity of fluorine is 3.40 eV. Find the dissociation energy neglecting any energy of repulsion. (c) The measured dissociation energy is 5.07 eV. Calculate the energy due to repulsion of the ions at the equilibrium separation.

24 ••• SSM Assume that the potential energy associated with the core repulsion of the two ions of a diatomic molecule with ionic bonding can be represented by a potential energy of the form $U_{rep} = C/r^n$, so the total potential energy is $U = U_e + U_{rep} + \Delta E$, where $U_e = -ke^2/r$. ΔE is the energy of the two ions at infinite separation less the energy of the two neutral atoms at infinite separation (see Figure 37-1). Use $\dfrac{dU}{dr} = 0$ at $r = r_0$ to show that $n = \dfrac{|U_e(r_0)|}{U_{rep}(r_0)}$.

25 ••• (a) Find U_{rep} at $r = r_0$ for NaCl. (b) Assume $U_{rep} = C/r^n$ and find C and n for NaCl. (See Problem 24.)

Energy Levels of Spectra of Diatomic Molecules

26 • ISOLVE The characteristic rotational energy E_{0r} for the rotation of the N_2 molecule is 2.48×10^{-4} eV. From this, find the separation distance of the 2 nitrogen atoms.

27 • SSM ISOLVE ✓ The separation of the two oxygen atoms in a molecule of O_2 is actually slightly greater than the 0.1 nm used in Example 37-3, and the characteristic energy of

rotation E_{0r} is 1.78×10^{-4} eV rather than the result obtained in that example. Use this value to calculate the separation distance of the two oxygen atoms.

28 •• Show that the reduced mass is smaller than either mass in a diatomic molecule and calculate it for (a) H_2, (b) N_2, (c) CO, and (d) HCl. Express your answers in unified mass units.

29 •• The CO molecule has a binding energy of approximately 11 eV. Find the vibrational quantum number v that would cause the molecule to have this much energy, and thus cause it to "shake" apart.

30 •• SSM ISOLVE ✓ The equilibrium separation between the nuclei of the LiH molecule is 0.16 nm. Determine the energy separation between the $\ell = 3$ and $\ell = 2$ rotational levels of this diatomic molecule.

31 •• SSM Derive Equations 37-14 and 37-15 for the moment of inertia in terms of the reduced mass of a diatomic molecule.

32 •• ISOLVE ✓ Use the separation of the K^+ and Cl^- ions given in Problem 20 and the reduced mass of KCl to calculate the characteristic rotational energy E_{0r}.

33 •• ISOLVE The central frequency for the absorption band of HCl shown in Figure 37-17 is at $f = 8.66 \times 10^{13}$ Hz, and the absorption peaks are separated by about $f = 6 \times 10^{11}$ Hz. Use this information to find (a) the lowest (zero-point) vibrational energy for HCl, (b) the moment of inertia of HCl, and (c) the equilibrium separation of the atoms.

34 •• ISOLVE Calculate the effective force constant for HCl from its reduced mass and the fundamental vibrational frequency obtained from Figure 37-17.

35 •• To see how the population of rotational states of the oxygen molecule depends on the angular momentum quantum number ℓ, use a spreadsheet program or graphing calculator to graph the function $(2\ell + 1)e^{-E_\ell/kT}$, where $E_\ell = \ell(\ell + 1)E_{0r}$ for values of $0 \le \ell \le 10$ at $T = 100K$, 200K, 300K, and 500K.

36 •• SSM Two objects of mass m_1 and m_2 are attached to a spring of force constant k and equilibrium length r_0. (a) Show that when m_1 is moved a distance Δr_1 from the center of mass, the force exerted by the spring is

$$F = -k\left(\frac{m_1 + m_2}{m_2}\right)\Delta r_1$$

(b) Show that the frequency of oscillation is $f = \left(\frac{1}{2\pi}\right)\sqrt{k/\mu}$, where μ is the reduced mass.

37 ••• Calculate the reduced mass for the $H^{35}Cl$ and $H^{37}Cl$ molecules and the fractional difference $\Delta\mu/\mu$. Show that the mixture of isotopes in HCl leads to a fractional difference in the frequency of a transition from one rotational state to another given by $\Delta f/f = -\Delta\mu/\mu$. Compute $\Delta f/f$ and compare your result with Figure 37-17.

General Problems

38 • Show that when one atom in a diatomic molecule is much more massive than the other the reduced mass is approximately equal to the mass of the lighter atom.

39 •• The equilibrium separation between the nuclei of the CO molecule is 0.113 nm. Determine the energy difference between the $\ell = 2$ and $\ell = 1$ rotational energy levels of this molecule.

40 •• **SSM** The effective force constant for the HF molecule is 970 N/m. Find the frequency of vibration for this molecule.

41 •• **iSOLVE** The frequency of vibration of the NO molecule is 5.63×10^{13} Hz. Find the effective force constant for NO.

42 •• The effective force constant of the hydrogen bond in the H_2 molecule is 580 N/m. Obtain the energies of the four lowest vibrational levels of the H_2, HD, and D_2 molecules and the wavelengths of photons resulting from transitions between adjacent vibrational levels of these molecules.

43 •• The potential energy between two atoms in a molecule can often be described rather well by the Lenard–Jones potential, which can be written as

$$U = U_0\left[\left(\frac{a}{r}\right)^{12} - 2\left(\frac{a}{r}\right)^6\right],$$

where U_0 and a are constants. Find the interatomic separation r_0 in terms of a for which the potential energy is a minimum. Find the corresponding value of U_{min}. Use Figure 37-4 to obtain numerical values of r_0 and U_0 for the H_2 molecule, and express your answers in nanometers and electron volts.

44 •• In this problem, you are to find how the van der Waals force between a polar molecule and a nonpolar molecule depends on the distance between the molecules. Let the dipole moment of the polar molecule be in the x direction and the nonpolar molecule be a distance x away. (a) How does the electric field due to an electric dipole depend on distance x? (b) Use the facts that (1) the potential energy of an electric dipole of moment $\vec{p}$ in an electric field $\vec{E}$ is $U = -\vec{p} \cdot \vec{E}$, and (2) the induced dipole moment of the nonpolar molecule is

proportional to E, to find how the potential energy of interaction of the two molecules depends on separation distance. (c) Using $F_x = -dU/dx$, find the x dependence of the force between the two molecules.

45 •• Find the dependence of the force on separation distance between two polar molecules. (See Problem 44.)

46 •• Use the infrared absorption spectrum of HCl in Figure 37-17 to obtain (a) the characteristic rotational energy E_{0r} (in eV) and (b) the vibrational frequency f and the vibrational energy hf (in eV).

47 •• **SSM** For a molecule such as CO, which has a permanent electric dipole moment, radiative transitions obeying the selection rule $\Delta\ell = \pm 1$ between two rotational energy levels of the same vibrational level are allowed. (That is, the selection rule $\Delta v = \pm 1$ does not hold.) (a) Find the moment of inertia of CO and calculate the characteristic rotational energy E_{0r} (in eV). (b) Make an energy-level diagram for the rotational levels for $\ell = 0$ to $\ell = 5$ for some vibrational level. Label the energies in electron volts, starting with $E = 0$ for $\ell = 0$. Indicate on your diagram the transitions that obey $\Delta\ell = -1$, and calculate the energy of the photon emitted. (c) Find the wavelength of the photons emitted for each transition in (b). In what region of the electromagnetic spectrum are these photons?

48 ••• **SSM** Use the results of Problem 24 to calculate the vibrational frequency of the LiCl molecule. The dissociation energy of LiCl is 4.86 eV, and the equilibrium separation is 0.202 nm. The electron affinity of chlorine is 3.62 eV, and the ionization energy of lithium is 5.39 eV. To do this, expand the potential about $r = r_0$, where r_0 is the equilibrium separation, in a Taylor series. Retain only the term proportional to $(r - r_0)^2$. Recall that the potential energy of a simple harmonic oscillator is given by $U_{SHO} = \frac{1}{2}m\omega^2x^2$. What is the wavelength resulting from transitions between adjacent harmonic oscillator levels of this molecule?

Solids

SILICON INGOT SILICON IS A SEMICONDUCTOR, AND SLICES OF INGOTS LIKE THIS ARE USED TO PRODUCE TRANSISTORS AND OTHER ELECTRONIC DEVICES. TO MAKE A TRANSISTOR, ATOMS OF ARSENIC AND GALLIUM ARE INJECTED INTO THE SILICON.

 **Do you know how many atoms of arsenic it takes to increase the charge-carrier density by a factor of 5 million? For more on this topic, see Example 38-7.**

The first microscopic model of electric conduction in metals was proposed by Paul K. Drude in 1900 and developed by Hendrik A. Lorentz about 1909. This model successfully predicts that the current is proportional to the potential drop (Ohm's law) and relates the resistivity ρ of conductors to the mean speed v_{av} and the mean free path[†] λ of the free electrons within the conductor. However, when v_{av} and λ are interpreted classically, there is a disagreement between the calculated values and the measured values of the resistivity, and a similar disagreement between the predicted temperature dependence and the observed temperature dependence. Thus, the classical theory fails to adequately describe the resistivity of metals. Furthermore, the classical theory says nothing about the most striking property of solids, namely that some materials are conductors, others are insulators, and still others are semiconductors, which are materials whose resistivity falls between that of conductors and insulators.

---

† The mean free path is the average distance traveled between collisions.

When v_{av} and λ are interpreted using quantum theory, both the magnitude and the temperature dependence of the resistivity are correctly predicted. In addition, quantum theory allows us to determine if a material will be a conductor, an insulator, or a semiconductor.

➤ **In this chapter, we use our understanding of quantum mechanics to discuss the structure of solids and solid-state semiconducting devices. Much of our discussion will be qualitative because, as in atomic physics, the quantum-mechanical calculations are very difficult.**

38-1 The Structure of Solids

The three phases of matter we observe—gas, liquid, and solid—result from the relative strengths of the attractive forces between molecules and the thermal energy of the molecules. Molecules in the gas phase have a relatively high thermal kinetic energy, and such molecules have little influence on one another except during their frequent but brief collisions. At sufficiently low temperatures (depending on the type of molecule), van der Waals forces will cause practically every substance to condense into a liquid and then into a solid. In liquids, the molecules are close enough—and their kinetic energy is low enough—that they can develop a temporary **short-range order.** As thermal kinetic energy is further reduced, the molecules form solids, which are characterized by a lasting order.

If a liquid is cooled slowly so that the kinetic energy of its molecules is reduced slowly, the molecules (or atoms or ions) may arrange themselves in a regular crystalline array, producing the maximum number of bonds and leading to a minimum potential energy. However, if the liquid is cooled rapidly so that its internal energy is removed before the molecules have a chance to arrange themselves, the solid formed is often not crystalline but instead resembles a snapshot of the liquid. Such a solid is called an **amorphous solid.** It displays short-range order but not the **long-range order** (over many molecular diameters) that is characteristic of a crystal. Glass is a typical amorphous solid. A characteristic result of the long-range ordering of a crystal is that it has a well-defined melting point, whereas an amorphous solid merely softens as its temperature is increased. Many materials may solidify into either an amorphous state or a crystalline state depending on how the materials are prepared; others exist only in one form or the other.

Most common solids are polycrystalline; that is, they consist of many single crystals that meet at grain boundaries. The size of a single crystal is typically a fraction of a millimeter. However, large single crystals do occur naturally and can be produced artificially. The most important property of a single crystal is the symmetry and regularity of its structure. It can be thought of as having a single unit structure that is repeated throughout the crystal. This smallest unit of a crystal is called the **unit cell;** its structure depends on the type of bonding—ionic, covalent, metallic, hydrogen, van der Waals—between the atoms, ions, or molecules. If more than one kind of atom is present, the structure will also depend on the relative sizes of the atoms.

Figure 38-1 shows the structure of the ionic crystal sodium chloride (NaCl). The Na^+ and Cl^- ions are spherically symmetric, and the Cl^- ion is approximately twice as large as the Na^+ ion. The minimum potential energy for this crystal occurs when an ion of either kind has six nearest neighbors of the other kind. This structure is called *face-centered-cubic* (fcc). Note that the Na^+ and Cl^- ions in solid NaCl are *not* paired into NaCl molecules.

The net attractive part of the potential energy of an ion in a crystal can be written

$$U_{att} = -\alpha \frac{ke^2}{r} \qquad 38\text{-}1$$

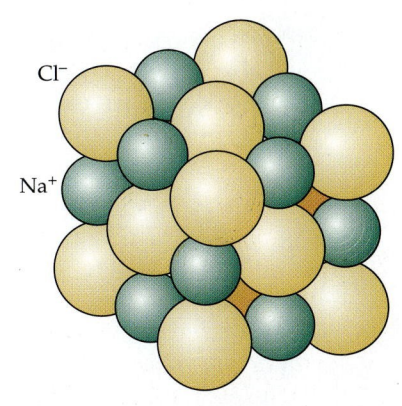

Cl^-

Na^+

FIGURE 38-1 Face-centered-cubic structure of the NaCl crystal.

where r is the separation distance between neighboring ions (0.281 nm for the Na^+ and Cl^- ions in crystalline NaCl), and α, called the **Madelung constant,** depends on the geometry of the crystal. If only the six nearest neighbors of each ion were important, α would be six. However, in addition to the six neighbors of the opposite charge at a distance r, there are twelve ions of the same charge at a distance $\sqrt{2}r$, eight ions of opposite charge at a distance $\sqrt{3}r$, and so on. The Madelung constant is thus an infinite sum:

$$\alpha = 6 - \frac{12}{\sqrt{2}} + \frac{8}{\sqrt{3}} - \dots \qquad\qquad 38\text{-}2$$

The result for face-centered-cubic structures is $\alpha = 1.7476.$[†]

[†] A large number of terms are needed to calculate the Madelung constant accurately because the sum converges very slowly.

Crystal structure. (*a*) The hexagonal symmetry of a snowflake arises from a hexagonal symmetry in its lattice of hydrogen atoms and oxygen atoms. (*b*) NaCl (salt) crystals, magnified approximately thirty times. The crystals are built up from a cubic lattice of sodium and chloride ions. In the absence of impurities, an exact cubic crystal is formed. This (false-color) scanning electron micrograph shows that in practice the basic cube is often disrupted by dislocations, giving rise to crystals with a wide variety of shapes. The underlying cubic symmetry, though, remains evident. (*c*) A crystal of quartz (SiO_2, silicon dioxide), the most abundant and widespread mineral on the earth. If molten quartz solidifies without crystallizing, glass is formed. (*d*) A soldering iron tip, ground down to reveal the copper core within its iron sheath. Visible in the iron is its underlying microcrystalline structure.

(*a*)

(*b*)

(*c*)

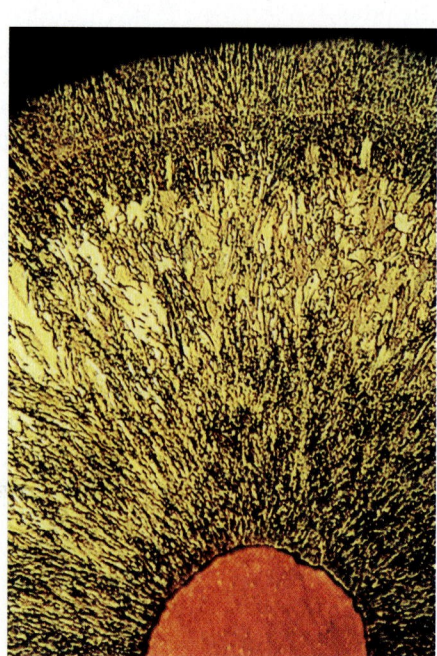

(*d*)

When Na^+ and Cl^- ions are very close together, they repel each other because of the overlap of their electrons and the exclusion-principle repulsion discussed in Section 37-1. A simple empirical expression for the potential energy associated with this repulsion that works fairly well is

$$U_{rep} = \frac{A}{r^n}$$

where A and n are constants. The total potential energy of an ion is then

$$U = -\alpha \frac{ke^2}{r} + \frac{A}{r^n} \qquad \text{38-3}$$

The equilibrium separation $r = r_0$ is that at which the force $F = -dU/dr$ is zero. Differentiating and setting $dU/dr = 0$ at $r = r_0$, we obtain

$$A = \frac{\alpha k e^2 r_0^{n-1}}{n} \qquad \text{38-4}$$

The total potential energy can thus be written

$$U = -\alpha \frac{ke^2}{r_0} \left[\frac{r_0}{r} - \frac{1}{n} \left(\frac{r_0}{r} \right)^n \right] \qquad \text{38-5}$$

At $r = r_0$, we have

$$U(r_0) = -\alpha \frac{ke^2}{r_0} \left(1 - \frac{1}{n} \right) \qquad \text{38-6}$$

If we know the equilibrium separation r_0, the value of n can be found approximately from the *dissociation energy* of the crystal, which is the energy needed to break up the crystal into atoms.

SEPARATION DISTANCE BETWEEN Na^+ AND Cl^- IN NaCl **EXAMPLE 38-1**

Calculate the equilibrium spacing r_0 for NaCl from the measured density of NaCl, which is $\rho = 2.16 \text{ g/cm}^3$.

PICTURE THE PROBLEM We consider each ion to occupy a cubic volume of side r_0. The mass of 1 mol of NaCl is 58.4 g, which is the sum of the atomic masses of sodium and chlorine. There are $2N_A$ ions in 1 mol of NaCl, where $N_A = 6.02 \times 10^{23}$ is Avogadro's number.

1. The volume v per mol of NaCl equals the number of ions times the volume per ion:

$$v = 2N_A r_0^3$$

2. Relate r_0 to the density ρ:

$$\rho = \frac{M}{v} = \frac{M}{2N_A r_0^3}$$

3. Solve for r_0^3 and substitute the known values:

$$r_0^3 = \frac{M}{2N_A \rho} = \frac{58.4 \text{ g}}{2(6.02 \times 10^{23})(2.16 \text{ g/cm}^3)}$$

$$= 2.25 \times 10^{-23} \text{ cm}^3$$

so

$$r_0 = 2.82 \times 10^{-8} \text{ cm} = \boxed{0.282 \text{ nm}}$$

The measured dissociation energy of NaCl is 770 kJ/mol. Using 1 eV = 1.602×10^{-19} J, and the fact that 1 mol of NaCl contains N_A pairs of ions, we can express the dissociation energy in electron volts per ion pair. The conversion between electron volts per ion pair and kilojoules per mole is

$$1\frac{eV}{\text{ion pair}} \times \frac{6.022 \times 10^{23} \text{ ion pairs}}{1 \text{ mol}} \times \frac{1.602 \times 10^{-19} \text{ J}}{1 \text{ eV}}$$

The result is

$$1\frac{eV}{\text{ion pair}} = 96.47 \frac{kJ}{\text{mol}} \qquad\qquad 38\text{-}7$$

Thus, 770 kJ/mol = 7.98 eV per ion pair. Substituting −7.98 eV for $U(r_0)$, 0.282 nm for r_0, and 1.75 for α in Equation 38-6, we can solve for n. The result is $n = 9.35 \approx 9$.

Most ionic crystals, such as LiF, KF, KCl, KI, and AgCl, have a face-centered-cubic structure. Some elemental solids that have fcc structure are silver, aluminum, gold, calcium, copper, nickel, and lead.

Figure 38-2 shows the structure of CsCl, which is called the *body-centered-cubic* (bcc) structure. In this structure, each ion has eight nearest neighbor ions of the opposite charge. The Madelung constant for these crystals is 1.7627. Elemental solids with bcc structure include barium, cesium, iron, potassium, lithium, molybdenum, and sodium.

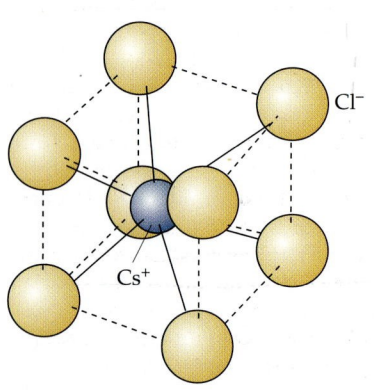

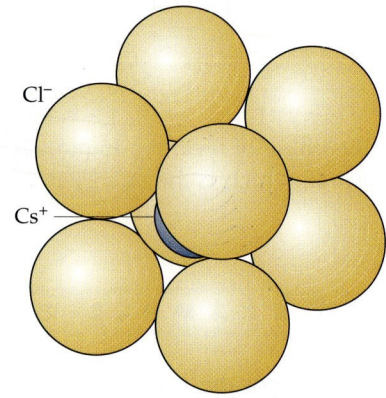

FIGURE 38-2 Body-centered-cubic structure of the CsCl crystal.

Figure 38-3 shows another important crystal structure: the *hexagonal close-packed* (hcp) structure. This structure is obtained by stacking identical spheres, such as bowling balls. In the first layer, each ball touches six others; thus, the name *hexagonal*. In the next layer, each ball fits into a triangular depression of the first layer. In the third layer, each ball fits into a triangular depression of the second layer, so it lies directly over a ball in the first layer. Elemental solids with hcp structure include beryllium, cadmium, cerium, magnesium, osmium, and zinc.

In some solids with covalent bonding, the crystal structure is determined by the directional nature of the bonds. Figure 38-4 illustrates the diamond structure of carbon, in which each atom is bonded to four other atoms as a result of hybridization, which is discussed in Section 38-2. This is also the structure of germanium and silicon.

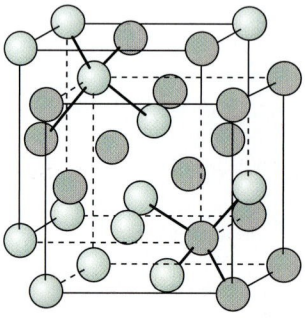

FIGURE 38-3 Hexagonal close-packed crystal structure.

FIGURE 38-4 Diamond crystal structure. This structure can be considered to be a combination of two interpenetrating face-centered-cubic structures.

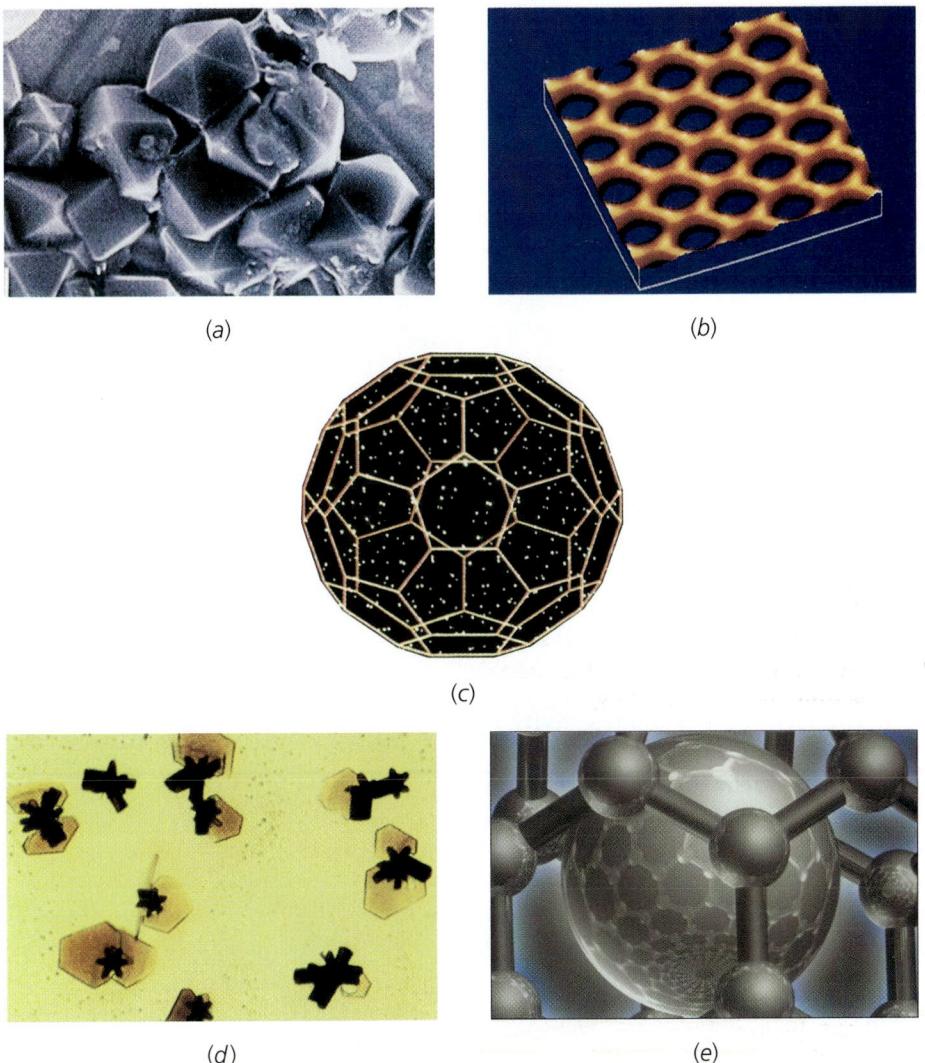

(a)

(b)

(c)

(d)

(e)

Carbon exists in three well-defined crystalline forms: diamond, graphite, and fullerenes (short for "buckminsterfullerenes"), the third of which was predicted and discovered less than two decades ago. The forms differ in how the carbon atoms are packed together in a lattice. A fourth form of carbon, in which no well-defined crystalline form exists, is common charcoal. (*a*) Synthetic diamonds, magnified approximately 75,000 times. In diamond, each carbon atom is centered in a tetrahedron of four other carbon atoms. The strength of these bonds accounts for the hardness of a diamond. (*b*) An atomic-force micrograph of graphite. In graphite, carbon atoms are arranged in sheets, with each sheet made up of atoms in hexagonal rings. The sheets slide easily across one another, a property that allows graphite to function as a lubricant. (*c*) A single sheet of carbon rings can be closed on itself if certain rings are allowed to be pentagonal, instead of hexagonal. A computer-generated image of the smallest such structure, C_{60}, is shown here. Each of the sixty vertices corresponds to a carbon atom; twenty of the faces are hexagons and twelve of the faces are pentagons. The same geometric pattern is encountered in a soccer ball. (*d*) Fullerene crystals, in which C_{60} molecules are close-packed. The smaller crystals tend to form thin brownish platelets; larger crystals are usually rod-like in shape. Fullerenes exist in which more than sixty carbon atoms appear. In the crystals shown here, about one-sixth of the molecules are C_{70}. (*e*) Carbon nanotubes have very interesting electrical properties. A single graphite sheet is a semimetal, which means that it has properties intermediate between those of semiconductors and those of metals. When a graphite sheet is rolled into a nanotube, not only do the carbon atoms have to line up around the circumference of the tube, but the wavefunctions of the electrons must also match up. This boundary-matching requirement places restrictions on these wavefunctions, which affects the motion of the electrons. Depending on exactly how the tube is rolled up, the nanotube can be either a semiconductor or a metal.

38-2 A Microscopic Picture of Conduction

We consider a metal as a regular three-dimensional lattice of ions filling some volume V and containing a large number N of electrons that are free to move throughout the whole metal. Experimentally, the number of free electrons in a metal is approximately one electron to four electrons per atom. In the absence of an electric field, the free electrons move about the metal randomly, much the way gas molecules move about in a container. We will often refer to these free electrons in a metal as an electron gas.

The current in a conducting wire segment is proportional to the voltage drop across the segment:

$$I = \frac{V}{R}, \quad \text{or } (V = IR)$$

The resistance R is proportional to the length L of the wire segment and inversely proportional to the cross-sectional area A:

$$R = \rho \frac{L}{A}$$

where ρ is the resistivity. Substituting $\rho L/A$ for R, and EL for V, we can write the current in terms of the electric field strength E and the resistivity. We have

$$I = \frac{V}{R} = \frac{EL}{\rho L/A} = \frac{1}{\rho}EA$$

Dividing both sides by the area A gives $I/A = (1/\rho)E$, or $J = (1/\rho)E$, where $J = I/A$ is the magnitude of the **current density** vector $\vec{J}$. The current density vector is defined as

$$\vec{J} = qn\vec{v}_{d}$$

38-8

DEFINITION—CURRENT DENSITY

where q, n, and $\vec{v}_d$ are the charge, the number density, and the drift velocity of the charge carrier. (This follows from Equation 25-3). In vector form, the relation between the current density and the electric field is

$$\vec{J} = \frac{1}{\rho}\vec{E}$$

38-9

This relation is the point form of Ohm's law. The reciprocal of the resistivity is called the **conductivity.**

According to Ohm's law, the resistivity is independent of both the current density and the electric field $\vec{E}$. Combining Equation 38-8 and Equation 38-9 gives

$$-en_e\vec{v}_d = \frac{1}{\rho}\vec{E}$$

38-10

where $-e$ has been substituted for q. According to Equation 38-10, the drift velocity $\vec{v}_d$ is proportional to $\vec{E}$.

In the presence of an electric field, a free electron experiences a force $-e\vec{E}$. If this were the only force acting, the electron would have a constant acceleration $-e\vec{E}/m_e$. However, Equation 38-10 implies a steady-state situation with a constant drift velocity that is proportional to the field $\vec{E}$. In the microscopic model, it is assumed that a free electron is accelerated for a short time and then makes a collision with a lattice ion. The velocity of the electron immediately after the collision is completely unrelated to the drift velocity. The justification for this assumption is that the magnitude of the drift velocity is extremely small compared with the random thermal speeds of the electrons.

For a typical electron, its velocity a time t after its last collision is $\vec{v}_0 - (e\vec{E}/m_e)t$, where $\vec{v}_0$ is its velocity immediately after that collision. Since the direction of $\vec{v}_0$ is random, it does not contribute to the average velocity of the electrons. Thus, the average velocity or drift velocity of the electrons is

$$\vec{v}_d = -\frac{e\vec{E}}{m_e}\tau$$

38-11

where τ is the average time since the last collision. Substituting for $\vec{v}_d$ in Equation 38-10, we obtain

$$-n_e e\left(-\frac{e\vec{E}}{m_e}\tau\right) = \frac{1}{\rho}\vec{E}$$

so

$$\rho = \frac{m_e}{n_e e^2 \tau} \qquad \text{38-12}$$

The time τ, called the **collision time,** is also the average time between collisions.[†] The average distance an electron travels between collisions is $v_{av}\tau$, which is called the mean free path λ:

$$\lambda = v_{av}\tau \qquad \text{38-13}$$

where v_{av} is the mean speed of the electrons. (The mean speed is many orders of magnitude greater than the drift speed.) In terms of the mean free path and the mean speed, the resistivity is

$$\rho = \frac{m_e v_{av}}{n_e e^2 \lambda} \qquad \text{38-14}$$

RESISTIVITY IN TERMS OF V_{AV} AND λ

According to Ohm's law, the resistivity ρ is independent of the electric field $\vec{E}$. Since m_e, n_e, and e are constants, the only quantities that could possibly depend on $\vec{E}$ are the mean speed v_{av} and the mean free path λ. Let us examine these quantities to see if they can possibly depend on the applied field $\vec{E}$.

Classical Interpretation of v_{av} and λ

Classically, at $T = 0$ all the free electrons in a conductor should have zero kinetic energy. As the conductor is heated, the lattice ions acquire an average kinetic energy of $\frac{3}{2}kT$, which is imparted to the electron gas by the collisions between the electrons and the ions. (This is a result of the equipartition theorem studied in Chapters 17 and 18.) The electron gas would then have a Maxwell–Boltzmann distribution just like a gas of molecules. In equilibrium, the electrons would be expected to have a mean kinetic energy of $\frac{3}{2}kT$, which at ordinary temperatures ($\sim$300 K) is approximately 0.04 eV. At $T = 300$ K, their root-mean-square (rms) speed,[‡] which is slightly greater than the mean speed, is

$$v_{av} \approx v_{rms} = \sqrt{\frac{3kT}{m_e}} = \sqrt{\frac{3(1.38 \times 10^{-23}\,\text{J/K})(300\,\text{K})}{9.11 \times 10^{-31}\,\text{kg}}} \qquad \text{38-15}$$

$$= 1.17 \times 10^5\,\text{m/s}$$

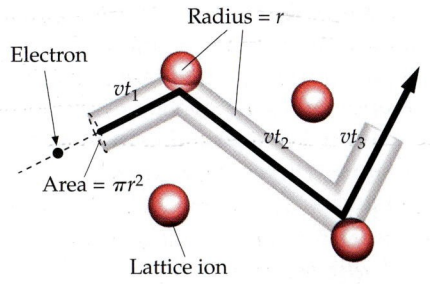

Note that this is about nine orders of magnitude greater than the typical drift speed of 3.5×10^{-5} m/s, which was calculated in Example 25-1. The very small drift speed caused by the electric field therefore has essentially no effect on the very large mean speed of the electrons, so v_{av} in Equation 38-14 cannot depend on the electric field $\vec{E}$.

The mean free path is related classically to the size of the lattice ions in the conductor and to the number of ions per unit volume. Consider one electron moving with speed v through a region of stationary ions that are assumed to be hard spheres (Figure 38-5). The size of the electron is assumed to be negligible.

FIGURE 38-5 Model of an electron moving through the lattice ions of a conductor. The electron, which is considered to be a point particle, collides with an ion if it comes within a distance r of the center of the ion, where r is the radius of the ion. If the electron speed is v, it collides in time t with all the ions whose centers are in the volume $\pi r^2 vt$. While this picture is in accord with the classical Drude model for conduction in metals, it is in conflict with the current quantum-mechanical model presented later in this chapter.

---

† It is tempting but incorrect to think that if τ is the average time between collisions, the average time since its last collision is $\frac{1}{2}\tau$ rather than τ. (If you find this confusing, you may take comfort in the fact that Drude used the incorrect result $\frac{1}{2}\tau$ in his original work.)
‡ See Equation 17-23.

The electron will collide with an ion if it comes within a distance r from the center of the ion, where r is the radius of the ion. In some time t_1, the electron moves a distance vt_1. If there is an ion whose center is in the cylindrical volume $\pi r^2 vt_1$, the electron will collide with the ion. The electron will then change directions and collide with another ion in time t_2 if the center of the ion is in the volume $\pi r^2 vt_2$. Thus, in the total time $t = t_1 + t_2 + \ldots$, the electron will collide with the number of ions whose centers are in the volume $\pi r^2 vt$. The number of ions in this volume is $n_{ion}\ \pi r^2 vt$, where n_{ion} is the number of ions per unit volume. The total path length divided by the number of collisions is the mean free path:

$$\lambda = \frac{vt}{n_{ion}\ \pi r^2 vt} = \frac{1}{n_{ion}\ \pi r^2} = \frac{1}{n_{ion} A} \qquad\qquad 38\text{-}16$$

where $A = \pi r^2$ is the cross-sectional area of a lattice ion.

Successes and Failures of the Classical Model

Neither n_{ion} nor r depends on the electric field $\vec{E}$, so λ also does not depend on $\vec{E}$. Thus, according to the classical interpretation of v_{av} and λ, neither depend on $\vec{E}$, so the resistivity ρ does not depend on $\vec{E}$, in accordance with Ohm's law. However, the classical theory gives an incorrect temperature dependence for the resistivity. Because λ depends only on the radius and the number density of the lattice ions, the only quantity in Equation 38-14 that depends on temperature in the classical theory is v_{av}, which is proportional to $\sqrt{T}$. But experimentally, ρ varies linearly with temperature. Furthermore, when ρ is calculated at $T = 300$ K using the Maxwell–Boltzmann distribution for v_{av} and Equation 38-16 for λ, the numerical result is about six times greater than the measured value.

The classical theory of conduction fails because electrons are not classical particles. The wave nature of the electrons must be considered. Because of the wave properties of electrons and the exclusion principle (to be discussed in the following section), the energy distribution of the free electrons in a metal is not even approximately given by the Maxwell–Boltzmann distribution. Furthermore, the collision of an electron with a lattice ion is not similar to the collision of a baseball with a tree. Instead, it involves the scattering of electron waves by the lattice. To understand the quantum theory of conduction, we need a qualitative understanding of the energy distribution of free electrons in a metal. This will also help us understand the origin of contact potentials between two dissimilar metals in contact and the contribution of free electrons to the heat capacity of metals.

38-3 The Fermi Electron Gas

We have used the term *electron gas* to describe the free electrons in a metal. Whereas the molecules in an ordinary gas, such as air, obey the classical Maxwell–Boltzmann energy distribution, the free electrons in a metal do not. Instead, they obey a quantum energy distribution called the Fermi–Dirac distribution. Because the behavior of this electron gas is so different from a gas of molecules, the electron gas is often called a **Fermi electron gas.** The main features of a Fermi electron gas can be understood by considering an electron in a metal to be a particle in a box, a problem whose one-dimensional version we studied extensively in Chapter 34. We discuss the main features of a Fermi electron gas semiquantitatively in this section and leave the details of the Fermi–Dirac distribution to Section 38-9.

Energy Quantization in a Box

In Chapter 34, we found that the wavelength associated with an electron of momentum p is given by the de Broglie relation:

$$\lambda = \frac{h}{p} \qquad\qquad 38\text{-}17$$

where h is Planck's constant. When a particle is confined to a finite region of space, such as a box, only certain wavelengths λ_n given by standing-wave conditions are allowed. For a one-dimensional box of length L, the standing-wave condition is

$$n\frac{\lambda_n}{2} = L \qquad\qquad 38\text{-}18$$

This results in the quantization of energy:

$$E_n = \frac{p_n^2}{2m} = \frac{(h/\lambda_n)^2}{2m} = \frac{h^2}{2m}\frac{1}{\lambda_n^2} = \frac{h^2}{2m}\frac{1}{(2L/n)^2}$$

or

$$E_n = n^2\frac{h^2}{8mL^2} \qquad\qquad 38\text{-}19$$

The wave function for the nth state is given by

$$\psi_n(x) = \sqrt{\frac{2}{L}}\sin\frac{n\pi x}{L} \qquad\qquad 38\text{-}20$$

The quantum number n characterizes the wave function for a particular state and the energy of that state. In three-dimensional problems, three quantum numbers arise, one associated with each dimension.

The Exclusion Principle

The distribution of electrons among the possible energy states is dominated by the exclusion principle, which states that no two electrons in an atom can be in the same quantum state; that is, they cannot have the same set of values for their quantum numbers. The exclusion principle applies to all "spin one-half" particles, which include electrons, protons, and neutrons. These particles have a *spin* quantum number m_s which has two possible values, $+\frac{1}{2}$ and $-\frac{1}{2}$. The quantum state of a particle is characterized by the spin quantum number m_s plus the quantum numbers associated with the spatial part of the wave function. Because the spin quantum numbers have just two possible values, the exclusion principle can be stated in terms of the spatial states:

There can be at most two electrons with the same set of values for their *spatial* quantum numbers.

EXCLUSION PRINCIPLE IN TERMS OF SPATIAL STATES

When there are more than two electrons in a system, such as an atom or metal, only two can be in the lowest energy state. The third and fourth electrons must go into the second-lowest state, and so on.

BOSON-SYSTEM ENERGY VERSUS FERMION-SYSTEM ENERGY   **EXAMPLE 38-2**

FIGURE 38-6

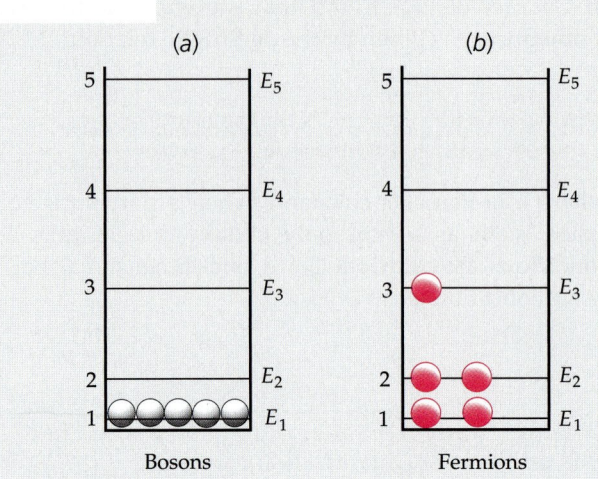

Compare the total energy of the ground state of five identical bosons of mass m in a one-dimensional box with that of five identical fermions of mass m in the same box.

PICTURE THE PROBLEM The ground state is the lowest possible energy state. The energy levels in a one-dimensional box are given by $E_n = n^2 E_1$, where $E_1 = h^2/(8mL^2)$. The lowest energy for five bosons occurs when all the bosons are in the state $n = 1$, as shown in Figure 38-6a. For fermions, the lowest state occurs with two fermions in the state $n = 1$, two fermions in the state $n = 2$, and one fermion in the state $n = 3$, as shown in Figure 38-6b.

1. The energy of five bosons in the state $n = 1$ is:

$$E = 5E_1$$

2. The energy of two fermions in the state $n = 1$, two fermions in the state $n = 2$, and one fermion in the state $n = 3$ is:

$$E = 2E_1 + 2E_2 + 1E_3 = 2E_1 + 2(2)^2 E_1 + 1(3)^2 E_1$$
$$= 2E_1 + 8E_1 + 9E_1 = 19E_1$$

3. Compare the total energies:

> The five identical fermions have 3.8 times the total energy of the five identical bosons.

REMARKS We see that the exclusion principle has a large effect on the total energy of a multiple-particle system.

The Fermi Energy

When there are many electrons in a box, at $T = 0$ the electrons will occupy the lowest energy states consistent with the exclusion principle. If we have N electrons, we can put two electrons in the lowest energy level, two electrons in the next lowest energy level, and so on. The N electrons thus fill the lowest $N/2$ energy levels (Figure 38-7). The energy of the last filled (or half-filled) level at $T = 0$ is called the Fermi energy E_F. If the electrons moved in a one-dimensional box, the Fermi energy would be given by Equation 38-19, with $n = N/2$:

$$E_F = \left(\frac{N}{2}\right)^2 \frac{h^2}{8m_e L^2} = \frac{h^2}{32m_e}\left(\frac{N}{L}\right)^2 \qquad 38-21$$

FERMI ENERGY AT $T = 0$ IN ONE DIMENSION

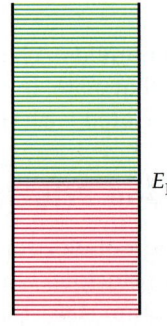

FIGURE 38-7 At $T = 0$ the electrons fill up the allowed energy states to the Fermi energy E_F. The levels are so closely spaced that they can be assumed to be continuous.

In a one-dimensional box, the Fermi energy depends on the number of free electrons per unit length of the box.

EXERCISE Suppose there is an ion, and therefore a free electron, every 0.1 nm in a one-dimensional box. Calculate the Fermi energy. *Hint:* Write Equation 38-21 as

$$E_F = \frac{(hc)^2}{32m_e c^2}\left(\frac{N}{L}\right)^2 = \frac{(1240 \text{ eV·nm})^2}{32(0.511 \text{ MeV})}\left(\frac{N}{L}\right)^2$$

$$E_f = \frac{(1240 \text{ eV·nm})^2}{32(511,000 \text{ eV})}\left(\frac{10 \text{ electrons}}{nm}\right)^2 = 9.403 \text{ eV}$$

(Answer $E_F = 9.4$ eV)

In our model of conduction, the free electrons move in a *three-dimensional* box of volume V. The derivation of the Fermi energy in three dimensions is somewhat difficult, so we will just give the result. In three dimensions, the Fermi energy at $T = 0$ is

$$E_F = \frac{h^2}{8m_e}\left(\frac{3N}{\pi V}\right)^{2/3}$$

38-22a

FERMI ENERGY AT $T = 0$ IN THREE DIMENSIONS

The Fermi energy depends on the number density of electrons N/V. Substituting numerical values for the constants gives

$$E_F = (0.365 \text{ eV·nm}^2)\left(\frac{N}{V}\right)^{2/3}$$

38-22b

FERMI ENERGY AT $T = 0$ IN THREE DIMENSIONS

THE FERMI ENERGY FOR COPPER

E X A M P L E   3 8 - 3

The number density for electrons in copper was calculated in Example 25-1 and found to be 84.7/nm³. Calculate the Fermi energy at $T = 0$ for copper.

1. The Fermi energy is given by Equation 38-22:

$$E_F = (0.365 \text{ eV·nm}^2)\left(\frac{N}{V}\right)^{2/3}$$

2. Substitute the given number density for copper:

$$E_F = (0.365 \text{ eV·nm}^2)(84.7/\text{nm}^3)^{2/3}$$

$$= \boxed{7.04 \text{ eV}}$$

REMARKS Note that the Fermi energy is much greater than kT at ordinary temperatures. For example, at $T = 300$ K, kT is only about 0.026 eV.

EXERCISE Use Equation 38-22b to calculate the Fermi energy at $T = 0$ for gold, which has a free-electron number density of 59.0/nm³. (*Answer*   5.53 eV)

$$E_F = (0.365 \text{ eV·nm}^2)\left(\frac{59 \text{ electrons}}{\text{nm}^3}\right)^{\frac{2}{3}}$$

Table 38-1 lists the free-electron number densities and Fermi energies at $T = 0$ for several metals.

The average energy of a free electron can be calculated from the complete energy distribution of the electrons, which is discussed in Section 38-9. At $T = 0$, the average energy turns out to be

$$E_{av} = \tfrac{3}{5} E_F$$

38-23

AVERAGE ENERGY OF ELECTRONS IN A FERMI GAS AT $T = 0$

For copper, E_{av} is approximately 4 eV. This average energy is huge compared with typical thermal energies of about $kT \approx 0.026$ eV at a normal temperature of $T = 300$ K. This result is very different from the classical Maxwell–Boltzmann distribution result that at $T = 0$, $E = 0$, and that at some temperature T, E is of the order of kT.

TABLE 38-1

Free-Electron Number Densities[†] and Fermi Energies at $T = 0$ for Selected Elements

	Element	N/V, electrons/nm³	E_F, eV
Al	Aluminum	181	11.7
Ag	Silver	58.6	5.50
Au	Gold	59.0	5.53
Cu	Copper	84.7	7.04
Fe	Iron	170	11.2
K	Potassium	14.0	2.11
Li	Lithium	47.0	4.75
Mg	Magnesium	86.0	7.11
Mn	Manganese	165	11.0
Na	Sodium	26.5	3.24
Sn	Tin	148	10.2
Zn	Zinc	132	9.46

† Number densities are measured using the Hall effect, discussed in Section 26-4.

The Fermi Factor at $T = 0$

The probability of an energy state being occupied is called the **Fermi factor,** $f(E)$. At $T = 0$ all the states below E_F are filled, whereas all those above this energy are empty, as shown in Figure 38-8. Thus, at $T = 0$ the Fermi factor is simply

$$f(E) = 1, \qquad E < E_F$$
$$f(E) = 0, \qquad E > E_F \qquad\qquad 38\text{-}24$$

The Fermi Factor for $T > 0$

At temperatures greater than $T = 0$, some electrons will occupy higher energy states because of thermal energy gained during collisions with the lattice. However, an electron cannot move to a higher or lower state unless it is unoccupied. Since the kinetic energy of the lattice ions is of the order of kT, electrons cannot gain much more energy than kT in collisions with the lattice ions. Therefore, only those electrons with energies within about kT of the Fermi energy can gain energy as the temperature is increased. At 300 K, kT is only 0.026 eV, so the exclusion principle prevents all but a very few electrons near the top of the energy distribution from gaining energy through random collisions with the lattice ions. Figure 38-9 shows the Fermi factor for some temperature T. Since for $T > 0$ there is no distinct energy that separates filled levels from unfilled levels, the definition of the Fermi energy must be slightly modified. At temperature T, the Fermi energy is defined to be that energy for which the probability of being occupied is $\frac{1}{2}$. For all but extremely high temperatures, the difference between the Fermi energy at temperature T and the Fermi energy at temperature $T = 0$ is very small. The **Fermi temperature** T_F is defined by

$$kT_F = E_F \qquad\qquad 38\text{-}25$$

For temperatures much lower than the Fermi temperature, the average energy of the lattice ions will be much less than the Fermi energy, and the electron energy distribution will not differ greatly from that at $T = 0$.

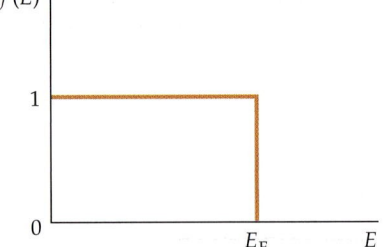

FIGURE 38-8 Fermi factor versus energy at $T = 0$.

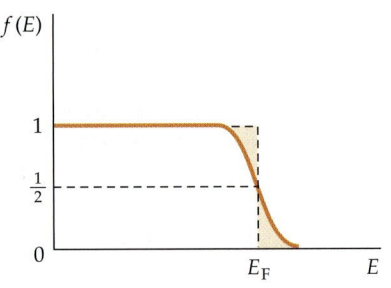

FIGURE 38-9 The Fermi factor for some temperature T. Some electrons with energies near the Fermi energy are excited, as indicated by the shaded regions. The Fermi energy is that value of E for which $f(E) = \frac{1}{2}$.

THE FERMI TEMPERATURE FOR COPPER **EXAMPLE 38-4**

Find the Fermi temperature for copper.

Use $E_F = 7.04$ eV and $k = 8.62 \times 10^{-5}$ eV/K in Equation 38-25: $\quad T_F = \dfrac{E_F}{k} = \dfrac{7.04 \text{ eV}}{8.62 \times 10^{-5} \text{ eV/K}} = \boxed{81{,}700 \text{ K}}$

REMARKS We can see from this example that the Fermi temperature of copper is much greater than any temperature T for which copper remains a solid.

Because an electric field in a conductor accelerates all of the conduction electrons together, the exclusion principle does not prevent the free electrons in filled states from participating in conduction. Figure 38-10 shows the Fermi factor in one dimension versus *velocity* for an ordinary temperature. The factor is

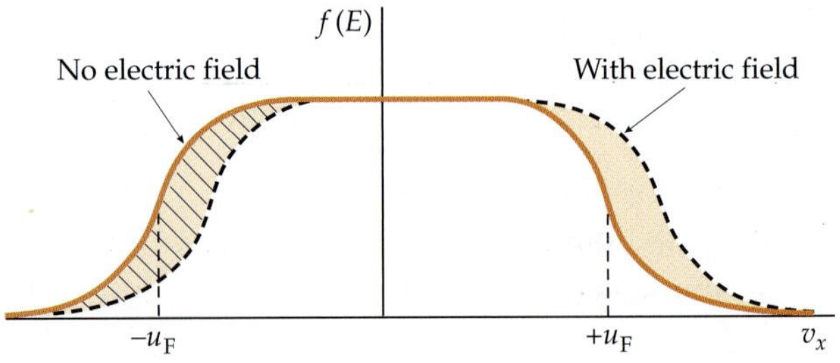

FIGURE 38-10 Fermi factor versus velocity in one dimension with no electric field (solid) and with an electric field in the $-x$ direction (dashed). The difference is greatly exaggerated.

approximately 1 for speeds v_x in the range $-u_F < v_x < u_F$, where the Fermi speed u_F is related to the Fermi energy by $E_F = \frac{1}{2}mu_F^2$. Then

$$u_F = \sqrt{\frac{2E_F}{m_e}}$$

38-26

THE FERMI SPEED FOR COPPER

E X A M P L E   3 8 - 5

Calculate the Fermi speed for copper.

Use Equation 38-26 with $E_F = 7.04$ eV:

$$u_F = \sqrt{\frac{2(7.04 \text{ eV})}{9.11 \times 10^{-31} \text{ kg}} \left(\frac{1.60 \times 10^{-19} \text{ J}}{1 \text{ eV}}\right)} = \boxed{1.57 \times 10^6 \text{ m/s}}$$

The dashed curve in Figure 38-10 shows the Fermi factor after the electric field has been acting for some time t. Although all of the free electrons have their velocities shifted in the direction opposite to the electric field, the net effect is equivalent to shifting only the electrons near the Fermi energy.

Contact Potential

When two different metals are placed in contact, a potential difference $V_{contact}$ called the **contact potential** develops between them. The contact potential depends on the work functions of the two metals, ϕ_1 and ϕ_2 (we encountered work functions when the photoelectric effect was introduced in Chapter 34), and the Fermi energies of the two metals. When the metals are in contact, the total energy of the system is lowered if electrons near the boundary move from the metal with the higher Fermi energy into the metal with the lower Fermi energy until the Fermi energies of the two metals are the same, as shown in Figure 38-11. When equilibrium is established, the metal with the lower initial Fermi energy is negatively charged and the other is positively charged, so that between them there is a potential difference $V_{contact}$ given by

$$V_{contact} = \frac{\phi_1 - \phi_2}{e}$$

38-27

Table 38-2 lists the work functions for several metals.

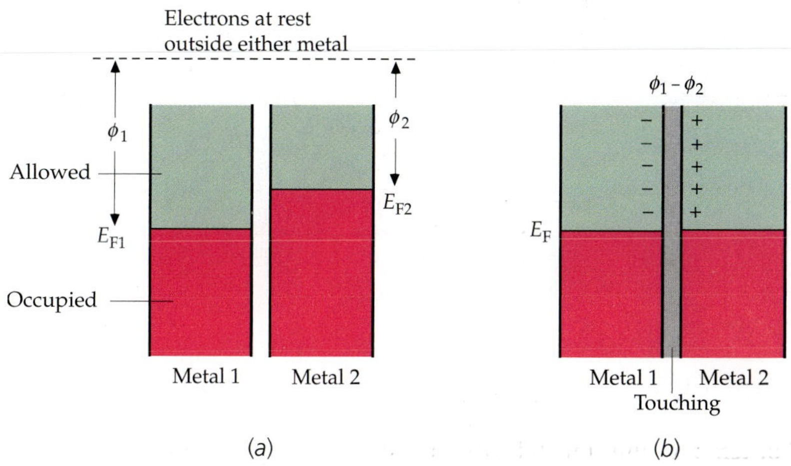

FIGURE 38-11 (*a*) Energy levels for two different metals with different Fermi energies and work functions. (*b*) When the metals are in contact, electrons flow from the metal that initially has the higher Fermi energy to the metal that initially has the lower Fermi energy until the Fermi energies are equal.

TABLE 38-2

Work Functions for Some Metals

	Metal	ϕ, eV		Metal	ϕ, eV
Ag	Silver	4.7	K	Potassium	2.1
Au	Gold	4.8	Mn	Manganese	3.8
Ca	Calcium	3.2	Na	Sodium	2.3
Cu	Copper	4.1	Ni	Nickel	5.2

CONTACT POTENTIAL BETWEEN SILVER AND TUNGSTEN **EXAMPLE 38-6**

The threshold wavelength for the photoelectric effect is 271 nm for tungsten and 262 nm for silver. What is the contact potential developed when silver and tungsten are placed in contact?

PICTURE THE PROBLEM The contact potential is proportional to the difference in the work functions for the two metals. The work function ϕ can be found from the given threshold wavelengths using $\phi = hc/\lambda_t$ (Equation 34-4).

1. The contact potential is given by Equation 38-27:

$$V_{contact} = \frac{\phi_1 - \phi_2}{e}$$

2. The work function is related to the threshold wavelength (Equation 34-4):

$$\phi = \frac{hc}{\lambda_t}$$

3. Substitute $\lambda_t = 271$ nm for tungsten:

$$\phi_W = \frac{hc}{\lambda_t} = \frac{1240 \text{ eV·nm}}{271 \text{ nm}} = 4.58 \text{ eV}$$

4. Substitute $\lambda_t = 262$ nm for silver:

$$\phi_{Ag} = \frac{1240 \text{ eV·nm}}{262 \text{ nm}} = 4.73 \text{ eV}$$

5. The contact potential is thus:

$$V_{contact} = \frac{\phi_{Ag} - \phi_W}{e} = 4.73 \text{ V} - 4.58 \text{ V}$$

$$= \boxed{0.15 \text{ V}}$$

Heat Capacity Due to Electrons in a Metal

The quantum-mechanical modification of the electron distribution in metals allows us to understand why the contribution of the electron gas to the heat capacity of a metal is much less that of the ions. According to the classical equipartition theorem, the energy of the lattice ions in n moles of a solid is $3nRT$, and thus the molar specific heat is $c' = 3R$, where R is the universal gas constant (see Section 18-7). In a metal, there is a free electron gas containing a number of electrons approximately equal to the number of lattice ions. If these electrons obey the classical equipartition theorem, they should have an energy of $\frac{3}{2}nRT$ and contribute an additional $\frac{3}{2}R$ to the molar specific heat. But measured heat capacities of metals are just slightly greater than those of insulators. We can understand this because at some temperature T, only those electrons with energies near the Fermi energy can be excited by random collisions with the lattice ions. The number of these electrons is of the order of $(kT/E_F)N$, where N is the total number of electrons. The energy of these electrons is increased from that at $T = 0$ by an amount that is of the order of kT. So the total increase in thermal energy is of the order of $(kT/E_F)N \times kT$. We can thus express the energy of N electrons at temperature T as

$$E = NE_{av}(0) + \alpha N \frac{kT}{E_F} kT \qquad \text{38-28}$$

where α is some constant that we expect to be of the order of 1 if our reasoning is correct. The calculation of α is quite difficult. The result is $\alpha = \pi^2/4$. Using this result and writing E_F in terms of the Fermi temperature, $E_F = kT_F$, we obtain the following for the contribution of the electron gas to the heat capacity at constant volume:

$$C_V = \frac{dE}{dT} = 2\alpha N k \frac{kT}{E_F} = \frac{\pi^2}{2} n R \frac{T}{T_F}$$

where we have written Nk in terms of the gas constant R ($Nk = nR$). The molar specific heat at constant volume is then

$$c'_V = \frac{\pi^2}{2} R \frac{T}{T_F} \qquad\qquad 38\text{-}29$$

We can see that because of the large value of T_F, the contribution of the electron gas is a small fraction of R at ordinary temperatures. Because $T_F = 81,700$ K for copper, the molar specific heat of the electron gas at $T = 300$ K is

$$c'_V = \frac{\pi^2}{2} \left(\frac{300 \text{ K}}{81,700} \right) R \approx 0.02 \, R$$

which is in good agreement with experiment.

38-4 Quantum Theory of Electrical Conduction

We can use Equation 38-14 for the resistivity if we use the Fermi speed u_F in place of v_{av}:

$$\rho = \frac{m_e u_F}{n_e e^2 \lambda} \qquad\qquad 38\text{-}30$$

We now have two problems. First, since the Fermi speed u_F is approximately independent of temperature, the resistivity given by Equation 38-30 is independent of temperature unless the mean free path depends on it. The second problem concerns magnitudes. As mentioned earlier, the classical expression for resistivity using v_{av} calculated from the Maxwell–Boltzmann distribution gives values that are about 6 times too large at $T = 300$ K. Since the Fermi speed u_F is about 16 times the Maxwell–Boltzmann value of v_{av}, the magnitude of ρ predicted by Equation 38-30 will be approximately 100 times greater than the experimentally determined value. The resolution of both of these problems lies in the calculation of the mean free path λ.

The Scattering of Electron Waves

In Equation 38-16 for the classical mean free path $\lambda = 1/(n_{ion}A)$, the quantity $A = \pi r^2$ is the area of the lattice ion as seen by an electron. In the quantum calculation, the mean free path is related to the scattering of electron waves by the crystal lattice. Detailed calculations show that, for a *perfectly* ordered crystal, $\lambda = \infty$; that is, there is no scattering of the electron waves. The scattering of electron waves arises because of *imperfections* in the crystal lattice, which have nothing to do with the actual cross-sectional area A of the lattice ions. According to the quantum theory of electron scattering, A depends merely on *deviations* of the lattice ions from a perfectly ordered array and not on the size of the ions. The most common causes of such deviations are thermal vibrations of the lattice ions or impurities.

We can use $\lambda = 1/(n_{ion}A)$ for the mean free path if we reinterpret the area A. Figure 38-12 compares the classical picture and the quantum picture of this area. In the quantum picture, the lattice ions are points that have no size but present an

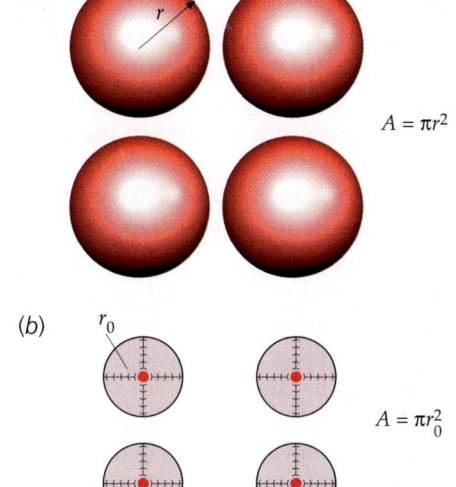

(a)

$A = \pi r^2$

(b) r_0

$A = \pi r_0^2$

FIGURE 38-12 (*a*) Classical picture of the lattice ions as spherical balls of radius *r* that present an area πr^2 to the electrons. (*b*) Quantum-mechanical picture of the lattice ions as points that are vibrating in three dimensions. The area presented to the electrons is πr_0^2, where r_0 is the amplitude of oscillation of the ions.

area $A = \pi r_0^2$, where r_0 is the amplitude of thermal vibrations. In Chapter 14, we saw that the energy of vibration in simple harmonic motion is proportional to the square of the amplitude, which is πr_0^2. Thus, the effective area A is proportional to the energy of vibration of the lattice ions. From the equipartition theorem,[†] we know that the average energy of vibration is proportional to kT. Thus, A is proportional to T, and λ is proportional to $1/T$. Then the resistivity given by Equation 38-14 is proportional to T, in agreement with experiment.

The effective area A due to thermal vibrations can be calculated, and the results give values for the resistivity that are in agreement with experiment. At $T = 300$ K, for example, the effective area turns out to be about 100 times smaller than the actual area of a lattice ion. We see, therefore, that the free-electron model of metals gives a good account of electrical conduction if the classical mean speed v_{av} is replaced by the Fermi speed u_F and if the collisions between electrons and the lattice ions are interpreted in terms of the scattering of electron waves, for which only deviations from a perfectly ordered lattice are important.

The presence of impurities in a metal also causes deviations from perfect regularity in the crystal lattice. The effects of impurities on resistivity are approximately independent of temperature. The resistivity of a metal containing impurities can be written $\rho = \rho_t + \rho_i$, where ρ_t is the resistivity due to the thermal motion of the lattice ions and ρ_i is the resistivity due to impurities. Figure 38-13 shows typical resistance curves versus temperature curves for metals with impurities. As the absolute temperature approaches zero, ρ_t approaches zero, and the resistivity approaches the constant ρ_i due to impurities.

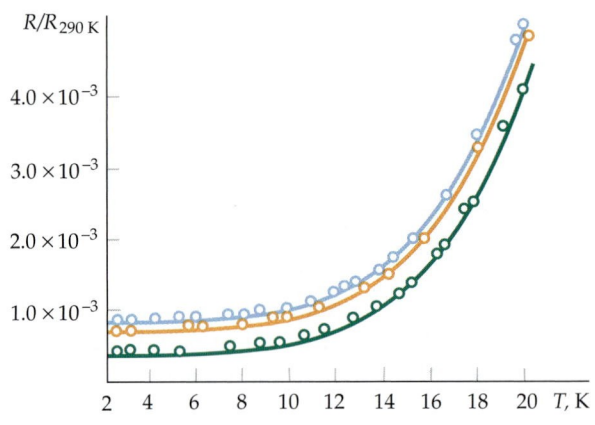

FIGURE 38-13 Relative resistance versus temperature for three samples of sodium. The three curves have the same temperature dependence but different magnitudes because of differing amounts of impurities in the samples.

38-5 Band Theory of Solids

Resistivities vary enormously between insulators and conductors. For a typical insulator, such as quartz, $\rho \sim 10^{16}$ Ω·m, whereas for a typical conductor, $\rho \sim 10^{-8}$ Ω·m. The reason for this enormous variation is the variation in the number density of free electrons n_e. To understand this variation, we consider the effect of the lattice on the electron energy levels.

We begin by considering the energy levels of the individual atoms as they are brought together. The allowed energy levels in an isolated atom are often far apart. For example, in hydrogen, the lowest allowed energy $E_1 = -13.6$ eV is 10.2 eV below the next lowest allowed energy $E_2 = (-13.6 \text{ eV})/4 = -3.4$ eV.[‡] Let us consider two identical atoms and focus our attention on one particular energy level. When the atoms are far apart, the energy of a particular level is the same for each atom. As the atoms are brought closer together, the energy level for each atom changes because of the influence of the other atom. As a result, the level splits into two levels of slightly different energies for the two-atom system. If we bring three atoms close together, a particular energy level splits into three separate levels of slightly different energies. Figure 38-14 shows the energy splitting of two energy levels for six atoms as a function of the separation of the atoms.

If we have N identical atoms, a particular energy level in the isolated atom splits into N different, closely spaced energy levels when the atoms are close together. In a macroscopic solid, N is very large—of the order of 10^{23}—so each energy level splits into a very large

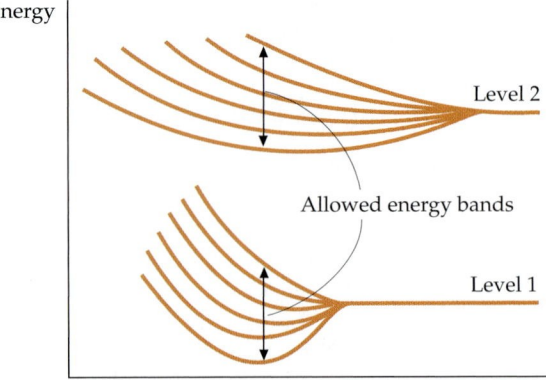

FIGURE 38-14 Energy splitting of two energy levels for six atoms as a function of the separation of the atoms. When there are many atoms, each level splits into a near-continuum of levels called a band.

---

† The equipartition theorem does hold for the lattice ions, which obey the Maxwell–Boltzmann energy distribution.
‡ The energy levels in hydrogen are discussed in Chapter 36.

number of levels called a **band.** The levels are spaced almost continuously within the band. There is a separate band of levels for each particular energy level of the isolated atom. The bands may be widely separated in energy, they may be close together, or they may even overlap, depending on the kind of atom and the type of bonding in the solid.

The lowest energy bands, corresponding to the lowest energy levels of the atoms in the lattice, are filled with electrons that are bound to the individual atoms. The electrons that can take part in conduction occupy the higher energy bands. The highest energy band that contains electrons is called the **valence band.** The valence band may be completely filled with electrons or only partially filled, depending on the kind of atom and the type of bonding in the solid.

We can now understand why some solids are conductors and why others are insulators. If the valence band is only partially full, there are many available empty energy states in the band, and the electrons in the band can easily be raised to a higher energy state by an electric field. Accordingly, this material is a good conductor. If the valence band is full and there is a large energy gap between it and the next available band, a typical applied electric field will be too weak to excite an electron from the upper energy levels of the filled band across the large gap into the energy levels of the empty band, so the material is an insulator. The lowest band in which there are unoccupied states is called the **conduction band.** In a conductor, the valence band is only partially filled, so the valence band is also the conduction band. An energy gap between allowed bands is called a **forbidden energy band.**

The band structure for a conductor, such as copper, is shown in Figure 38-15a. The lower bands (not shown) are filled with the inner electrons of the atoms. The valence band is only about half full. When an electric field is established in the conductor, the electrons in the conduction band are accelerated, which means that their energy is increased. This is consistent with the Pauli exclusion principle because there are many empty energy states just above those occupied by electrons in this band. These electrons are thus the conduction electrons.

Figure 38-15b shows the band structure for magnesium, which is also a conductor. In this case, the highest occupied band is full, but there is an empty band above it that overlaps it. The two bands thus form a combined valence–conduction band that is only partially filled.

Figure 38-15c shows the band structure for a typical insulator. At $T = 0$ K, the valence band is completely full. The next energy band containing empty energy states, the conduction band, is separated from the valence band by a large energy gap. At $T = 0$, the conduction band is empty. At ordinary temperatures, a few electrons can be excited to states in this band, but most cannot be excited to states because the energy gap is large compared with the energy an electron might obtain by thermal excitation. Very few electrons can be thermally excited to the nearly empty conduction band, even at fairly high temperatures. When an electric field of ordinary magnitude is established in the solid, electrons cannot be

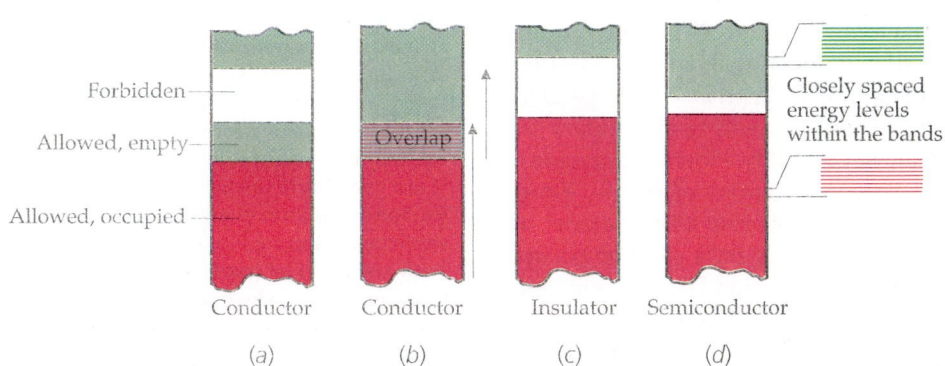

FIGURE 38-15 Four possible band structures for a solid. (*a*) A typical conductor. The valence band is only partially full, so electrons can be easily excited to nearby energy states. (*b*) A conductor in which the allowed energy bands overlap. (*c*) A typical insulator. There is a forbidden band with a large energy gap between the filled valence band and the conduction band. (*d*) A semiconductor. The energy gap between the filled valence band and the conduction band is very small, so some electrons are excited to the conduction band at normal temperatures, leaving holes in the valence band.

accelerated because there are no empty energy states at nearby energies. We describe this by saying that there are no free electrons. The small conductivity that is observed is due to the very few electrons that are thermally excited into the nearly empty conduction band. When an electric field applied to an insulator is sufficiently strong to cause an electron to be excited across the energy gap to the empty band, dielectric breakdown occurs.

In some materials, the energy gap between the filled valence band and the empty conduction band is very small, as shown in Figure 38-15d. At $T = 0$, there are no electrons in the conduction band and the material is an insulator. However, at ordinary temperatures, there are an appreciable number of electrons in the conduction band due to thermal excitation. Such a material is called an **intrinsic semiconductor.** For typical intrinsic semiconductors, such as silicon and germanium, the energy gap is only about 1 eV. In the presence of an electric field, the electrons in the conduction band can be accelerated because there are empty states nearby. Also, for each electron in the conduction band there is a vacancy, or hole, in the nearly filled valence band. In the presence of an electric field, electrons in this band can also be excited to a vacant energy level. This contributes to the electric current and is most easily described as the motion of a hole in the direction of the field and opposite to the motion of the electrons. The hole thus acts like a positive charge. To visualize the conduction of holes, think of a two-lane, one-way road with one lane full of parked cars and the other lane empty. If a car moves out of the filled lane into the empty lane, it can move ahead freely. As the other cars move up to occupy the space left, the empty space propagates backward in the direction opposite the motion of the cars. Both the forward motion of the car in the nearly empty lane and the backward propagation of the empty space contribute to a net forward propagation of the cars.

An interesting characteristic of semiconductors is that the resistivity of the material decreases as the temperature increases, which is contrary to the case for normal conductors. The reason is that as the temperature increases, the number of free electrons increases because there are more electrons in the conduction band. The number of holes in the valence band also increases, of course. In semiconductors, the effect of the increase in the number of charge carriers, both electrons and holes, exceeds the effect of the increase in resistivity due to the increased scattering of the electrons by the lattice ions due to thermal vibrations. Semiconductors therefore have a negative temperature coefficient of resistivity.

38-6 Semiconductors

The semiconducting property of intrinsic semiconductors materials makes them useful as a basis for electronic circuit components whose resistivity can be controlled by application of an external voltage or current. Most such *solid-state devices,* however, such as the semiconductor diode and the transistor, make use of **impurity semiconductors,** which are created through the controlled addition of certain impurities to intrinsic semiconductors. This process is called **doping.** Figure 38-16a is a schematic illustration of silicon doped with a small amount of arsenic so that the arsenic atoms replace a few of the silicon atoms in the crystal lattice. The conduction band of pure silicon is virtually empty at ordinary temperatures, so pure silicon is a poor conductor of electricity. However, arsenic has five valence electrons rather than the four valence electrons of silicon. Four of these electrons take part in bonds with the four neighboring silicon atoms, and the fifth electron is very loosely bound to the atom. This extra electron occupies an energy level that is just slightly below the conduction band in the solid, and it is easily excited into the conduction band, where it can contribute to electrical conduction.

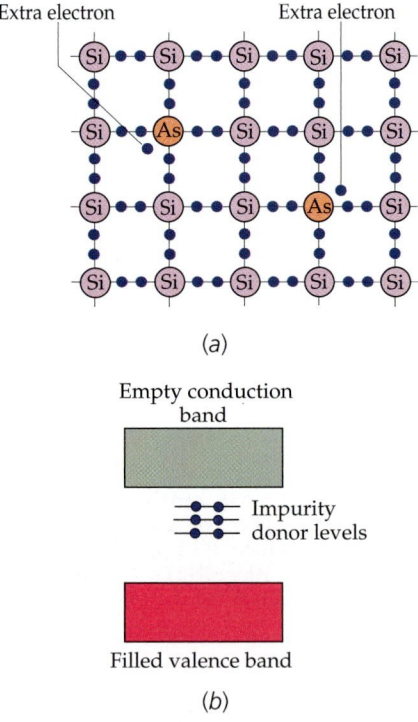

(a)

(b)

FIGURE 38-16 (a) A two-dimensional schematic illustration of silicon doped with arsenic. Because arsenic has five valence electrons, there is an extra, weakly bound electron that is easily excited to the conduction band, where it can contribute to electrical conduction. (b) Band structure of an *n*-type semiconductor, such as silicon doped with arsenic. The impurity atoms provide filled energy levels that are just below the conduction band. These levels donate electrons to the conduction band.

The effect on the band structure of a silicon crystal achieved by doping it with arsenic is shown in Figure 38-16b. The levels shown just below the conduction band are due to the extra electrons of the arsenic atoms. These levels are called **donor levels** because they donate electrons to the conduction band without leaving holes in the valence band. Such a semiconductor is called an *n*-type **semiconductor** because the major charge carriers are negative electrons. The conductivity of a doped semiconductor can be controlled by controlling the amount of impurity added. The addition of just one part per million can increase the conductivity by several orders of magnitude.

Another type of impurity semiconductor can be made by replacing a silicon atom with a gallium atom, which has three valence electrons (Figure 38-17a). The gallium atom accepts electrons from the valence band to complete its four covalent bonds, thus creating a hole in the valence band. The effect on the band structure of silicon achieved by doping it with gallium is shown in Figure 38-17b. The empty levels shown just above the valence band are due to the holes from the ionized gallium atoms. These levels are called **acceptor levels** because they accept electrons from the filled valence band when these electrons are thermally excited to a higher energy state. This creates holes in the valence band that are free to propagate in the direction of an electric field. Such a semiconductor is called a *p*-type **semiconductor** because the charge carriers are positive holes. The fact that conduction is due to the motion of positive holes can be verified by the Hall effect. (The Hall effect is discussed in chapter 26.)

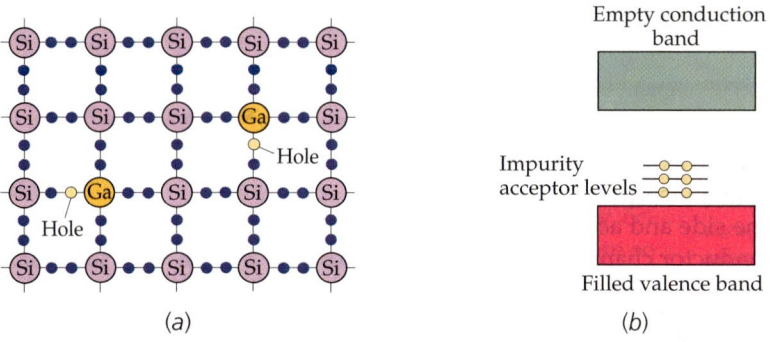

(a)

(b)

FIGURE 38-17 (a) A two-dimensional schematic illustration of silicon doped with gallium. Because gallium has only three valence electrons, there is a hole in one of its bonds. As electrons move into the hole the hole moves about, contributing to the conduction of electrical current. (b) Band structure of a *p*-type semiconductor, such as silicon doped with gallium. The impurity atoms provide empty energy levels just above the filled valence band that accept electrons from the valence band.

Synthetic crystal silicon is produced beginning with a raw material containing silicon (for instance, common beach sand), separating out the silicon, and melting it. From a seed crystal, the molten silicon grows into a cylindrical crystal, such as the one shown here. The crystals (typically about 1.3 m long) are formed under highly controlled conditions to ensure that they are flawless and the crystals are then sliced into thousands of thin wafers onto which the layers of an integrated circuit are etched.

NUMBER DENSITY OF FREE ELECTRONS IN ARSENIC-DOPED SILICON   **EXAMPLE 38-7**   **Try It Yourself**

The number of free electrons in pure silicon is approximately 10^{10} electrons/cm³ at ordinary temperatures. If one silicon atom out of every million atoms is replaced by an arsenic atom, how many free electrons per cubic centimeter are there? (The density of silicon is 2.33 g/cm³ and its molar mass is 28.1 g/mol.)

PICTURE THE PROBLEM The number of silicon atoms per cubic centimeter, n_{Si} can be found from $n_{Si} = N_A \rho / M$. Then, since each arsenic atom contributes one free electron, the number of electrons contributed by the arsenic atoms is $10^{-6} n_S$.

➤ In this chapter, we concentrate on the special theory (often referred to as *special relativity*). General relativity will be discussed briefly near the end of the chapter.

39-1 Newtonian Relativity

Newton's first law does not distinguish between a particle at rest and a particle moving with constant velocity. If there is no net external force acting, the particle will remain in its initial state, either at rest or moving with its initial velocity. A particle at rest relative to you is moving with constant velocity relative to an observer who is moving with constant velocity relative to you. How might we distinguish whether you and the particle are at rest and the second observer is moving with constant velocity, or the second observer is at rest and you and the particle are moving?

Let us consider some simple experiments. Suppose we have a railway boxcar moving along a straight, flat track with a constant velocity v. We note that a ball at rest in the boxcar remains at rest. If we drop the ball, it falls straight down, relative to the boxcar, with an acceleration g due to gravity. Of course, when viewed from the track the ball moves along a parabolic path because it has an initial velocity v to the right. No mechanics experiment that we can do—measuring the period of a pendulum, observing the collisions between two objects, or whatever—will tell us whether the boxcar is moving and the track is at rest or the track is moving and the boxcar is at rest. If we have a coordinate system attached to the track and another attached to the boxcar, Newton's laws hold in either system.

A set of coordinate systems at rest relative to each other is called a *reference frame*. A reference frame in which Newton's laws hold is called an *inertial reference frame*.[†] All reference frames moving at constant velocity relative to an inertial reference frame are also inertial reference frames. If we have two inertial reference frames moving with constant velocity relative to each other, there are no mechanics experiments that can tell us which is at rest and which is moving or if they are both moving. This result is known as the principle of **Newtonian relativity:**

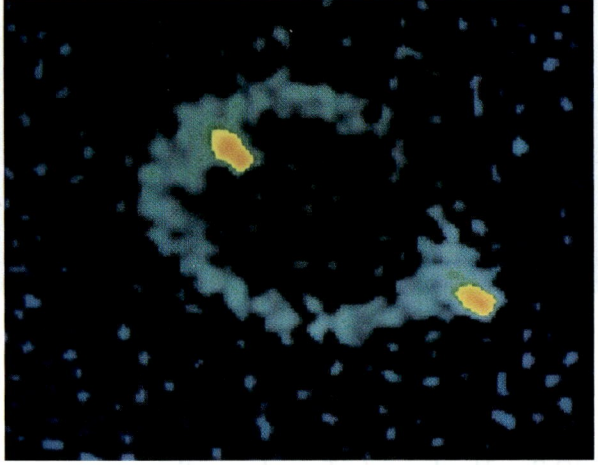

This ring-like structure of the radio source MG1131 + 0456 is thought to be due to *gravitational lensing*, first proposed by Albert Einstein in 1936, in which a source is imaged into a ring by a large, massive object in the foreground.

> Absolute motion cannot be detected.

PRINCIPLE OF NEWTONIAN RELATIVITY

This principle was well known by Galileo, Newton, and others in the seventeenth century. By the late nineteenth century, however, this view had changed. It was then generally thought that Newtonian relativity was not valid and that absolute motion could be detected in principle by a measurement of the speed of light.

Ether and the Speed of Light

We saw in Chapter 15 that the velocity of a wave depends on the properties of the medium in which the wave travels and not on the velocity of the source of the waves. For example, the velocity of sound relative to still air depends on the temperature of the air. Light and other electromagnetic waves (radio, X rays, etc.) travel through a vacuum with a speed $c \approx 3 \times 10^8$ m/s that is predicted by James Clerk Maxwell's equations for electricity and magnetism. But what is this speed

---

[†] Reference frames were first discussed in Section 2-1. Inertial reference frames were also discussed in Section 4-1.

relative to? What is the equivalent of still air for a vacuum? A proposed medium for the propagation of light was called the *ether;* it was thought to pervade all space. The velocity of light relative to the ether was assumed to be c, as predicted by Maxwell's equations. The velocity of any object relative to the ether was considered its absolute velocity.

Albert Michelson, first in 1881 and then again with Edward Morley in 1887, set out to measure the velocity of the earth relative to the ether by an ingenious experiment in which the velocity of light relative to the earth was compared for two light beams, one in the direction of the earth's motion relative to the sun and the other perpendicular to the direction of the earth's motion. Despite painstakingly careful measurements, they could detect no difference. The experiment has since been repeated under various conditions by a number of people, and no difference has ever been found. The absolute motion of the earth relative to the ether cannot be detected.

39-2 Einstein's Postulates

In 1905, at the age of 26, Albert Einstein published a paper on the electrodynamics of moving bodies.[†] In this paper, he postulated that absolute motion cannot be detected by any experiment. That is, there is no ether. The earth can be considered to be at rest and the velocity of light will be the same in any direction.[‡] His theory of special relativity can be derived from two postulates. Simply stated, these postulates are as follows:

> Postulate 1: Absolute uniform motion cannot be detected.
>
> Postulate 2: The speed of light is independent of the motion of the source.

EINSTEIN'S POSTULATES

Postulate 1 is merely an extension of the Newtonian principle of relativity to include all types of physical measurements (not just those that are mechanical). Postulate 2 describes a common property of all waves. For example, the speed of sound waves does not depend on the motion of the sound source. The sound waves from a car horn travel through the air with the same velocity independent of whether the car is moving or not. The speed of the waves depends only on the properties of the air, such as its temperature.

Although each postulate seems quite reasonable, many of the implications of the two postulates together are quite surprising and contradict what is often called common sense. For example, one important implication of these postulates is that every observer measures the same value for the speed of light independent of the relative motion of the source and the observer. Consider a light source S and two observers, R_1 at rest relative to S and R_2 moving toward S with speed v, as shown in Figure 39-1a. The speed of light measured by R_1 is $c = 3 \times 10^8$ m/s. What is the speed measured by R_2? The answer is *not* $c + v$. By postulate 1, Figure 39-1a is equivalent to Figure 39-1b, in which R_2 is at rest and the source S and R_1 are moving with speed v. That is, since absolute motion cannot be detected, it is not possible to say which is really moving and which is at rest. By postulate 2, the speed of light from a moving source is independent of the

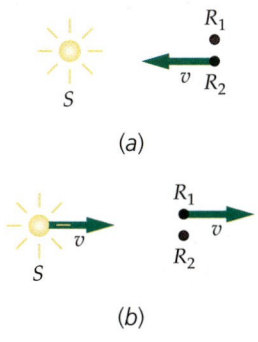

(a)

(b)

FIGURE 39-1 (*a*) A stationary light source S and a stationary observer R_1, with a second observer R_2 moving toward the source with speed v. (*b*) In the reference frame in which the observer R_2 is at rest, the light source S and observer R_1 move to the right with speed v. If absolute motion cannot be detected, the two views are equivalent. Since the speed of light does not depend on the motion of the source, observer R_2 measures the same value for that speed as observer R_1.

[†] *Annalen der Physik,* vol. 17, 1905, p. 841. For a translation from the original German, see W. Perrett and G. B. Jeffery (trans.), *The Principle of Relativity: A Collection of Original Memoirs on the Special and General Theory of Relativity* by H. A. Lorentz, A. Einstein, H. Minkowski, and W. Weyl, Dover, New York, 1923.

[‡] Einstein did not set out to explain the results of the Michelson–Morley experiment. His theory arose from his considerations of the theory of electricity and magnetism and the unusual property of electromagnetic waves that they propagate in a vacuum. In his first paper, which contains the complete theory of special relativity, he made only a passing reference to the Michelson–Morley experiment, and in later years he could not recall whether he was aware of the details of this experiment before he published his theory.

motion of the source. Thus, looking at Figure 39-1b, we see that R_2 measures the speed of light to be c, just as R_1 does. This result is often considered as an alternative to Einstein's second postulate:

> Postulate 2 (alternate): Every observer measures the same value c for the speed of light.

This result contradicts our intuitive ideas about relative velocities. If a car moves at 50 km/h away from an observer and another car moves at 80 km/h in the same direction, the velocity of the second car relative to the first car is 30 km/h. This result is easily measured and conforms to our intuition. However, according to Einstein's postulates, if a light beam is moving in the direction of the cars, observers in both cars will measure the same speed for the light beam. Our intuitive ideas about the combination of velocities are approximations that hold only when the speeds are very small compared with the speed of light. Even in an airplane moving with the speed of sound, to measure the speed of light accurately enough to distinguish the difference between the results c and $c + v$, where v is the speed of the plane, would require a measurement with six-digit accuracy.

39-3 The Lorentz Transformation

Einstein's postulates have important consequences for measuring time intervals and space intervals, as well as relative velocities. Throughout this chapter, we will be comparing measurements of the positions and times of events (such as lightning flashes) made by observers who are moving relative to each other. We will use a rectangular coordinate system xyz with origin O, called the S reference frame, and another system $x'y'z'$ with origin O', called the S' frame, that is moving with a constant velocity $\vec{v}$ relative to the S frame. Relative to the S' frame, the S frame is moving with a constant velocity $-\vec{v}$. For simplicity, we will consider the S' frame to be moving along the x axis in the positive x direction relative to S. In each frame, we will assume that there are as many observers as are needed who are equipped with measuring devices, such as clocks and metersticks, that are identical when compared at rest (see Figure 39-2).

We will use Einstein's postulates to find the general relation between the coordinates x, y, and z and the time t of an event as seen in reference frame S and the coordinates x', y', and z' and the time t' of the same event as seen in reference frame S', which is moving with uniform velocity relative to S. We assume that the origins are coincident at time $t = t' = 0$. The classical relation, called the **Galilean transformation**, is

$$x = x' + vt', \qquad y = y', \qquad z = z', \qquad t = t' \qquad \text{39-1a}$$

GALILEAN TRANSFORMATION

The inverse transformation is

$$x' = x - vt, \qquad y' = y, \qquad z' = z, \qquad t' = t \qquad \text{39-1b}$$

These equations are consistent with experimental observations as long as v is much less than c. They lead to the familiar classical addition law for velocities. If a particle has velocity $u_x = dx/dt$ in frame S, its velocity in frame S' is

$$u'_x = \frac{dx'}{dt'} = \frac{dx'}{dt} = \frac{dx}{dt} - v = u_x - v \qquad \text{39-2}$$

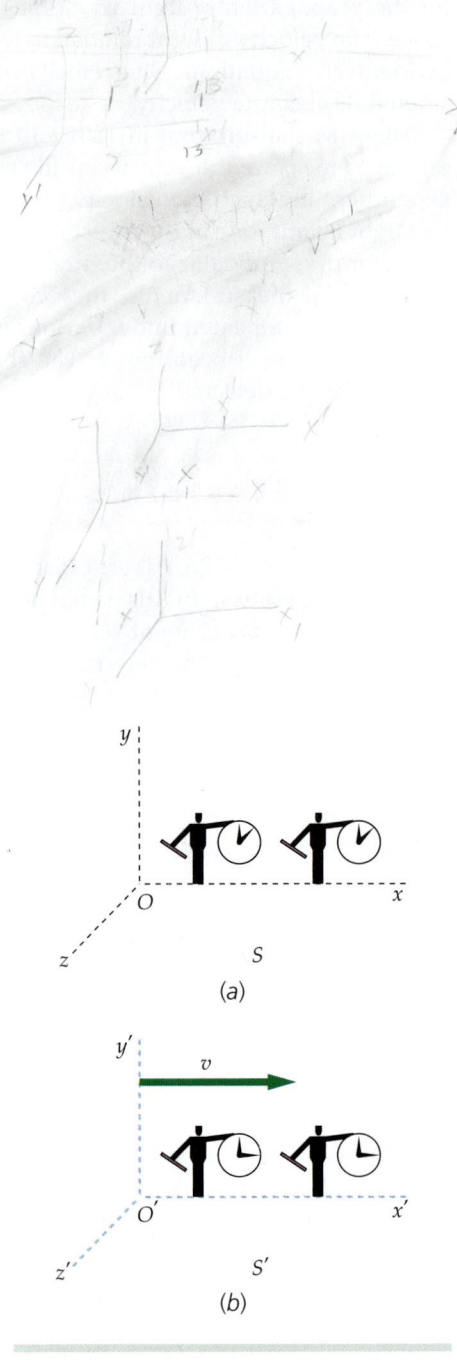

FIGURE 39-2 Coordinate reference frames S and S' moving with relative speed v. In each frame, there are observers with metersticks and clocks that are identical when compared at rest.

If we differentiate this equation again, we find that the acceleration of the particle is the same in both frames:

$$a_x = \frac{du_x}{dt} = \frac{du'_x}{dt'} = a'_x$$

It should be clear that the Galilean transformation is not consistent with Einstein's postulates of special relativity. If light moves along the x axis with speed $u'_x = c$ in S', these equations imply that the speed in S' is $u_x = c + v$ rather than $u_x = c$, which is consistent with Einstein's postulates and with experiment. The classical transformation equations must therefore be modified to make them consistent with Einstein's postulates. We will give a brief outline of one method of obtaining the relativistic transformation.

We assume that the relativistic transformation equation for x is the same as the classical equation (Equation 39-1a) except for a constant multiplier on the right side. That is, we assume the equation is of the form

$$x = \gamma(x' + vt') \qquad\qquad 39\text{-}3$$

where γ is a constant that can depend on v and c but not on the coordinates. The inverse transformation must look the same except for the sign of the velocity:

$$x' = \gamma(x - vt) \qquad\qquad 39\text{-}4$$

Let us consider a light pulse that starts at the origin of S at $t = 0$. Since we have assumed that the origins are coincident at $t = t' = 0$, the pulse also starts at the origin of S' at $t' = 0$. Einstein's postulates require that the equation for the x component of the wave front of the light pulse is $x = ct$ in frame S and $x' = ct'$ in frame S'. Substituting ct for x and ct' for x' in Equation 39-3 and Equation 39-4, we obtain

$$ct = \gamma(ct' + vt') = \gamma(c + v)t' \qquad\qquad 39\text{-}5$$

and

$$ct' = \gamma(ct - vt) = \gamma(c - v)t \qquad\qquad 39\text{-}6$$

We can eliminate the ratio t'/t from these two equations and determine γ. Thus,

$$\gamma = \frac{1}{\sqrt{1 - \dfrac{v^2}{c^2}}} \qquad\qquad 39\text{-}7$$

Note that γ is always greater than 1, and that when v is much less than c, $\gamma \approx 1$. The relativistic transformation for x and x' is therefore given by Equation 39-3 and Equation 39-4, with γ given by Equation 39-7. We can obtain equations for t and t' by combining Equation 39-3 with the inverse transformation given by Equation 39-4. Substituting $x = \gamma(x' + vt')$ for x in Equation 39-4, we obtain

$$x' = \gamma\big[\gamma(x' + vt') - vt\big] \qquad\qquad 39\text{-}8$$

which can be solved for t in terms of x' and t'. The complete relativistic transformation is

$$x = \gamma(x' + vt'), \qquad y = y', \qquad z = z'$$ 39-9

$$t = \gamma\left(t' + \frac{vx'}{c^2}\right)$$ 39-10

LORENTZ TRANSFORMATION

The inverse transformation is

$$x' = \gamma(x - vt), \qquad y' = y, \qquad z' = z$$ 39-11

$$t' = \gamma\left(t - \frac{vx}{c^2}\right)$$ 39-12

The transformation described by Equation 39-9 through Equation 39-12 is called the **Lorentz transformation.** It relates the space and time coordinates x, y, z, and t of an event in frame S to the coordinates x', y', z', and t' of the same event as seen in frame S', which is moving along the x axis with speed v relative to frame S.

We will now look at some applications of the Lorentz transformation.

Time Dilation

Consider two events that occur at a single point x'_0 at times t'_1 and t'_2 in frame S'. We can find the times t_1 and t_2 for these events in S from Equation 39-10. We have

$$t_1 = \gamma\left(t'_1 + \frac{vx'_0}{c^2}\right)$$

and

$$t_2 = \gamma\left(t'_2 + \frac{vx'_0}{c^2}\right)$$

so

$$t_2 - t_1 = \gamma(t'_2 - t'_1)$$

The time between events that happen at the *same place* in a reference frame is called **proper time** t_p. In this case, the time interval $t'_2 - t'_1$ measured in frame S' is proper time. The time interval Δt measured in any other reference frame is always longer than the proper time. This expansion is called **time dilation:**

$$\Delta t = \gamma \, \Delta t_p$$ 39-13

TIME DILATION

SPATIAL SEPARATION AND TEMPORAL SEPARATION OF TWO EVENTS

EXAMPLE 39-1

Two events occur at the same point x'_0 at times t'_1 and t'_2 in frame S', which is traveling at speed v relative to frame S. (*a*) What is the spatial separation of these events in frame S? (*b*) What is the temporal separation of these events in frame S?

PICTURE THE PROBLEM The spatial separation in S is $x_2 - x_1$, where x_2 and x_1 are the coordinates of the events in S, which are found using Equation 39-9.

(a) 1. The position x_1 in S is given by Equation 39-9 with $x'_1 = x'_0$:

$$x_1 = \gamma(x'_0 + vt'_1)$$

2. Similarly, the position x_2 in S is given by:

$$x_2 = \gamma(x'_0 + vt'_2)$$

3. Subtract to find the spatial separation:

$$\Delta x = x_2 - x_1 = \gamma v(t'_2 - t'_1) = \boxed{\frac{v(t'_2 - t'_1)}{\sqrt{1 - (v^2/c^2)}}}$$

(b) Using the time dilation formula, relate the two time intervals. The two events occur at the same place in S', so the proper time between the two events is $\Delta t_p = t'_2 - t'_1$:

$$\Delta t = t_2 - t_1 = \gamma(t'_2 - t'_1) = \boxed{\frac{(t'_2 - t'_1)}{\sqrt{1 - (v^2/c^2)}}}$$

REMARKS Dividing the Part (a) result by the Part (b) result gives $\Delta x/\Delta t = v$. The spatial separation of these two events in S is the distance a fixed point, such as x'_0 in S', moves in S during the time interval between the events in S.

We can understand time dilation directly from Einstein's postulates without using the Lorentz transformation. Figure 39-3a shows an observer A' a distance D from a mirror. The observer and the mirror are in a spaceship that is at rest in frame S'. The observer explodes a flash gun and measures the time interval $\Delta t'$ between the original flash and his seeing the return flash from the mirror. Because light travels with speed c, this time is

$$\Delta t' = \frac{2D}{c}$$

We now consider these same two events, the original flash of light and the receiving of the return flash, as observed in reference frame S, in which observer A' and the mirror are moving to the right with speed v, as shown in Figure 39-3b.

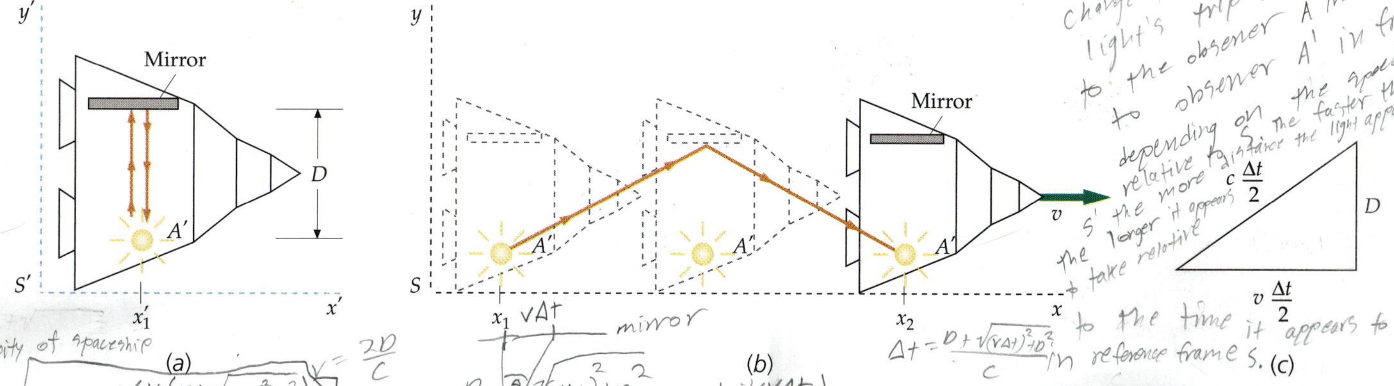

[handwritten margin notes, right side: since the observer A' in frame S is moving relative to frame S, sees the light travel $2D$ in time t, $c' = \frac{2D}{t}$. However, the observer in frame S sees the light travel further. Since the speed of light is constant: to not change the speed of light, the light's trip must appear longer to the observer A in frame S than to observer A' in frame S', depending on the speed of S' relative to S. The faster the speed of the more distance the light appears to travel, the longer it appears + take relative to the time it appears to take in reference frame S.]

[handwritten margin notes, left side: v = velocity of spaceship. as speed of spacecraft increases $\theta \to 90°$ but will never reach $90°$ because $\tan(90°)$ is undefined) and therefore the speed of light is unattainable. We cannot solve for v if $\theta = 90°$]

[boxed handwritten: $\theta = \tan^{-1}\left(\frac{V(D + \sqrt{(V\Delta t)^2 + D^2})}{CD}\right)$]

[handwritten near figure a: $= \frac{2D}{C}$]

[handwritten near figure b: $D \quad \theta \quad \sqrt{(V\Delta t)^2 + D^2}$ mirror $\quad \theta = \tan^{-1}\left(\frac{V\Delta t}{D}\right)$]

[handwritten near figure c: $\Delta t = \frac{D + \sqrt{(V\Delta t)^2 + D^2}}{c}$]

FIGURE 39-3 (a) Observer A' and the mirror are in a spaceship at rest in frame S'. The time it takes for the light pulse to reach the mirror and return is measured by A' to be $2D/c$. (b) In frame S, the spaceship is moving to the right with speed v. If the speed of light is the same in both frames, the time it takes for the light to reach the mirror and return is longer than $2D/c$ in S because the distance traveled is greater than $2D$. (c) A right triangle for computing the time Δt in frame S.

The events happen at two different places x_1 and x_2 in frame S. During the time interval Δt (as measured in S) between the original flash and the return flash, observer A' and his spaceship have moved a horizontal distance $v\,\Delta t$. In Figure 39-3b, we can see that the path traveled by the light is longer in S than in S'. However, by Einstein's postulates, light travels with the same speed c in frame S as it does in frame S'. Because light travels farther in S at the same speed, it takes longer in S to reach the mirror and return. The time interval in S is thus longer than it is in S'. From the triangle in Figure 39-3c, we have

$$\left(\frac{c\,\Delta t}{2}\right)^2 = D^2 + \left(\frac{v\,\Delta t}{2}\right)^2$$

or

$$\Delta t = \frac{2D}{\sqrt{c^2 - v^2}} = \frac{2D}{c} \frac{1}{\sqrt{1 - (v^2/c^2)}}$$

Using $\Delta t' = 2D/c$, we obtain

$$\Delta t = \frac{\Delta t'}{\sqrt{1 - (v^2/c^2)}} = \gamma \Delta t'$$

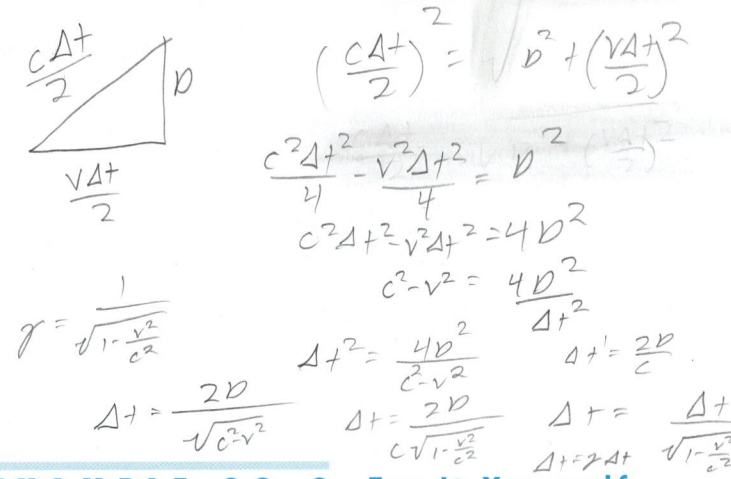

HOW LONG IS A ONE-HOUR NAP?   **EXAMPLE 39-2**   **Try It Yourself**

Astronauts in a spaceship traveling at $v = 0.6c$ relative to the earth sign off from space control, saying that they are going to nap for 1 h and then call back. How long does their nap last as measured on the earth?

PICTURE THE PROBLEM Because the astronauts go to sleep and wake up at the same place in their reference frame, the time interval for their nap of 1 h as measured by them is proper time. In the earth's reference frame, they move a considerable distance between these two events. The time interval measured in the earth's frame (using two clocks located at those events) is longer by the factor γ.

Cover the column to the right and try these on your own before looking at the answers.

Steps	Answers
1. Relate the time interval measured on the earth Δt to the proper time Δt_p.	$\Delta t = \gamma \Delta t_p$
2. Calculate γ for $v = 0.6c$.	$\gamma = 1.25$
3. Substitute to calculate the time of the nap in the earth's frame.	$\Delta t = \gamma \Delta t_p = \boxed{1.25\ \text{h}}$

EXERCISE If the spaceship is moving at $v = 0.8c$, how long would a 1 h nap last as measured on the earth? (*Answer* 1.67 h)

Length Contraction

A phenomenon closely related to time dilation is **length contraction.** The length of an object measured in the reference frame in which the object is at rest is called its **proper length** L_p. In a reference frame in which the object is moving, the measured length is shorter than its proper length. Consider a rod at rest in frame S' with one end at x_2' and the other end at x_1'. The length of the rod in this frame is its proper length $L_p = x_2' - x_1'$. Some care must be taken to find the length of the rod in frame S. In this frame, the rod is moving to the right with speed v, the speed of frame S'. The length of the rod in frame S is defined as $L = x_2 - x_1$, where x_2 is the position of one end at some time t_2, and x_1 is the position of the other end *at the same time* $t_1 = t_2$ as measured in frame S. Equation 39-11 is convenient to use to calculate $x_2 - x_1$ at some time t because it relates x and x' to t, whereas Equation 39-9 is not convenient because it relates x and x' to t':

$$x_2' = \gamma(x_2 - vt_2)$$

and

$$x_1' = \gamma(x_1 - vt_1)$$

Since $t_2 = t_1$, we obtain

$$x_2' - x_1' = \gamma(x_2 - x_1)$$

$$x_2 - x_1 = \frac{1}{\gamma}(x_2' - x_1') = (x_2' - x_1')\sqrt{1 - \frac{v^2}{c^2}}$$

or

$$L = \frac{1}{\gamma}L_\text{p} = L_\text{p}\sqrt{1 - \frac{v^2}{c^2}} \qquad\qquad 39\text{-}14$$

LENGTH CONTRACTION

Thus, the length of a rod is smaller when it is measured in a frame in which it is moving. Before Einstein's paper was published, Hendrik A. Lorentz and George F. FitzGerald tried to explain the null result of the Michelson–Morley experiment by assuming that distances in the direction of motion contracted by the amount given in Equation 39-14. This length contraction is now known as the **Lorentz–FitzGerald contraction.**

THE LENGTH OF A MOVING METERSTICK   **E X A M P L E   3 9 - 3**

A stick that has a proper length of 1 m moves in a direction along its length with speed v relative to you. The length of the stick as measured by you is 0.914 m. What is the speed v?

PICTURE THE PROBLEM Since both L and L_p are given, we can find v directly from Equation 39-14.

1. Equation 39-14 relates the lengths L and L_p and the speed v:

$$L = L_\text{p}\sqrt{1 - \frac{v^2}{c^2}}$$

2. Solve for v:

$$v = c\sqrt{1 - \frac{L^2}{L_\text{p}^2}} = c\sqrt{1 - \frac{(0.914\ \text{m})^2}{(1\ \text{m})^2}} = \boxed{0.406c}$$

An interesting example of time dilation or length contraction is afforded by the appearance of muons as secondary radiation from cosmic rays. Muons decay according to the statistical law of radioactivity:

$$N(t) = N_0 e^{-t/\tau} \qquad\qquad 39\text{-}15$$

where N_0 is the original number of muons at time $t = 0$, $N(t)$ is the number remaining at time t, and τ is the mean lifetime, which is approximately 2 μs for muons at rest. Since muons are created (from the decay of pions) high in the atmosphere, usually several thousand meters above sea level, few muons should reach sea level. A typical muon moving with speed $0.9978c$ would travel only about 600 m in 2 μs. However, the lifetime of the muon measured in the earth's reference frame is increased by the factor $1/\sqrt{1 - (v^2/c^2)}$, which is 15 for this particular speed. The mean lifetime measured in the earth's reference frame is therefore 30 μs, and a muon with speed $0.9978c$ travels approximately 9000 m in this time. From the muon's point of view, it lives only 2 μs, but the atmosphere is rushing past it with a speed of $0.9978c$. The distance of 9000 m in the earth's frame is thus contracted to only 600 m in the muon's frame, as indicated in Figure 39-4.

It is easy to distinguish experimentally between the classical and relativistic predictions of the observation of muons at sea level. Suppose that we observe 10^8 muons at an altitude of 9000 m in some time interval with a muon detector. How many would we expect to observe at sea level in the same time interval? According to the non-relativistic prediction, the time it takes for these muons to travel 9000 m is $(9000\text{ m})/(0.998c) \approx 30\ \mu\text{s}$, which is 15 lifetimes. Substituting $N_0 = 10^8$ and $t = 15\tau$ into Equation 39-15, we obtain

$$N = 10^8 e^{-15} = 30.6$$

We would thus expect all but about 31 of the original 100 million muons to decay before reaching sea level.

According to the relativistic prediction, the earth must travel only the contracted distance of 600 m in the rest frame of the muon. This takes only $2\ \mu\text{s} = 1\tau$. Therefore, the number of muons expected at sea level is

$$N = 10^8 e^{-1} = 3.68 \times 10^7$$

Thus, relativity predicts that we would observe 36.8 million muons in the same time interval. Experiments of this type have confirmed the relativistic predictions.

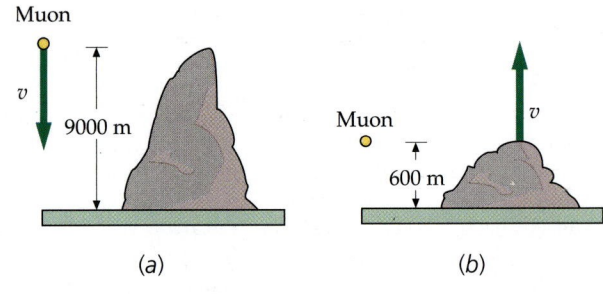

FIGURE 39-4 Although muons are created high above the earth and their mean lifetime is only about 2 μs when at rest, many appear at the earth's surface. (*a*) In the earth's reference frame, a typical muon moving at 0.998c has a mean lifetime of 30 μs and travels 9000 m in this time. (*b*) In the reference frame of the muon, the distance traveled by the earth is only 600 m in the muon's lifetime of 2 μs.

The Relativistic Doppler Effect

For light or other electromagnetic waves in a vacuum, a distinction between motion of source and receiver cannot be made. Therefore, the expressions we derived in Chapter 15 for the Doppler effect cannot be correct for light. The reason is that in that derivation, we assumed the time intervals in the reference frames of the source and receiver to be the same.

Consider a source moving toward a receiver with velocity v, relative to the receiver. If the source emits N electromagnetic waves in a time Δt_R (measured in the frame of the receiver), the first wave will travel a distance $c\,\Delta t_R$ and the source will travel a distance $v\,\Delta t_R$ measured in the frame of the receiver. The wavelength will be

$$\lambda' = \frac{c\,\Delta t_R - v\,\Delta t_R}{N}$$

The frequency f' observed by the receiver will therefore be

$$f' = \frac{c}{\lambda'} = \frac{c}{c - v}\frac{N}{\Delta t_R} = \frac{1}{1 - (v/c)}\frac{N}{\Delta t_R}$$

If the frequency of the source is f_0, it will emit $N = f_0\,\Delta t_S$ waves in the time Δt_S measured by the source. Then

$$f' = \frac{1}{1 - (v/c)}\frac{N}{\Delta t_R} = \frac{1}{1 - (v/c)}\frac{f_0\,\Delta t_S}{\Delta t_R} = \frac{f_0}{1 - (v/c)}\frac{\Delta t_S}{\Delta t_R}$$

Here Δt_S is the proper time interval (the first wave and the Nth wave are emitted at the same place in the source's reference frame). Times Δt_S and Δt_R are related by Equation 39-13 for time dilation:

$$\Delta t_R = \gamma\,\Delta t_S = \frac{\Delta t_S}{\sqrt{1 - (v^2/c^2)}}$$

Thus, when the source and the receiver are moving toward one another we obtain

$$f' = \frac{f_0}{1 - (v/c)} \frac{1}{\gamma} = \frac{\sqrt{1 - (v/c)^2}}{1 - (v/c)} f_0 = \sqrt{\frac{1 + (v/c)}{1 - (v/c)}} f_0, \quad \text{approaching} \quad 39\text{-}16a$$

This differs from our classical equation only in the time-dilation factor. It is left as a problem (Problem 27) for you to show that the same results are obtained if the calculations are done in the reference frame of the source.

When the source and the receiver are moving away from one another, the same analysis shows that the observed frequency is given by

$$f' = \frac{\sqrt{1 - (v/c)^2}}{1 + (v/c)} f_0 = \sqrt{\frac{1 - (v/c)}{1 + (v/c)}} f_0, \quad \text{receding} \qquad\qquad 39\text{-}16b$$

An application of the relativistic Doppler effect is the **redshift** observed in the light from distant galaxies. Because the galaxies are moving away from us, the light they emit is shifted toward the longer red wavelengths. The speed of the galaxies relative to us can be determined by measuring this shift.

CONVINCING THE JUDGE   **E X A M P L E 3 9 - 4**   **Put It in Context**

As part of a community volunteering option on your campus, you are spending the day shadowing two police officers. You have just had the excitement of pulling over a car that went through a red light. The driver claims that the red light looked green because the car was moving toward the stoplight, which shifted the wavelength of the observed light. You quickly do some calculations to see if the driver has a reasonable case or not.

PICTURE THE PROBLEM We can use the Doppler shift formula for approaching objects in Equation 39-16a. This will tell us the velocity, but we need to know the frequencies of the light. We can make good guesses for the wavelengths of red light and green light and use the definition of the speed of a wave $c = f\lambda$ to determine the frequencies.

1. The observer is approaching the light source, so we use the Doppler formula (Equation 39-16a) for approaching sources:

$$f' = \sqrt{\frac{1 + (v/c)}{1 - (v/c)}} f_0$$

2. Substitute c/λ for f, then simplify:

$$\frac{c}{\lambda'} = \sqrt{\frac{1 + (v/c)}{1 - (v/c)}} \frac{c}{\lambda_0}$$

$$\left(\frac{\lambda_0}{\lambda'}\right)^2 = \frac{1 + (v/c)}{1 - (v/c)}$$

3. Cross multiply and solve for v/c:

$$(\lambda_0)^2\left(1 - \frac{v}{c}\right) = (\lambda')^2\left(1 + \frac{v}{c}\right)$$

$$(\lambda_0)^2 - (\lambda')^2 = \left[(\lambda_0)^2 + (\lambda')^2\right]\left(\frac{v}{c}\right)$$

$$\frac{v}{c} = \frac{(\lambda_0)^2 - (\lambda')^2}{(\lambda_0)^2 + (\lambda')^2} = \frac{1 - (\lambda'/\lambda_0)^2}{1 + (\lambda'/\lambda_0)^2}$$

train. We can understand this by considering the motion of C' as seen in frame S (Figure 39-6). By the time the light from the front flash reaches C', C' has moved some distance toward the front flash and some distance away from the back flash. Thus, the light from the back flash has not yet reached C', as indicated in the figure. Observer C' must therefore conclude that the events are not simultaneous and that the front of the train was struck before the back. Furthermore, all observers in S' on the train will agree with C' when they have corrected for the time it takes the light to reach them.

Figure 39-7 shows the events of the lightning bolts as seen in the reference frame of the train (S'). In this frame the platform is moving, so the distance between the burns on the platform is contracted. The platform is shorter than it is in S, and, since the train is at rest, the train is longer than its contracted length in S. When the lightning bolt strikes the front of the train at A', the front of the train is at point A, and the back of the train has not yet reached point B. Later, when the lightning bolt strikes the back of the train at B', the back has reached point B on the platform.

The time discrepancy of two clocks that are synchronized in frame S as seen in frame S' can be found from the Lorentz transformation equations. Suppose we have clocks at points x_1 and x_2 that are synchronized in S. What are the times t_1 and t_2 on these clocks as observed from frame S' at a time t'_0? From Equation 39-12, we have

$$t'_0 = \gamma\left(t_1 - \frac{vx_1}{c^2}\right)$$

and

$$t'_0 = \gamma\left(t_2 - \frac{vx_2}{c^2}\right)$$

$$\gamma\left(t_1 - \frac{vx_1}{c^2}\right) = \gamma\left(t_2 - \frac{vx_2}{c^2}\right)$$

$$t_2 - t_1 = \frac{vx_2}{c^2} - \frac{vx_1}{c^2}$$

$$t_2 - t_1 = \frac{v}{c^2}(x_2 - x_1)$$

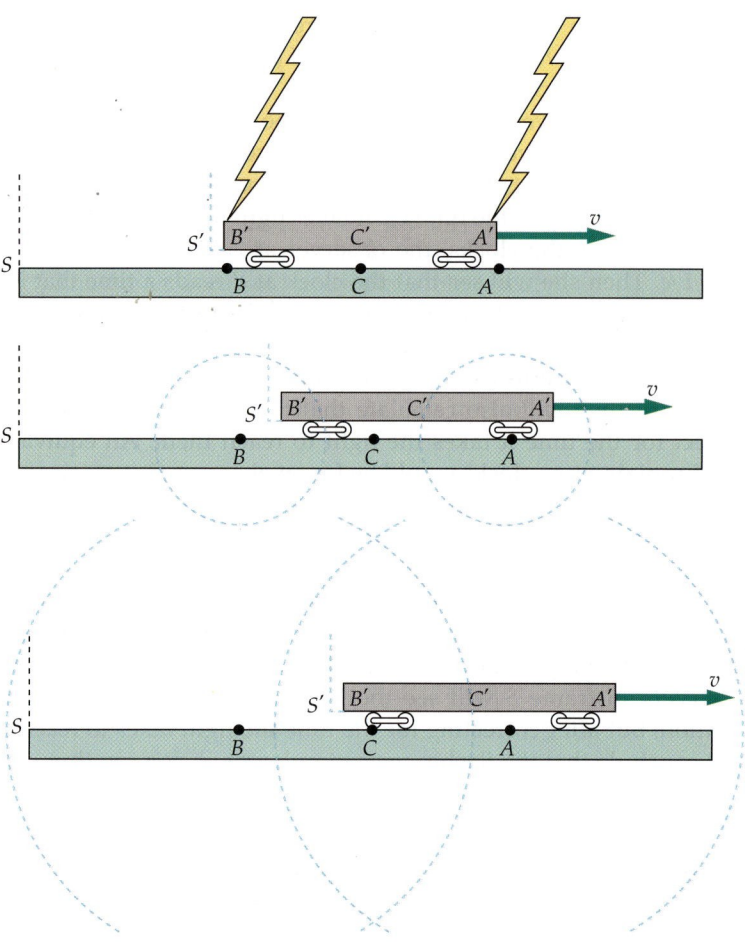

FIGURE 39-6 In frame S attached to the platform, the light from the lightning bolt at the front of the train reaches observer C', standing on the train at its midpoint, before the light from the bolt at the back of the train. Since C' is midway between the events (which occur at the front and rear of the train), these events are not simultaneous for him.

FIGURE 39-7 The lightning bolts of Figure 39-5 as seen in frame S' of the train. In this frame, the distance between A and B on the platform is less than $L_{\text{p,platform}}$, and the proper length of the train $L_{\text{p,train}}$ is longer than $L_{\text{p,platform}}$. The first lightning bolt strikes the front of the train when A' and A are coincident. The second bolt strikes the rear of the train when B' and B are coincident.

(a)

(b)

Let pl
frame S
neglect t
moving
respectiv
coasts in
celerates
Ulysses
then coas
decelerat
(and dec
coasting
$L_p = 8$ li
  It is ea
Accordin
for an ec
interval i
shorter b
Ulysses's

$\Delta t' =$

Since the
12 y for th
  From
contracte

$L' = \frac{L}{}$

At $v = 0.8$
  The re
twin age
and Hom
$3/5(6$ y)
trip? This
point of
happens
detail, we
in the stu
can get so
each othe
send a si
counting
1 per yea
Equation
for the cas

$f' = \frac{}{}$

When the
  Consid
it takes hi
frame), h
2 signals.

Then

$$t_2 - t_1 = \frac{v}{c^2}(x_2 - x_1)$$

Note that the chasing clock (at x_2) leads the other (at x_1) by an amount that is proportional to their proper separation $L_p = x_2 - x_1$.

> If two clocks are synchronized in the frame in which they are both at rest, in a frame in which they are moving along the line through both clocks, the chasing clock leads (shows a later time) by an amount
> *trails ?*
>
> $$\Delta t_S = L_p \frac{v}{c^2} \qquad\qquad 39\text{-}17$$
>
> where L_p is the proper distance between the clocks.

CHASING CLOCK SHOWS LATER TIME

A numerical example should help clarify time dilation, clock synchronization, and the internal consistency of these results.

SYNCHRONIZING CLOCKS **EXAMPLE 39-6**

An observer in a spaceship has a flashgun and a mirror, as shown in Figure 39-3. The distance from the gun to the mirror is 15 light-minutes (written $15c\cdot$min) and the spaceship, at rest in frame S', travels with speed $v = 0.8c$ relative to a very long space platform that is at rest in frame S. The platform has two synchronized clocks, one clock at the position x_1 of the spaceship when the observer explodes the flashgun, and the other clock at the position x_2 of the spaceship when the light returns to the gun from the mirror. Find the time intervals between the events (exploding the flashgun and receiving the return flash from the mirror) (a) in the frame of the spaceship and (b) in the frame of the platform. (c) Find the distance traveled by the spaceship and (d) the amount by which the clocks on the platform are out of synchronization according to observers on the spaceship.

(a) 1. In the spaceship, the light travels from the gun to the mirror and back, a total distance $D = 30\ c\cdot$min. The time required is D/c:
$$\Delta t' = \frac{D}{c} = \frac{30\ c\cdot\text{min}}{c} = 30\ \text{min}$$

2. Since these events happen at the same place in the spaceship, the time interval is proper time:
$$\Delta t_p = \boxed{30\ \text{min}}$$

(b) 1. In frame S, the time between the events is longer by the factor γ:
$$\Delta t = \gamma\,\Delta t_p = \gamma(30\ \text{min})$$

2. Calculate γ:
$$\gamma = \frac{1}{\sqrt{1 - (v^2/c^2)}} = \frac{1}{\sqrt{1 - (0.8)^2}} = \frac{1}{\sqrt{0.36}} = \frac{5}{3}$$

3. Use this value of γ to calculate the time between the events as observed in frame S:
$$\Delta t = \gamma\,\Delta t_p = \tfrac{5}{3}(30\ \text{min}) = \boxed{50\ \text{min}}$$

(c) In frame S, the distance traveled by the spaceship is $v\,\Delta t$:
$$x_2 - x_1 = v\,\Delta t = (0.8c)(50\ \text{min}) = \boxed{40\ c\cdot\text{min}}$$

(d) 1. The amc
of synch
between

2. The Part
clocks on

REMARKS Ol
running slow
whereas the tir

Figure 39-8
viewed from t
platform is tra
speed 0.8c. The
which coincide
flashgun is ex
point x_2, which
when the retur
the mirror. We
x_1 reads 12:00
light flash. The
synchronized i
the clock at x_2,
at x_1, leads by 3
12:32 to an obse
reads 12:50. Th
according to ob
trip that takes 3
factor 30/18 = !
Every obser
According to ol
interval in S' (3
by the factor 5/
time that is too
clocks in S are
second clock lea

The Twin Pa

Homer and Uly:
beyond the sola
are together aga
answer is that F
variations, has l
very few who di
cause of the seer
result in their ag
roles is noted. T
strong but incor:
case with some i
lations easy.

to receive 3 signals per year. In the 6 y it takes him to return he receives 18 signals, giving a total of 20 for the trip. He accordingly expects his twin to have aged 20 years.

We now consider the situation from Homer's point of view. He receives signals at the rate of $\frac{1}{3}$ signal per year not only for the 10 y it takes Ulysses to reach the planet but also for the time it takes for the last signal sent by Ulysses before he turns around to get back to the earth. (He cannot know that Ulysses has turned around until the signals begin reaching him with increased frequency.) Since the planet is 8 light-years away, there is an additional 8 y of receiving signals at the rate of $\frac{1}{3}$ signal per year. During the first 18 y, Homer receives 6 signals. In the final 2 y before Ulysses arrives, Homer receives 6 signals, or 3 per year. (The first signal sent after Ulysses turns around takes 8 y to reach the earth, whereas Ulysses, traveling at 0.8c, takes 10 y to return and therefore arrives just 2 y after Homer begins to receive signals at the faster rate.) Thus, Homer expects Ulysses to have aged 12 y. In this analysis, the asymmetry of the twins' roles is apparent. When they are together again, both twins agree that the one who has been accelerated will be younger than the one who stayed home.

The predictions of the special theory of relativity concerning the twin paradox have been tested using small particles that can be accelerated to such large speeds that γ is appreciably greater than 1. Unstable particles can be accelerated and trapped in circular orbits in a magnetic field, for example, and their lifetimes can then be compared with those of identical particles at rest. In all such experiments, the accelerated particles live longer on the average than the particles at rest, as predicted. These predictions have also been confirmed by the results of an experiment in which high-precision atomic clocks were flown around the world in commercial airplanes, but the analysis of this experiment is complicated due to the necessity of including gravitational effects treated in the general theory of relativity.

39-5 The Velocity Transformation

We can find how velocities transform from one reference frame to another by differentiating the Lorentz transformation equations. Suppose a particle has velocity $u'_x = dx'/dt'$ in frame S', which is moving to the right with speed v relative to frame S. The particle's velocity in frame S is

$$u_x = \frac{dx}{dt}$$

From the Lorentz transformation equations (Equation 39-9 and Equation 39-10), we have

$$dx = \gamma(dx' + v\,dt')$$

and

$$dt = \gamma\left(dt' + \frac{v\,dx'}{c^2}\right)$$

The velocity in S is thus

$$u_x = \frac{dx}{dt} = \frac{\gamma(dx' + v\,dt')}{\gamma\left(dt' + \dfrac{v\,dx'}{c^2}\right)} = \frac{\dfrac{dx'}{dt'} + v}{1 + \dfrac{v}{c^2}\dfrac{dx'}{dt'}} = \frac{u'_x + v}{1 + \dfrac{v\,u'_x}{c^2}}$$

If a particle has components of velocity along the y or z axes, we can use the same relation between dt and dt', with $dy = dy'$ and $dz = dz'$, to obtain

$$u_y = \frac{dy}{dt} = \frac{dy'}{\gamma\left(dt' + \frac{v\,dx'}{c^2}\right)} = \frac{\frac{dy'}{dt'}}{\gamma\left(1 + \frac{v}{c^2}\frac{dx'}{dt'}\right)} = \frac{u_y'}{\gamma\left(1 + \frac{vu_x'}{c^2}\right)}$$

and

$$u_z = \frac{u_z'}{\gamma\left(1 + \frac{vu_x'}{c^2}\right)}$$

The complete relativistic velocity transformation is

$$u_x = \frac{u_x' + v}{1 + \frac{vu_x'}{c^2}} \qquad \text{39-18}a$$

$$u_y = \frac{u_y'}{\gamma\left(1 + \frac{vu_x'}{c^2}\right)} \qquad \text{39-18}b$$

Why only dependent on x values?

$$u_z = \frac{u_z'}{\gamma\left(1 + \frac{vu_x'}{c^2}\right)} \qquad \text{39-18}c$$

RELATIVISTIC VELOCITY TRANSFORMATION

The inverse velocity transformation equations are

$$u_x' = \frac{u_x - v}{1 - \frac{vu_x}{c^2}} \qquad \text{39-19}a$$

$u_x' + v = u_x\left(1 + \frac{vu_x'}{c^2}\right)$

$u_x' = u_x + \frac{vu_xu_x'}{c^2} - v \implies u_x' = \frac{u_x - v}{\left(1 - \frac{vu_x}{c^2}\right)}$

$u_x'\left(1 - \frac{vu_x}{c^2}\right) = u_x - v$

$$u_y' = \frac{u_y}{\gamma\left(1 - \frac{vu_x}{c^2}\right)} \qquad \text{39-19}b$$

$$u_z' = \frac{u_z}{\gamma\left(1 - \frac{vu_x}{c^2}\right)} \qquad \text{39-19}c$$

These equations differ from the classical and intuitive result $u_x = u_x' + v$, $u_y = u_y'$, and $u_z = u_z'$ because the denominators in the equations are not equal to 1. When v and u_x' are small compared with the speed of light c, $\gamma \approx 1$ and $vu_x'/c^2 \ll 1$. Then the relativistic and classical expressions are the same.

RELATIVE VELOCITY AT NONRELATIVISTIC SPEEDS   **E X A M P L E   3 9 - 7**

A supersonic plane moves away from you along the x axis with speed 1000 m/s (about 3 times the speed of sound) relative to you. A second plane moves along the x axis away from you, and away from the first plane, at speed 500 m/s relative to the first plane. How fast is the second plane moving relative to you?

As in classical mechanics, we will define kinetic energy as the work done by the net force in accelerating a particle from rest to some final velocity u_f. Considering one dimension only, we have

$$K = \int_{u=0}^{u=u_f} F_{net}\, ds = \int_0^{u_f} \frac{dp}{dt}\, ds = \int_0^{u_f} u\, dp = \int_0^{u_f} u\, d\left(\frac{mu}{\sqrt{1-(u^2/c^2)}}\right) \qquad 39\text{-}21$$

where we have used $u = ds/dt$. It is left as a problem (Problem 37) for you to show that

$$d\left(\frac{mu}{\sqrt{1-(u^2/c^2)}}\right) = m\left(1 - \frac{u^2}{c^2}\right)^{-3/2} du$$

If we substitute this expression into the integrand in Equation 39-21, we obtain

$$K = \int_0^{u_f} u\, d\left(\frac{mu}{\sqrt{1-(u^2/c^2)}}\right) = \int_0^{u_f} m\left(1 - \frac{u^2}{c^2}\right)^{-3/2} u\, du$$

$$= mc^2\left(\frac{1}{\sqrt{1-(u_f^2/c^2)}} - 1\right)$$

or

$$K = \frac{mc^2}{\sqrt{1-(u^2/c^2)}} - mc^2 \qquad\qquad 39\text{-}22$$

RELATIVISTIC KINETIC ENERGY

(In this expression the final speed u_f is arbitrary, so the subscript f is not needed.)

The expression for kinetic energy consists of two terms. The first term depends on the speed of the particle. The second, mc^2, is independent of the speed. The quantity mc^2 is called the **rest energy** E_0 of the particle. The rest energy is the product of the mass and c^2:

$$E_0 = mc^2 \qquad\qquad 39\text{-}23$$

REST ENERGY

The total **relativistic energy** E is then defined to be the sum of the kinetic energy and the rest energy:

$$E = K + mc^2 = \frac{mc^2}{\sqrt{1-(u^2/c^2)}} \qquad\qquad 39\text{-}24$$

RELATIVISTIC ENERGY

Thus, the work done by an unbalanced force increases the energy from the rest energy mc^2 to the final energy $mc^2/\sqrt{1-(u^2/c^2)} = m_{rel}c^2$, where $m_{rel} = m/\sqrt{1-(u^2/c^2)}$ is the relativistic mass. We can obtain a useful expression for the velocity of a particle by multiplying Equation 39-20 for the relativistic momentum by c^2 and comparing the result with Equation 39-24 for the relativistic energy. We have

$$pc^2 = \frac{mc^2 u}{\sqrt{1-(u^2/c^2)}} = Eu$$

or

$$\frac{u}{c} = \frac{pc}{E} \qquad\qquad\qquad 39\text{-}25$$

Energies in atomic and nuclear physics are usually expressed in units of electron volts (eV) or mega-electron volts (MeV):

$$1 \text{ eV} = 1.602 \times 10^{-19} \text{ J}$$

A convenient unit for the masses of atomic particles is eV/c^2 or MeV/c^2, which is the rest energy of the particle divided by c^2. The rest energies of some elementary particles and light nuclei are given in Table 39-1.

TABLE 39-1

Rest Energies of Some Elementary Particles and Light Nuclei

Particle	Symbol	Rest energy, MeV
Photon	γ	0
Electron (positron)	e or e^- (e^+)	0.5110
Muon	$\mu^\pm$	105.7
Pion	π^0	135
	$\pi^\pm$	139.6
Proton	p	938.280
Neutron	n	939.573
Deuteron	^{2}H or d	1875.628
Triton	^{3}H or t	2808.944
Helium-3	^{3}He	2808.41
Alpha particle	^{4}He or α	3727.409

TOTAL ENERGY, KINETIC ENERGY, AND MOMENTUM **EXAMPLE 39-11**

An electron (rest energy 0.511 MeV) moves with speed $u = 0.8c$. Find (a) its total energy, (b) its kinetic energy, and (c) the magnitude of its momentum.

(a) The total energy is given by Equation 39-24:

$$E = \frac{mc^2}{\sqrt{1 - (u^2/c^2)}} = \frac{0.511 \text{ MeV}}{\sqrt{1 - 0.64}} = \frac{0.511 \text{ MeV}}{0.6} = \boxed{0.852 \text{ MeV}}$$

(b) The kinetic energy is the total energy minus the rest energy:

$$K = E - mc^2 = 0.852 \text{ MeV} - 0.511 \text{ MeV} = \boxed{0.341 \text{MeV}}$$

(c) The magnitude of the momentum is found from Equation 39-20. We can simplify by multiplying both numerator and denominator by c^2 and using the Part (a) result:

$$p = \frac{mu}{\sqrt{1 - (u^2/c^2)}}$$

$$= \frac{mc^2}{\sqrt{1 - (u^2/c^2)}} \frac{u}{c^2} = (0.852 \text{ MeV}) \frac{0.8c}{c^2} = \boxed{0.681 \text{ MeV}/c}$$

REMARKS The technique used to solve Part (c) (multiplying numerator and denominator by c^2) is equivalent to using Equation 39-25.

The expression for kinetic energy given by Equation 39-22 does not look much like the classical expression $\frac{1}{2}mu^2$. However, when u is much less than c, we can approximate $1/\sqrt{1 - (u^2/c^2)}$ using the binomial expansion

$$(1 + x)^n = 1 + nx + n(n - 1)\frac{x^2}{2} + \cdots \approx 1 + nx \qquad\qquad 39\text{-}26$$

Then

$$\frac{1}{\sqrt{1 - (u^2/c^2)}} = \left(1 - \frac{u^2}{c^2}\right)^{-1/2} \approx 1 + \frac{1}{2}\frac{u^2}{c^2}$$

From this result, when u is much less than c, the expression for relativistic kinetic energy becomes

$$K = mc^2\left[\frac{1}{\sqrt{1 - (u^2/c^2)}} - 1\right] \approx mc^2\left(1 + \frac{1}{2}\frac{u^2}{c^2} - 1\right) = \frac{1}{2}mu^2$$

Thus, at low speeds, the relativistic expression is the same as the classical expression.

We note from Equation 39-24 that as the speed u approaches the speed of light c, the energy of the particle becomes very large because $1/\sqrt{1 - (u^2/c^2)}$ becomes very large. At $u = c$, the energy becomes infinite. For u greater than c, $\sqrt{1 - (u^2/c^2)}$ is the square root of a negative number and is therefore imaginary. A simple interpretation of the result that it takes an infinite amount of energy to accelerate a particle to the speed of light is that no particle that is ever at rest in any inertial reference frame can travel as fast or faster than the speed of light c. As we noted in Example 39-8, if the speed of a particle is less than c in one reference frame, it is less than c in all other reference frames moving relative to that frame at speeds less than c.

In practical applications, the momentum or energy of a particle is often known rather than the speed. Equation 39-20 for the relativistic momentum and Equation 39-24 for the relativistic energy can be combined to eliminate the speed u. The result is

$$E^2 = p^2c^2 + (mc^2)^2 \qquad\qquad 39\text{-}27$$

RELATION FOR TOTAL ENERGY, MOMENTUM, AND REST ENERGY

This useful equation can be conveniently remembered from the right triangle shown in Figure 39-12. If the energy of a particle is much greater than its rest energy mc^2, the second term on the right side of Equation 39-27 can be neglected, giving the useful approximation

$$E \approx pc, \qquad \text{for } E \gg mc^2 \qquad\qquad 39\text{-}28$$

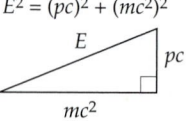

$E^2 = (pc)^2 + (mc^2)^2$

FIGURE 39-12 Right triangle to remember Equation 39-27.

Equation 39-28 is an exact relation between energy and momentum for particles with no mass, such as photons.

EXERCISE A proton (mass 938 MeV/c^2) has a total energy of 1400 MeV. Find (a) $1/\sqrt{1 - (u^2/c^2)}$, (b) the momentum of the proton, and (c) the speed u of the proton. (*Answer* (a) 1.49, (b) $p = 1.04 \times 10^3$ MeV/c, and (c) $u = 0.74c$)

Mass and Energy

Einstein considered Equation 39-23 relating the energy of a particle to its mass to be the most significant result of the theory of relativity. Energy and inertia, which were formerly two distinct concepts, are related through this famous equation. As discussed in Chapter 7, the conversion of rest energy to kinetic energy with a corresponding decrease in mass is a common occurrence in radioactive decay and nuclear reactions, including nuclear fission and nuclear fusion. We illustrated this in Section 7-3 with the deuteron, whose mass is $2.22 \text{ MeV}/c^2$ less than the mass of its parts, a proton and a neutron. When a neutron and a proton combine to form a deuteron, 2.22 MeV of energy is released. The breaking up of a deuteron into a neutron and a proton requires 2.22 MeV of energy input. The proton and the neutron are thus bound together in a deuteron by a binding energy of 2.22 MeV. Any stable composite particle, such as a deuteron or a helium nucleus (2 neutrons plus 2 protons), that is made up of other particles has a mass and rest energy that are less than the sum of the masses and rest energies of its parts. The difference in rest energy is the binding energy of the composite particle. The binding energies of atoms and molecules are of the order of a few electron volts, which leads to a negligible difference in mass between the composite particle and its parts. The binding energies of nuclei are of the order of several MeV, which leads to a noticeable difference in mass. Some very heavy nuclei, such as radium, are radioactive and decay into a lighter nucleus plus an alpha particle. In this case, the original nucleus has a rest energy greater than that of the decay particles. The excess energy appears as the kinetic energy of the decay products.

To further illustrate the interrelation of mass and energy, we consider a perfectly inelastic collision of two particles. Classically, kinetic energy is lost in such a collision. Relativistically, this loss in kinetic energy shows up as an increase in rest energy of the system; that is, the total energy of the system is conserved. Consider a particle of mass m_1 moving with initial speed u_1 that collides with a particle of mass m_2 moving with initial speed u_2. The particles collide and stick together, forming a particle of mass M that moves with speed u_f, as shown in Figure 39-13. The initial total energy of particle 1 is

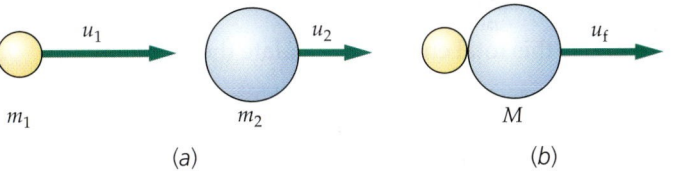

(a) (b)

FIGURE 39-13 A perfectly inelastic collision between two particles. One particle of mass m_1 collides with another particle of mass m_2. After the collision, the particles stick together, forming a composite particle of mass M that moves with speed u_f so that relativistic momentum is conserved. Kinetic energy is lost in this process. If we assume that the total energy is conserved, the loss in kinetic energy must equal c^2 times the increase in the mass of the system.

$$E_1 = K_1 + m_1 c^2$$

where K_1 is its initial kinetic energy. Similarly the initial total energy of particle 2 is

$$E_2 = K_2 + m_2 c^2$$

The total initial energy of the system is

$$E_i = E_1 + E_2 = K_1 + m_1 c^2 + K_2 + m_2 c^2 = K_i + M_i c^2$$

where $K_i = K_1 + K_2$ and $M_i = m_1 + m_2$ are the initial kinetic energy and initial mass of the system. The final total energy of the system is

$$E_f = K_f + M_f c^2$$

If we set the final total energy equal to the initial total energy, we obtain

$$K_f + M_f c^2 = K_i + M_i c^2$$

Rearranging gives $K_f - K_i = -(M_f - M_i)c^2$, which can be expressed

$$\Delta K + (\Delta M)c^2 = 0 \qquad\qquad 39\text{-}29$$

where $\Delta M = M_f - M_i$ is the change in mass of the system.

3. We use the value for p to solve for $E_{R,f}$:

$$E^2_{R,f} = p^2c^2 + (m_R c^2)^2 = (5.00 \times 10^2 \text{ kg})^2 c^4 + (10^6 \text{ kg})^2 c^4 = (1.00 \times 10^{12} \text{ kg}^2)c^4$$

so

$$E_{R,f} = (1.00 \times 10^6 \text{ kg})c^2$$

4. Using our Part (*b*), step 1 result, we solve for u_R:

$$u_R = \frac{pc^2}{E_{R,f}} = \frac{(5.00 \times 10^2 \text{ kg})c^3}{(1.00 \times 10^6 \text{ kg})c^2} = \boxed{5.00 \times 10^{-4}c = 1.50 \times 10^5 \text{ m/s}}$$

(*c*) Equate the magnitude of the classical expressions for the momentum of the rocket and burned fuel and solve for u_R:

$$m_R u_R = m_F u_F$$

$$u_R = \frac{m_F}{m_R}u_F = \frac{10^3 \text{ kg}}{10^6 \text{ kg}}0.5c = \boxed{1.5 \times 10^5 \text{ m/s}}$$

REMARK If carried out to five figures, the relativistic calculation gives $u_R = 4.9994 \times 10^4\,c$ for the final speed of the rocket. However, the classical calculation gives $u_R = 5.0000 \times 10^4\,c$. These two values differ by less than one part in 8000.

EXERCISE If the matter being ejected were a 1×10^3-kg rigid block launched by a spring with one end attached to the rocket, would the rest mass of the block change or would the rest mass of the spring change? (*Answer* Only the rest mass of the spring would change.)

39-8 General Relativity

The generalization of the theory of relativity to noninertial reference frames by Einstein in 1916 is known as the general theory of relativity. It is much more difficult mathematically than the special theory of relativity, and there are fewer situations in which it can be tested. Nevertheless, its importance calls for a brief qualitative discussion.

The basis of the general theory of relativity is the **principle of equivalence:**

A homogeneous gravitational field is completely equivalent to a uniformly accelerated reference frame.

PRINCIPLE OF EQUIVALENCE

This principle arises in Newtonian mechanics because of the apparent identity of gravitational mass and inertial mass. In a uniform gravitational field, all objects fall with the same acceleration $\vec{g}$ independent of their mass because the gravitational force is proportional to the (gravitational) mass, whereas the acceleration varies inversely with the (inertial) mass. Consider a compartment in space undergoing a uniform acceleration $\vec{a}$, as shown in Figure 39-14*a*. No mechanics experiment can be performed *inside* the compartment that will distinguish whether the compartment is actually accelerating in space or is at rest (or is moving with uniform velocity) in the presence of a uniform gravitational field $\vec{g} = -\vec{a}$, as shown in Figure 39-14*b*. If objects are dropped in the compartment, they will fall to the floor with an acceleration $\vec{g} = -\vec{a}$. If people stand on a spring scale, it will read their weight of magnitude ma.

Einstein assumed that the principle of equivalence applies to all physics and not just to mechanics. In effect, he assumed that there is no experiment of any kind that can distinguish uniformly accelerated motion from the presence of a gravitational field.

One consequence of the principle of equivalence—the deflection of a light beam in a gravitational field—was one of the first to be tested experimentally. In a region

(*a*)

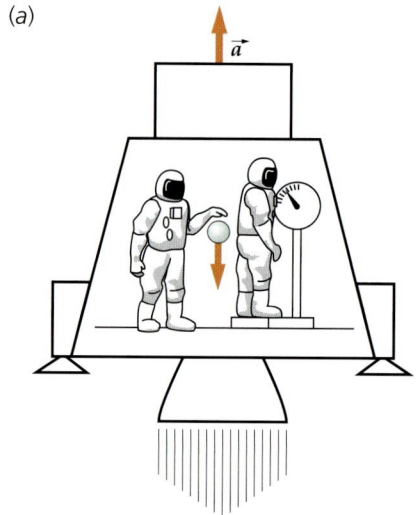

(*b*)

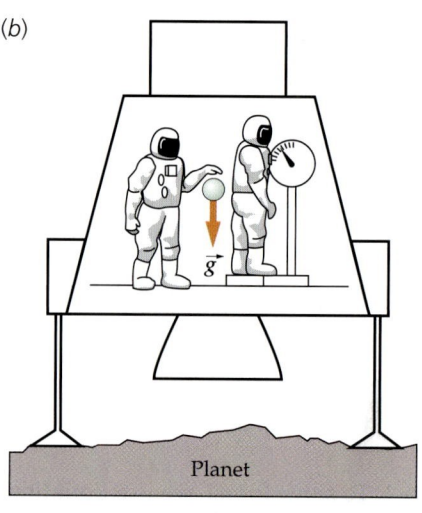

Planet

FIGURE 39-14 The results of experiments in a uniformly accelerated reference frame (*a*) cannot be distinguished from those in a uniform gravitational field (*b*) if the acceleration $\vec{a}$ and the gravitational field $\vec{g}$ have the same magnitude.

with no gravitational field, a light beam will travel in a straight line at speed c. The principle of equivalence tells us that a region with no gravitational field exists only in a compartment that is in free fall. Figure 39-15 shows a beam of light entering a compartment that is accelerating relative to a nearby reference frame in free fall. Successive positions of the compartment at equal time intervals are shown in Figure 39-15a. Because the compartment is accelerating, the distance it moves in each time interval increases with time. The path of the beam of light as observed from inside the compartment is therefore a parabola, as shown in Figure 39-15b. But according to the principle of equivalence, there is no way to distinguish between an accelerating compartment and one moving with uniform velocity in a uniform gravitational field. We conclude, therefore, that a beam of light will accelerate in a gravitational field, just like objects that have mass. For example, near the surface of the earth, light will fall with an acceleration of 9.81 m/s^2. This is difficult to observe because of the enormous speed of light. In a distance of 3000 km, which takes light about 0.01 s to traverse, a beam of light should fall approximately 0.5 mm. Einstein pointed out that the deflection of a light beam in a gravitational field might be observed when light from a distant star passes close to the sun, as illustrated in Figure 39-16. Because of the brightness of the sun, this cannot ordinarily be seen. Such a deflection was first observed in 1919 during an eclipse of the sun. This well-publicized observation brought instant worldwide fame to Einstein.

A second prediction from Einstein's theory of general relativity, which we will not discuss in detail, is the excess precession of the perihelion of the orbit of Mercury of about 0.01° per century. This effect had been known and unexplained for some time, so, in a sense, explaining it constituted an immediate success of the theory.

A third prediction of general relativity concerns the change in time intervals and frequencies of light in a gravitational field. In Chapter 11, we found that the gravitational potential energy between two masses M and m a distance r apart is

$$U = -\frac{GMm}{r}$$

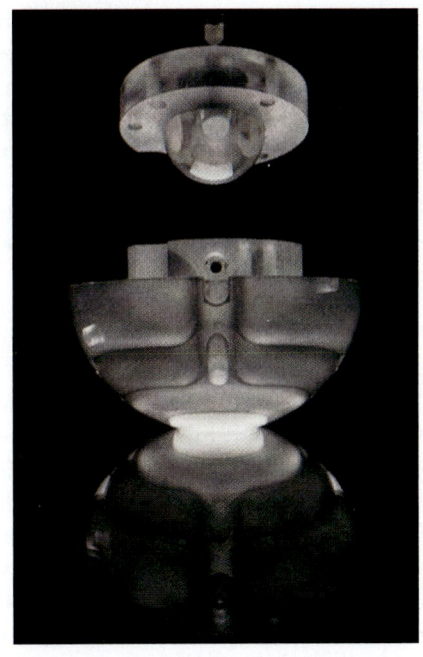

The quartz sphere in the top part of the container is probably the world's most perfectly round object. It is designed to spin as a gyroscope in a satellite orbiting the earth. General relativity predicts that the rotation of the earth will cause the axis of rotation of the gyroscope to precess in a circle at a rate of approximately 1 revolution in 100,000 years.

where G is the universal gravitational constant, and the point of zero potential energy has been chosen to be when the separation of the masses is infinite. The potential energy per unit mass near a mass M is called the *gravitational potential* ϕ:

$$\phi = -\frac{GM}{r} \qquad \text{39-30}$$

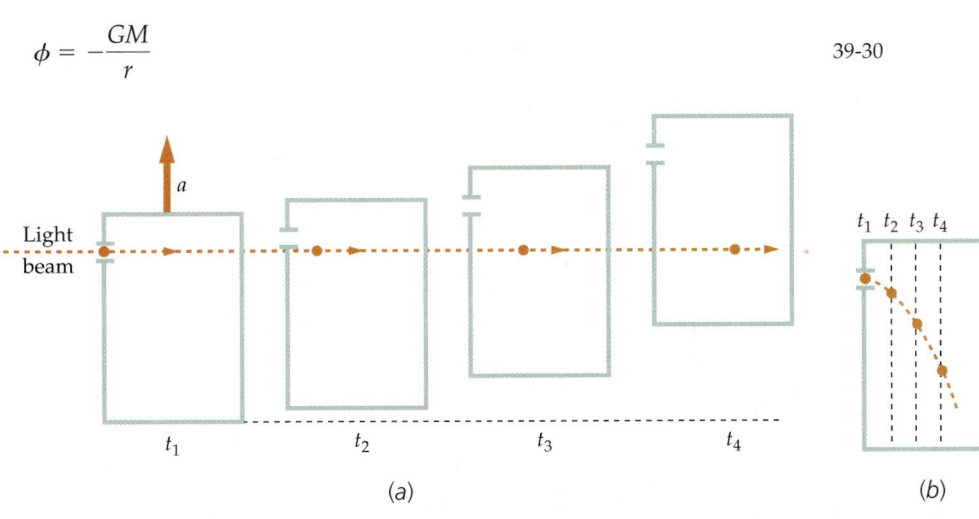

(a)

(b)

FIGURE 39-15 (a) A light beam moving in a straight line through a compartment that is undergoing uniform acceleration relative to a nearby reference frame in free fall. The position of the beam is shown at equally spaced times t_1, t_2, t_3, and t_4. (b) In the reference frame of the compartment, the light travels in a parabolic path as a ball would if it were projected horizontally. The vertical displacements are greatly exaggerated in Figure 39-15a and Figure 39-15b for emphasis.

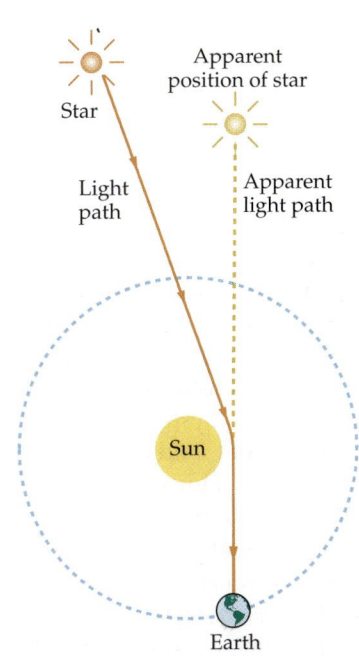

FIGURE 39-16 The deflection (greatly exaggerated) of a beam of light due to the gravitational attraction of the sun.

According to the general theory of relativity, clocks run more slowly in regions of lower gravitational potential. (Since the gravitational potential is negative, as can be seen from Equation 39-30, the nearer the mass the more negative, and therefore the lower the gravitational potential.) If Δt_1 is a time interval between two events measured by a clock where the gravitational potential is ϕ_1 and Δt_2 is the interval between the same events as measured by a clock where the gravitational potential is ϕ_2, general relativity predicts that the fractional difference between these times will be approximately[†]

$$\frac{\Delta t_2 - \Delta t_1}{\Delta t} = \frac{1}{c^2}(\phi_2 - \phi_1) \qquad \text{39-31}$$

A clock in a region of low gravitational potential will therefore run slower than a clock in a region of high potential. Since a vibrating atom can be considered to be a clock, the frequency of vibration of an atom in a region of low potential, such as near the sun, will be lower than the frequency of vibration of the same atom on the earth. This shift toward a lower frequency, and therefore a longer wavelength, is called the **gravitational redshift.**

As our final example of the predictions of general relativity, we mention **black holes,** which were first predicted by J. Robert Oppenheimer and Hartland Snyder in 1939. According to the general theory of relativity, if the density of an object such as a star is great enough, its gravitational attraction will be so great that once inside a critical radius, nothing can escape, not even light or other electromagnetic radiation. (The effect of a black hole on objects outside the critical radius is the same as that of any other mass.) A remarkable property of such an object is that nothing that happens inside it can be communicated to the outside. As sometimes occurs in physics, a simple but incorrect calculation gives the correct results for the relation between the mass and the critical radius of a black hole. In Newtonian mechanics, the speed needed for a particle to escape from the surface of a planet or a star of mass M and radius R is given by Equation 11-21:

$$v_e = \sqrt{\frac{2GM}{R}}$$

If we set the escape speed equal to the speed of light and solve for the radius, we obtain the critical radius R_S, called the **Schwarzschild radius:**

$$R_S = \frac{2GM}{c^2} \qquad \text{39-32}$$

For an object with a mass equal to five times that of our sun (theoretically the minimum mass for a black hole) to be a black hole, its radius would have to be approximately 15 km. Since no radiation is emitted from a black hole and its radius is expected to be small, the detection of a black hole is not easy. The best chance of detection occurs if a black hole is a close companion to a normal star in a binary star system. Then both stars revolve around their center of mass and the gravitational field of the black hole will pull gas from the normal star into the black hole. However, to conserve angular momentum, the gas does not go straight into the black hole. Instead, the gas orbits around the black hole in a disk, called an accretion disk, while slowly being pulled closer to the black hole. The gas in this disk emits X rays because the temperature of the gas being pulled inward reaches several millions of kelvins. The mass of a black-hole candidate can often be estimated. An estimated mass of at least five solar masses, along with the emission of X rays, establishes a strong inference that the candidate is, in fact, a black hole. In addition to the black holes just described, there are supermassive black holes that exist at the centers of galaxies. At the center of the Milky Way is a supermassive black hole with a mass of about two million solar masses.

This extremely accurate hydrogen maser clock was launched in a satellite in 1976, and its time was compared to that of an identical clock on the earth. In accordance with the prediction of general relativity, the clock on the earth, where the gravitational potential was lower, *lost* approximately 4.3×10^{-10} s each second compared with the clock orbiting the earth at an altitude of approximately 10,000 km.

[†] Since this shift is usually very small, it does not matter by which interval we divide on the left side of the equation.

Topic	Relevant Equations and Remarks	
1. Einstein's Postulates	The special theory of relativity is based on two postulates of Albert Einstein. All of the results of special relativity can be derived from these postulates.	
	Postulate 1: Absolute uniform motion cannot be detected.	
	Postulate 2: The speed of light is independent of the motion of the source.	
	An important implication of these postulates is	
	Postulate 2 (alternate): Every observer measures the same value c for the speed of light.	
2. The Lorentz Transformation	$x = \gamma(x' + vt'), \quad y = y', \quad z = z'$	39-9
	$t = \gamma\left(t' + \dfrac{vx'}{c^2}\right)$	39-10
	$\gamma = \dfrac{1}{\sqrt{1 - (v^2/c^2)}}$	39-7
Inverse transformation	$x' = \gamma(x - vt), \quad y' = y, \quad z' = z$	39-11
	$t' = \gamma\left(t - \dfrac{vx}{c^2}\right)$	39-12
3. Time Dilation	The time interval measured between two events that occur at the same point in space in some reference frame is called the proper time t_{p}. In another reference frame in which the events occur at different places, the time interval between the events is longer by the factor γ.	
	$\Delta t = \gamma \, \Delta t_{\mathrm{p}}$	39-13
4. Length Contraction	The length of an object measured in the reference frame in which the object is at rest is called its proper length L_{p}. When measured in another reference frame, the length of the object is	
	$L = \dfrac{L_{\mathrm{p}}}{\gamma}$	39-14
5. The Relativistic Doppler Effect	$f' = \dfrac{\sqrt{1 - (v^2/c^2)}}{1 - (v/c)} f_0, \quad$ approaching	39-16a
	$f' = \dfrac{\sqrt{1 - (v^2/c^2)}}{1 + (v/c)} f_0, \quad$ receding	39-16b
6. Clock Synchronization and Simultaneity	Two events that are simultaneous in one reference frame typically are not simultaneous in another frame that is moving relative to the first. If two clocks are synchronized in the frame in which they are at rest, they will be out of synchronization in another frame. In the frame in which they are moving, the chasing clock leads by an amount	
	$\Delta t_{\mathrm{S}} = L_{\mathrm{p}} \dfrac{v}{c^2}$	39-17
	where L_{p} is the proper distance between the clocks.	
7. The Velocity Transformation	$u_x = \dfrac{u'_x + v}{1 + (vu'_x/c^2)}$	39-18a

$$u_y = \frac{u_y'}{\gamma[1 + (vu_x'/c^2)]}$$ 39-18b

$$u_z = \frac{u_z'}{\gamma[1 + (vu_x'/c^2)]}$$ 39-18c

Inverse velocity transformation $$u_x' = \frac{u_x - v}{1 - (vu_x/c^2)}$$ 39-19a

$$u_y' = \frac{u_y}{\gamma[1 - (vu_x/c^2)]}$$ 39-19b

$$u_z' = \frac{u_z}{\gamma[1 - (vu_x/c^2)]}$$ 39-19c

8. Relativistic Momentum $$\vec{p} = \frac{m\vec{u}}{\sqrt{1 - (u^2/c^2)}}$$ 39-20

where m is the mass of the particle.

9. Relativistic Energy

Kinetic energy $$K = \frac{mc^2}{\sqrt{1 - (u^2/c^2)}} - mc^2 = \frac{mc^2}{\sqrt{1 - (u^2/c^2)}} - E_0$$ 39-22

Rest energy $$E_0 = mc^2$$ 39-23

Total energy $$E = K + E_0 = \frac{mc^2}{\sqrt{1 - (u^2/c^2)}}$$ 39-24

10. Useful Formulas for Speed, Energy, and Momentum $$\frac{u}{c} = \frac{pc}{E}$$ 39-25

$$E^2 = p^2c^2 + (mc^2)^2$$ 39-27

$$E \approx pc, \quad \text{for } E \gg mc^2$$ 39-28

PROBLEMS

- Single-concept, single-step, relatively easy
- •• Intermediate-level, may require synthesis of concepts
- ••• Challenging
- SSM Solution is in the *Student Solutions Manual*
- iSOLVE Problems available on iSOLVE online homework service
- iSOLVE✓ These "Checkpoint" online homework service problems ask students additional questions about their confidence level, and how they arrived at their answer.

In a few problems, you are given more data than you actually need; in a few other problems, you are required to supply data from your general knowledge, outside sources, or informed estimates.

Conceptual Problems

1 • SSM The approximate total energy of a particle of mass m moving at speed $u \ll c$ is (a) $mc^2 + \frac{1}{2}mu^2$. (b) $\frac{1}{2}mu^2$. (c) cmu. (d) mc^2. (e) $\frac{1}{2}cmu$.

2 • SSM A set of twins work in an office building. One twin works on the top floor and the other twin works in the basement. Considering general relativity, which twin will age more quickly? (a) They will age at the same rate. (b) The twin who works on the top floor will age more quickly. (c) The twin who works in the basement will age more quickly. (d) It depends on the speed of the office building. (e) None of these is correct.

3 • True or false:

(a) The speed of light is the same in all reference frames.

(b) Proper time is the shortest time interval between two events.

(c) Absolute motion can be determined by means of length contraction.

(d) The light-year is a unit of distance.

(e) Simultaneous events must occur at the same place.

(f) If two events are not simultaneous in one frame, they cannot be simultaneous in any other frame.

(g) If two particles are tightly bound together by strong attractive forces, the mass of the system is less than the sum of the masses of the individual particles when separated.

4 • An observer sees a system consisting of a mass oscillating on the end of a spring moving past at a speed u and notes that the period of the system is T. Another observer, who is moving with the mass–spring system, also measures its period. The second observer will find a period that is (a) equal to T. (b) less than T. (c) greater than T. (d) either (a) or (b) depending on whether the system was approaching or receding from the first observer. (e) There is not sufficient information to answer the question.

5 • The Lorentz transformation for y and z is the same as the classical result: $y = y'$ and $z = z'$. Yet the relativistic velocity transformation does not give the classical result $u_y = u_y'$ and $u_z = u_z'$. Explain.

Estimation and Approximation

6 •• The sun radiates energy at the rate of approximately 4×10^{26} W. Assume that this energy is produced by a reaction whose net result is the fusion of 4 H nuclei to form 1 He nucleus, with the release of 25 MeV for each He nucleus formed. Calculate the sun's loss of mass per day.

7 •• **SSM** The most distant galaxies that can be seen by the Hubble telescope are moving away from us with a redshift parameter of about $z = 5$. (See Problem 30 for a definition of z.) (a) What is the velocity of these galaxies relative to us (expressed as a fraction of the speed of light)? (b) *Hubble's law* states that the recession velocity is given by the expression $v = Hx$, where v is the velocity of recession, x is the distance, and H is the Hubble constant, $H = 75$ km/s/Mpc. (1 pc = 3.26 c·y.) Estimate the distance of such a galaxy using the information given.

Time Dilation and Length Contraction

8 • The proper mean lifetime of a muon is 2 μs. Muons in a beam are traveling through a laboratory at 0.95c. (a) What is their mean lifetime as measured in the laboratory? (b) How far do they travel, on average, before they decay?

9 •• In the Stanford linear collider, small bundles of electrons and positrons are fired at each other. In the laboratory's frame of reference, each bundle is approximately 1 cm long and 10 μm in diameter. In the collision region, each particle has an energy of 50 GeV, and the electrons and the positrons are moving in opposite directions. (a) How long and how wide is each bundle in its own reference frame? (b) What must be the minimum proper length of the accelerator for a bundle to have both its ends simultaneously in the accelerator in its own reference frame? (The ac-

tual length of the accelerator is less than 1000 m.) (c) What is the length of a positron bundle in the reference frame of the electron bundle?

10 •• **SSM** Unobtainium (Un) is an unstable particle that decays into normalium (Nr) and standardium (St) particles. (a) An accelerator produces a beam of Un that travels to a detector located 100 m away from the accelerator. The particles travel with a velocity of $v = 0.866c$. How long do the particles take (in the laboratory frame) to get to the detector? (b) By the time the particles get to the detector, half of the particles have decayed. What is the half-life of Un? (*Note:* Half-life as it would be measured in a frame moving with the particles.) (c) A new detector is going to be used, which is located 1000 m away from the accelerator. How fast should the particles be moving if half of the particles are to make it to the new detector?

11 •• Star A and Star B are at rest relative to the earth. Star A is 27 c·y from earth, and Star B is located beyond (behind) Star A as viewed from earth. (a) A spaceship is making a trip from earth to Star A at a speed such that the trip from earth to Star A takes 12 y according to clocks on the spaceship. At what speed, relative to earth, must the ship travel? (Assume that the times for acceleration are very short compared to the overall trip time.) (b) Upon reaching Star A, the ship speeds up and departs for Star B at a speed such that the gamma factor, γ, is twice that of Part (a). The trip from Star A to Star B takes 5 y (ship's time). How far, in c·y, is Star B from Star A in the rest frame of the earth and the two stars? (c) Upon reaching Star B, the ship departs for earth at the same speed as in Part (b). It takes it 10 y (ship's time) to return to earth. If you were born on earth the day the ship left earth (and you remain on earth), how old are you on the day the ship returns to earth?

12 • A spaceship travels to a star 35 c·y away at a speed of 2.7×10^8 m/s. How long does the spaceship take to get to the star (a) as measured on the earth and (b) as measured by a passenger on the spaceship?

13 • Use the binomial expansion equation

$$(1 + x)^n = 1 + nx + \frac{n(n-1)}{2}x^2 + \ldots \approx 1 + nx, \quad \text{for } x \ll 1$$

to derive the following results for the case when v is much less than c.

(a) $\gamma \approx 1 + \frac{1}{2}\frac{v^2}{c^2}$

(b) $\frac{1}{\gamma} \approx 1 - \frac{1}{2}\frac{v^2}{c^2}$

(c) $\gamma - 1 \approx 1 - \frac{1}{\gamma} \approx \frac{1}{2}\frac{v^2}{c^2}$

14 •• A clock on Spaceship A measures the time interval between two events, both of which occur at the location of the clock. You are on Spaceship B. According to your careful measurements, the time interval between the two events is 1 percent longer than that measured by the two clocks on Spaceship A. How fast is Ship A moving relative to Ship B. (Use one or more of the results of Problem 13.)

15 •• If a plane flies at a speed of 2000 km/h, how long must the plane fly before its clock loses 1 s because of time dilation? (Use one or more of the results of Problem 13.)

The Lorentz Transformation, Clock Synchronization, and Simultaneity

16 •• Show that when $v \ll c$ the transformation equations for x, t, and u reduce to the Galilean equations.

17 •• **SSM** **iSOLVE** A spaceship of proper length $L_p = 400$ m moves past a transmitting station at a speed of $0.76c$. At the instant that the nose of the spaceship passes the transmitter, clocks at the transmitter and in the nose of the spaceship are synchronized to $t = t' = 0$. The instant that the tail of the spaceship passes the transmitter a signal is sent and subsequently detected by the receiver in the nose of the spaceship. (a) When, according to the clock in the spaceship, is the signal sent? (b) When, according to the clock at the transmitter, is the signal received by the spaceship? (c) When, according to the clock in the spaceship, is the signal received? (d) Where, according to an observer at the transmitter, is the nose of the spaceship when the signal is received?

18 •• In frame S, event B occurs 2 μs after event A, which occurs at $x = 1.5$ km from event A. How fast must an observer be moving along the $+x$ axis so that events A and B occur simultaneously? Is it possible for event B to precede event A for some observer?

19 •• Observers in reference frame S see an explosion located at $x_1 = 480$ m. A second explosion occurs 5 μs later at $x_2 = 1200$ m. In reference frame S', which is moving along the $+x$ axis at speed v, the explosions occur at the same point in space. What is the separation in time between the two explosions as measured in S'?

20 ••• Two events in S are separated by a distance $D = x_2 - x_1$ and a time $T = t_2 - t_1$. (a) Use the Lorentz transformation to show that in frame S', which is moving with speed v relative to S, the time separation is $t_2' - t_1' = \gamma(T - vD/c^2)$. (b) Show that the events can be simultaneous in frame S' only if D is greater than cT. (c) If one of the events is the *cause* of the other, the separation D must be less than cT, since D/c is the smallest time that a signal can take to travel from x_1 to x_2 in frame S. Show that if D is less than cT, t_2' is greater than t_1' in all reference frames. This shows that if the cause precedes the effect in one frame, it must precede it in all reference frames. (d) Suppose that a signal could be sent with speed $c' > c$ so that in frame S the cause precedes the effect by the time $T = D/c'$. Show that there is then a reference frame moving with speed v less than c in which the effect precedes the cause.

21 ••• A rocket with a proper length of 700 m is moving to the right at a speed of $0.9c$. It has two clocks, one in the nose and one in the tail, that have been synchronized in the frame of the rocket. A clock on the ground and the nose clock on the rocket both read $t = 0$ as they pass. (a) At $t = 0$, what does the tail clock on the rocket read as seen by an observer on the ground? When the tail clock on the rocket passes the ground clock, (b) what does the tail clock read as seen by an observer on the ground, (c) what does the nose clock read as seen by an observer on the ground, and (d) what does the nose clock read as seen by an observer on the rocket? (e) At $t = 1$ h, as measured on the rocket, a light signal is sent from the nose of the rocket to an observer standing by the ground clock. What does the ground clock read when the observer receives this signal? (f) When the observer on the ground receives the signal, he sends a return signal to the nose of the rocket. When is this signal received at the nose of the rocket as seen on the rocket?

22 ••• **SSM** An observer in frame S standing at the origin observes two flashes of colored light separated spatially by $\Delta x = 2400$ m. A blue flash occurs first, followed by a red flash 5 μs later. An observer in S' moving along the x axis at speed v relative to S also observes the flashes 5 μs apart and with a separation of 2400 m, but the red flash is observed first. Find the magnitude and direction of v.

The Velocity Transformation

23 •• Show that if u_x' and v in Equation 39-18a are both positive and less than c, then u_x is positive and less than c. [Hint: Let $u_x' = (1 - \varepsilon_1)c$ and $v = (1 - \varepsilon_2)c$, where ε_1 and ε_2 are positive numbers that are less than 1.]

24 •• **SSM** A spaceship, at rest in a certain reference frame S, is given a speed increment of $0.50c$ (call this boost 1). Relative to its new rest frame, the spaceship is given a further $0.50c$ increment 10 seconds later (as measured in its new rest frame; call this boost 2). This process is continued indefinitely, at 10-s intervals, as measured in the rest frame of the ship. (Assume that the boost itself takes a very short time compared to 10 s.) (a) Using a spreadsheet program, calculate and graph the velocity of the spaceship in reference frame S as a function of the boost number for boost 1 to boost 10. (b) Graph the gamma factor the same way. (c) How many boosts does it take until the velocity of the ship in S is greater than $0.999c$? (d) How far has the spaceship moved after 5 boosts, as measured in reference frame S? What is the average speed of the spaceship (between boost 1 and boost 5) as measured in S?

The Relativistic Doppler Effect

25 • Sodium light of wavelength 589 nm is emitted by a source that is moving toward the earth with speed v. The wavelength measured in the frame of the earth is 547 nm. Find v.

26 • **iSOLVE** A distant galaxy is moving away from us at a speed of 1.85×10^7 m/s. Calculate the fractional redshift $(\lambda' - \lambda_0)/\lambda_0$ in the light from this galaxy.

27 •• Derive Equation 39-16a for the frequency received by an observer moving with speed v toward a stationary source of electromagnetic waves.

28 • Show that if v is much less than c, the Doppler shift is given approximately by

$$\Delta f/f \approx \pm v/c$$

29 •• **SSM** **iSOLVE** A clock is placed in a satellite that orbits the earth with a period of 90 min. By what time interval will this clock differ from an identical clock on the earth after 1 y? (Assume that special relativity applies and neglect general relativity.)

30 •• For light that is Doppler-shifted with respect to an observer, define the redshift parameter

$$z = \frac{f - f'}{f'}$$

where f is the frequency of the light measured in the rest frame of the emitter, and f' is the frequency measured in the rest frame of the observer. If the emitter is moving directly away from the observer, show that the relative velocity between the emitter and the observer is

$$v = c\left(\frac{u^2 - 1}{u^2 + 1}\right)$$

where $u = z + 1$.

31 • A light beam moves along the y' axis with speed c in frame S', which is moving to the right with speed v relative to frame S. (a) Find the x and y components of the velocity of the light beam in frame S. (b) Show that the magnitude of the velocity of the light beam in S is c.

32 • **ISOLVE** ✓ A spaceship is moving east at speed $0.90c$ relative to the earth. A second spaceship is moving west at speed $0.90c$ relative to the earth. What is the speed of one spaceship relative to the other spaceship?

33 •• **SSM** A particle moves with speed $0.8c$ along the x'' axis of frame S'', which moves with speed $0.8c$ along the x' axis relative to frame S'. Frame S' moves with speed $0.8c$ along the x axis relative to frame S. (a) Find the speed of the particle relative to frame S'. (b) Find the speed of the particle relative to frame S.

Relativistic Momentum and Relativistic Energy

34 • **SSM** A proton (rest energy 938 MeV) has a total energy of 2200 MeV. (a) What is its speed? (b) What is its momentum?

35 • If the kinetic energy of a particle equals twice its rest energy, what percentage error is made by using $p = mu$ for its momentum?

36 •• **ISOLVE** A particle with momentum of 6 MeV/c has total energy of 8 MeV. (a) Determine the mass of the particle. (b) What is the energy of the particle in a reference frame in which its momentum is 4 MeV/c? (c) What are the relative velocities of the two reference frames?

37 •• Show that

$$d\left(\frac{mu}{\sqrt{1 - (u^2/c^2)}}\right) = m\left(1 - \frac{u^2}{c^2}\right)^{-3/2} du$$

38 •• **ISOLVE** The K^0 particle has a mass of 497.7 MeV/c^2. It decays into a π^- and π^+, each with mass 139.6 MeV/c^2. Following the decay of a K^0, one of the pions is at rest in the laboratory. Determine the kinetic energy of the other pion and of the K^0 prior to the decay.

39 •• **SSM** Two protons approach each other head-on at $0.5c$ relative to reference frame S'. (a) Calculate the total kinetic energy of the two protons as seen in frame S'. (b) Calculate the total kinetic energy of the protons as seen in reference frame S, which is moving with speed $0.5c$ relative to S' so that one of the protons is at rest.

40 •• An antiproton has the same rest energy as a proton. It is created in the reaction $p + p \rightarrow p + p + p + \bar{p}$. In an experiment, protons at rest in the laboratory are bombarded with protons of kinetic energy K_L, which must be great enough so that kinetic energy equal to $2mc^2$ can be converted into the rest energy of the two particles. In the frame of the laboratory, the total kinetic energy cannot be converted into rest energy because of conservation of momentum. However, in the zero-momentum reference frame in which the two initial protons are moving toward each other with equal speed u, the total kinetic energy can be converted into rest energy. (a) Find the speed of each proton u so that the total kinetic energy in the zero-momentum frame is $2mc^2$. (b) Transform to the laboratory's frame in which one proton is at rest, and find the speed u' of the other proton. (c) Show that the kinetic energy of the moving proton in the laboratory's frame is $K_L = 6mc^2$.

41 ••• **ISOLVE** A particle of mass 1 MeV/c^2 and kinetic energy 2 MeV collides with a stationary particle of mass 2 MeV/c^2. After the collision, the particles stick together. Find (a) the speed of the first particle before the collision, (b) the total energy of the first particle before the collision, (c) the initial total momentum of the system, (d) the total kinetic energy after the collision, and (e) the mass of the system after the collision.

General Relativity

42 •• **SSM** Light traveling in the direction of increasing gravitational potential undergoes a frequency redshift. Calculate the shift in wavelength if a beam of light of wavelength $\lambda = 632.8$ nm is sent up a vertical shaft of height $L = 100$ m.

43 •• Let us revisit a problem from Chapter 3: Two cannons are pointed directly toward each other, as shown in Figure 39-17. When fired, the cannonballs will follow the trajectories shown. Point P is the point where the trajectories cross each other. Ignore the effects of air resistance. Using the principle of equivalence, show that if the cannons are fired simultaneously, the cannonballs will hit each other at point P.

FIGURE 39-17 Problem 43

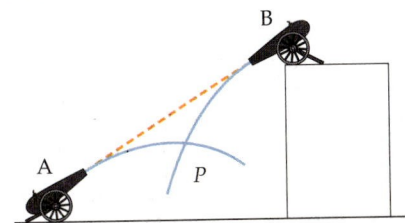

44 ••• A horizontal turntable rotates with angular speed ω. There is a clock at the center of the turntable and one at a distance r from the center. In an inertial reference frame, the clock at distance r is moving with speed $u = r\omega$. (a) Show that from time dilation according to special relativity, time intervals Δt_0 for the clock at rest and Δt_r for the moving clock are related by

$$\frac{\Delta t_r - \Delta t_0}{\Delta t_0} = -\frac{r^2\omega^2}{2c^2}, \quad \text{if } r\omega \ll c$$

(b) In a reference frame rotating with the table, both clocks are at rest. Show that the clock at distance r experiences a pseudo-force $F_r = mr\omega^2$ in this accelerated frame and that this is

Find the binding energy of the last neutron in ⁴He.

PICTURE THE PROBLEM The binding energy is c^2 times the difference in mass of ³He plus a neutron and ⁴He. We find these masses from Table 40-1 and convert to energy using Equation 40-3.

1. Add the mass of the neutron to that of ³He:

$$m_{^3He} + m_n = 3.016\ 029\ u + 1.008\ 665\ u$$
$$= 4.024\ 694\ u$$

2. Subtract the mass of ⁴He from the result:

$$\Delta m = (m_{^3He} + m_n) - m_{^4He}$$
$$= 4.024\ 694\ u - 4.002\ 603\ u = 0.022\ 091\ u$$

3. Multiply this mass difference by c^2 and convert to MeV:

$$E_b = (\Delta m)c^2$$
$$= (0.022\ 091\ u)c^2 \times \frac{931.5\ \text{MeV}/c^2}{1\ u}$$
$$= \boxed{20.58\ \text{MeV}}$$

Figure 40-3 shows the binding energy per nucleon E_b/A versus A. The mean value is approximately 8.3 MeV. The flatness of this curve for $A > 50$ shows that E_b is approximately proportional to A. This indicates that there is saturation of nuclear forces in the nucleus as would be the case if each nucleon were attracted only to its nearest neighbors. Such a situation also leads to a constant nuclear density consistent with the measurements of the radius. If, for example, there were no saturation and each nucleon bonded to each other nucleon, there would be $A - 1$ bonds for each nucleon and a total of $A(A - 1)$ bonds altogether. The total binding energy, which is a measure of the energy needed to break all these bonds, would then be proportional to $A(A - 1)$, and E_b/A would not be approximately constant. The steep rise in the curve for low A is due to the increase in the number of nearest neighbors and therefore to the increased number of bonds per nucleon. The gradual decrease at high A is due to the Coulomb repulsion of the protons, which increases as Z^2 and decreases the binding energy. Eventually, for very large A, this Coulomb repulsion becomes so great that a nucleus with A greater than approximately 300 is unstable and undergoes spontaneous fission.

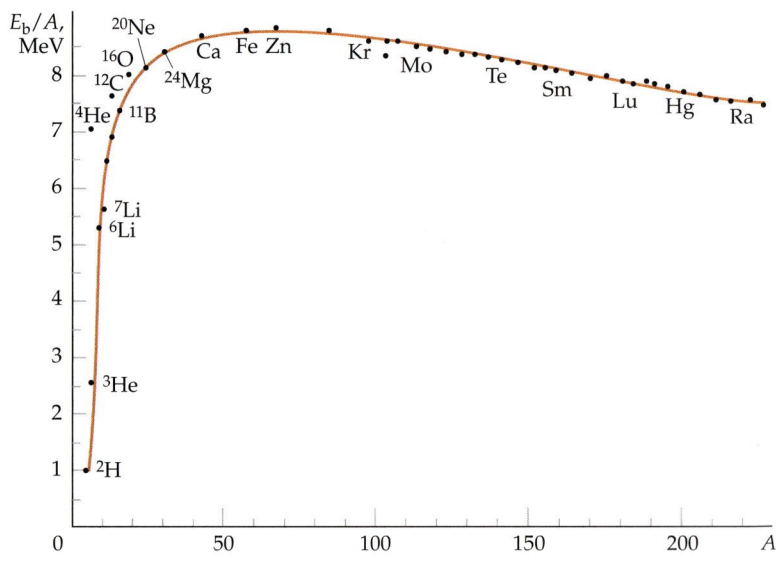

FIGURE 40-3 The binding energy per nucleon versus the mass number A. For nuclei with values of A greater than 50, the curve is approximately constant, indicating that the total binding energy is approximately proportional to A.

40-2 Radioactivity

Many nuclei are radioactive; that is, they decay into other nuclei by the emission of particles such as photons, electrons, neutrons, or α particles. The terms α decay, β decay, and γ decay were used before it was discovered that α particles are

^{4}He nuclei, β particles are either electrons (β^-) or positrons[†] (β^+), and γ-rays are photons. The rate of decay is not constant over time, but decreases exponentially. *This exponential time dependence is characteristic of all radioactivity and indicates that radioactive decay is a statistical process.* Because each nucleus is well shielded from others by the atomic electrons, pressure and temperature changes have little or no effect on the rate of radioactive decay or other nuclear properties.

Let N be the number of radioactive nuclei at some time t. If the decay of an individual nucleus is a random event, we expect the number of nuclei that decay in some time interval dt to be proportional to N and to dt. Because of these decays, the number N will decrease. The change in N is given by

$$dN = -\lambda N \, dt \qquad\qquad 40\text{-}4$$

where λ is a constant of proportionality called the **decay constant.** The rate of change of N, dN/dt, is proportional to N. This is characteristic of exponential decay. To solve Equation 40-4 for N, we first divide each side by N, thus separating the variables N and t:

$$\frac{dN}{N} = -\lambda \, dt$$

Integrating, we obtain

$$\int_{N_0}^{N'} \frac{dN}{N} = -\lambda \int_0^{t'} dt$$

or

$$\ln \frac{N'}{N_0} = -\lambda t' \qquad\qquad 40\text{-}5$$

where N' is the number of nuclei that remain at time t'. For convenience, we drop the primes from N' and t'. This introduces no ambiguity because the parameters N and t have been integrated out of the equation. Taking the exponential of each side, we obtain

$$\frac{N}{N_0} = e^{-\lambda t}$$

or

$$N = N_0 e^{-\lambda t} \qquad\qquad 40\text{-}6$$

The number of radioactive decays per second is called the decay rate R:

$$R = -\frac{dN}{dt} = \lambda N = \lambda N_0 e^{-\lambda t} = R_0 e^{-\lambda t} \qquad\qquad 40\text{-}7$$

DECAY RATE

where

$$R_0 = \lambda N_0 \qquad\qquad 40\text{-}8$$

---

† The positron is identical to an electron except it has a charge of $+e$.

is the rate of decay at time $t = 0$. The decay rate R is the quantity that is determined experimentally.

The average or **mean lifetime** τ is equal to the reciprocal of the decay constant (see Problem 42):

$$\tau = \frac{1}{\lambda} \qquad\qquad 40\text{-}9$$

The mean lifetime is analogous to the time constant in the exponential decrease in the charge on a capacitor in an RC circuit that we discussed in Section 25-6. After a time equal to the mean lifetime, the number of radioactive nuclei and the decay rate are each equal to $e^{-1} = 37$ percent of their original values. The **half-life** $t_{1/2}$ is defined as the time it takes for the number of nuclei and the decay rate to decrease by half. Setting $t = t_{1/2}$ and $N = N_0/2$ in Equation 40-6 gives

$$\frac{N_0}{2} = N_0 e^{-\lambda t_{1/2}} \qquad\qquad 40\text{-}10$$

or

$$e^{+\lambda t_{1/2}} = 2$$

Solving for $t_{1/2}$ gives

$$t_{1/2} = \frac{\ln 2}{\lambda} = \frac{0.693}{\lambda} = 0.693\tau \qquad\qquad 40\text{-}11$$

Figure 40-4 shows a plot of N versus t. If we multiply the numbers on the N axis by λ, this graph becomes a plot of R versus t. After each time interval of one half-life, both the number of nuclei left and the decay rate have decreased to half of their previous values. For example, if the decay rate is R_0 initially, it will be $\frac{1}{2}R_0$ after one half-life, $(\frac{1}{2})(\frac{1}{2})R_0$ after two half-lives, and so forth. After n half-lives, the decay rate will be

$$R = (\tfrac{1}{2})^n R_0 \qquad\qquad 40\text{-}12$$

The half-lives of radioactive nuclei vary from very small times (less than 1 μs) to very large times (up to 10^{16} y).

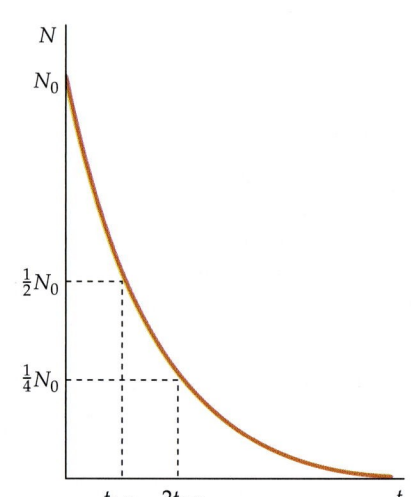

FIGURE 40-4 Exponential radioactive decay. After each half-life $t_{1/2}$, the number of nuclei remaining has decreased by one-half. The decay rate $R = \lambda N$ has the same time dependence, as does N.

COUNTING RATE FOR RADIOACTIVE DECAY   **EXAMPLE 40-2**

A radioactive source has a half-life of 1 min. At time $t = 0$, the radioactive source is placed near a detector, and the counting rate (the number of decay particles detected per unit time) is observed to be 2000 counts/s. Find the counting rate at times $t = 1$ min, $t = 2$ min, $t = 3$ min, and $t = 10$ min.

PICTURE THE PROBLEM The counting rate decreases by a factor of 2 each minute.

1. Since the half-life is 1 min, the counting rate will be half as great at $t = 1$ min as at $t = 0$:

$$r_1 = \tfrac{1}{2}r_0 = \tfrac{1}{2}(2000 \text{ counts/s})$$
$$= \boxed{1000 \text{ counts/s at 1 min}}$$

2. At $t = 2$ min, the rate is half that at 1 min. It decreases by one-half each minute:

$$r_2 = \tfrac{1}{2}r_1 = \tfrac{1}{2}(1000 \text{ counts/s})$$

$$= \boxed{500 \text{ counts/s at 2 min}}$$

$$r_3 = \tfrac{1}{2}r_2 = \tfrac{1}{2}(500 \text{ counts/s})$$

$$= \boxed{250 \text{ counts/s at 3 min}}$$

3. At $t = 10$ min, the rate will be $(\tfrac{1}{2})^{10}$ times the initial rate:

$$r_{10} = (\tfrac{1}{2})^{10}r_0 = (\tfrac{1}{2})^{10}(2000 \text{ counts/s})$$

$$= 1.95 \text{ counts/s}$$

$$\approx \boxed{2 \text{ counts/s at 10 min}}$$

DETECTION-EFFICIENCY CONSIDERATIONS   **EXAMPLE 40-3**

If the detection efficiency in Example 40-2 is 20 percent, (*a*) how many radioactive nuclei are there at time $t = 0$ and (*b*) at time $t = 1$ min? (*c*) How many nuclei decay in the first minute?

PICTURE THE PROBLEM The detection efficiency depends on the probability that a radioactive decay particle will enter the detector and the probability that upon entering the detector it will produce a count. If the efficiency is 20 percent, the decay rate must be five times the counting rate.

(*a*) 1. The number of radioactive nuclei is related to the decay rate R, and the decay constant λ:

$$R = \lambda N$$

2. The decay constant is related to the half-life:

$$\lambda = \frac{0.693}{t_{1/2}} = \frac{0.693}{1 \text{ min}} = 0.693 \text{ min}^{-1}$$

3. Because the detection efficiency is 20 percent, the decay rate is five times the counting rate. Calculate the initial decay rate:

$$R_0 = 5 \times 2000 \text{ counts/s}$$

$$= 10{,}000 \text{ s}^{-1}$$

4. Substitute to calculate the initial number of radioactive nuclei N_0 at $t = 0$:

$$N_0 = \frac{R_0}{\lambda} = \frac{10{,}000 \text{ s}^{-1}}{0.693 \text{ min}^{-1}} \times \frac{60 \text{ s}}{1 \text{ min}}$$

$$= \boxed{8.66 \times 10^5}$$

(*b*) At time $t = 1$ min $= t_{1/2}$, there are half as many radioactive nuclei as at $t = 0$:

$$N_1 = \tfrac{1}{2}(8.66 \times 10^5) = \boxed{4.33 \times 10^5}$$

(*c*) The number of nuclei that decay in the first minute is $N_0 - N_1$:

$$\Delta N = N_0 - N_1$$

$$= 8.66 \times 10^5 - 4.33 \times 10^5$$

$$= \boxed{4.33 \times 10^5}$$

The SI unit of radioactive decay is the **becquerel** (Bq), which is defined as one decay per second:

$$1 \text{ Bq} = 1 \text{ decay/s}$$

A historical unit that applies to all types of radioactivity is the **curie** (Ci), which is defined as

$$1 \text{ Ci} = 3.7 \times 10^{10} \text{ decays/s} = 3.7 \times 10^{10} \text{ Bq} \qquad \text{40-14}$$

The curie is the rate at which radiation is emitted by 1 g of radium. Since this is a very large unit, the millicurie (mCi) or microcurie (μCi) are often used.

Beta Decay

Beta decay occurs in nuclei that have too many neutrons or too few neutrons for stability. In β decay, A remains the same while Z either increases by 1 (β^- decay) or decreases by 1 (β^+ decay).

The simplest example of β decay is the decay of the free neutron into a proton plus an electron. (The half-life of a free neutron is about 10.8 min.) The energy of decay is 0.782 MeV, which is the difference between the rest energy of the neutron and the rest energy of the proton plus electron. More generally, in β^- decay, a nucleus of mass number A and atomic number Z decays into a nucleus, referred to as the **daughter nucleus,** of mass number A and atomic number $Z' = Z + 1$ with the emission of an electron. If the decay energy were shared by only the daughter nucleus and the emitted electron, the energy of the electron would be uniquely determined by the conservation of energy and momentum. Experimentally, however, the energies of the electrons emitted in the β^- decay of a nucleus are observed to vary from zero to the maximum energy available. A typical energy spectrum for these electrons is shown in Figure 40-5.

To explain the apparent nonconservation of energy in β decay, Wolfgang Pauli in 1930 suggested that a third particle, which he called the **neutrino,** is also emitted. Because the measured maximum energy of the emitted electrons is equal to the total available for the decay, the rest energy and therefore the mass of the neutrino was assumed to be zero. (It is now believed that the mass of the neutrino is very small but not zero.) In 1948, measurements of the momenta of the emitted electron and the recoiling nucleus showed that the neutrino was also needed for the conservation of linear momentum in β decay. The neutrino was first observed experimentally in 1957. It is now known that there are at least three kinds of neutrinos, one (v_e) associated with electrons, one (v_μ) associated with muons, and one (v_τ), which was first observed at Fermi National Laboratory in 2000, associated with the tau particle, τ. Moreover, each neutrino has an antiparticle, written $\bar{v}_e$, $\bar{v}_\mu$, and $\bar{v}_\tau$. It is the electron antineutrino that is emitted in the decay of a neutron, which is written[†]

$$n \rightarrow p + \beta^- + \bar{v}_e \qquad \text{40-15}$$

In β^+ decay, a proton changes into a neutron with the emission of a positron (and a neutrino). A free proton cannot decay by positron emission because of conservation of energy (the mass of the neutron plus the positron is greater than that of the proton); however, because of binding-energy effects, a proton inside a nucleus can decay. A typical β^+ decay is

$$^{13}_{7}\text{N} \rightarrow {}^{13}_{6}\text{C} + \beta^+ + v_e \qquad \text{40-16}$$

The electrons or the positrons emitted in β decay do not exist inside the nucleus. They are created in the process of decay, just as photons are created when an atom makes a transition from a higher energy state to a lower energy state.

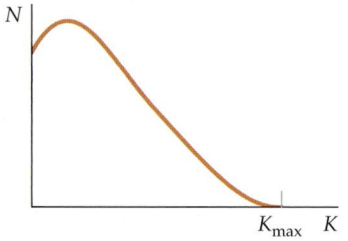

FIGURE 40-5 Number of electrons emitted in β^- decay versus kinetic energy. The fact that all the electrons do not have the same energy K_{max} suggests that another particle, one that shares the energy available for decay, is emitted.

---

† This reaction is also written $n \rightarrow p + e^- + \bar{v}_e$.

An important example of β decay is that of ^{14}C, which is used in radioactive carbon dating:

$$^{14}C \rightarrow {}^{14}N + \beta^- + \bar{v}_e \qquad\qquad\qquad 40\text{-}17$$

The half-life for this decay is 5730 y. The radioactive isotope ^{14}C is produced in the upper atmosphere in nuclear reactions caused by cosmic rays. The chemical behavior of carbon atoms with ^{14}C nuclei is the same as those with ordinary ^{12}C nuclei. For example, atoms with these nuclei combine with oxygen to form CO_2 molecules. Since living organisms continually exchange CO_2 with the atmosphere, the ratio of ^{14}C to ^{12}C in a living organism is the same as the equilibrium ratio in the atmosphere, which is about 1.3×10^{-12}. After an organism dies, it no longer absorbs ^{14}C from the atmosphere, so the ratio of ^{14}C to ^{12}C continually decreases due to the radioactive decay of ^{14}C. The number of ^{14}C decays per minute per gram of carbon in a living organism can be calculated from the known half-life of ^{14}C and the number of ^{14}C nuclei in a gram of carbon. The result is that there are approximately 15.0 decays per minute per gram of carbon in a living organism. Using this result and the measured number of decays per minute per gram of carbon in a nonliving sample of bone, wood, or other object containing carbon, we can determine the age of the sample. For example, if the measured rate were 7.5 decays per minute per gram, the age of the sample would be one half-life = 5730 years.

How Old Is the Artifact?   **EXAMPLE 40-4**   **Put It in Context**

You have a summer job working in an archeological research lab. Your supervisor calls to tell you that they found a new bone at their current dig and asks you to determine the age of the bone from a sample that is in the mail. When the bone arrives, you take a sample that contains 200 grams of carbon and you find a beta decay rate of 400 decays/min.

PICTURE THE PROBLEM We first obtain a rough estimate of the age of the bone. If the bone were from a living organism, we would expect the decay rate to be $[(15 \text{ decays/min})/\text{g}](200 \text{ g}) = 3000$ decays/min. Since 400/3000 is roughly 1/8 (actually 1/7.5), the sample must be approximately three half-lives old, which is about $3(5730 \text{ y}) = 17{,}190$ y. To find the age of the bone more accurately, we need to determine the number of half-lives of the bone. We can do this by using the equality $R_n = (1/2)^n R_0$ where R_n is the current decay rate, R_0 is the initial decay rate, and n is the number of half-lives. We can determine the initial decay rate by multiplying the decay rate per gram times the mass of the carbon of the sample.

1. Write the decay rate after n half-lives in terms of the initial decay rate:

$$R_n = \left(\tfrac{1}{2}\right)^n R_0$$

2. Calculate the initial decay rate (the decay for 200 g of carbon when the organism stopped breathing):

$$R_0 = [(15 \text{ decays/min})/\text{g}](200 \text{ g})$$
$$= 3000 \text{ decays/min}$$

3. Substitute the values for R_0 and R_n into the step 1 equation and solve for n:

$$R_n = \left(\tfrac{1}{2}\right)^n R_0$$

$$400 \,\frac{\text{decays}}{\text{min}} = \left(\tfrac{1}{2}\right)^n 3000 \,\frac{\text{decays}}{\text{min}}$$

$$\left(\tfrac{1}{2}\right)^n = \frac{400}{3000}$$

$$2^n = \frac{3000}{400} = 7.5$$

4. We solve for n by taking the logarithm of each side:

$$n \ln 2 = \ln 7.5$$

$$n = \frac{\ln 7.5}{\ln 2} = 2.91$$

5. The age of the bone is $nt_{1/2}$:

$$t = nt_{1/2} = 2.91(5730 \text{ y}) = \boxed{1.67 \times 10^4 \text{ y}}$$

EXERCISE Picture the Problem of Example 40-4 states "Since 400/3000 is roughly 1/8 (actually 1/7.5), the sample must be approximately three half-lives old, ..." Explain. [*Answer* It is because $\frac{1}{7.5} \approx \frac{1}{8} = (\frac{1}{2})^3$, so $n = 3$.]

Gamma Decay

In γ decay, a nucleus in an excited state decays to a lower-energy state by the emission of a photon. This is the nuclear counterpart of spontaneous emission of photons by atoms and molecules. Unlike β decay or α decay, neither the mass number A nor the atomic number Z change during γ decay. Since the spacing of the nuclear energy levels is of the order of 1 MeV (as compared with spacing of the order of 1 eV in atoms), the wavelengths of the emitted photons are of the order of 1 pm (1 pm = 10^{-12} m):

$$\lambda = \frac{hc}{E} \approx \frac{1240 \text{ eV·nm}}{1 \text{ MeV}} = 0.00124 \text{ nm} = 1.24 \text{ pm}$$

The mean lifetime for γ decay is often very short. Usually it is observed only because it follows either α decay or β decay. For example, if a radioactive parent nucleus decays by β decay to an excited state of the daughter nucleus, the daughter nucleus then decays to its ground state by γ emission. Direct measurements of mean lifetimes as short as approximately 10^{-11} s are possible. Measurements of mean lifetimes shorter than 10^{-11} s are difficult, but they can sometimes be made by indirect methods.

A few γ emitters have very long lifetimes, of the order of hours. Nuclear energy states that have such long lifetimes are called **metastable states.**

Alpha Decay

All very heavy nuclei ($Z > 83$) are theoretically unstable via α decay because the mass of the original radioactive nucleus is greater than the sum of the masses of the decay products—an α particle and the daughter nucleus. Consider the decay of ^{232}Th ($Z = 90$) into ^{228}Ra ($Z = 88$) plus an α particle. This is written as

$$^{232}\text{Th} \rightarrow {}^{228}\text{Ra} + \alpha = {}^{228}\text{Ra} + {}^4\text{He} \qquad \qquad \text{40-18}$$

The mass of the ^{232}Th atom is 232.038 050 u. The mass of the daughter atom ^{228}Ra is 228.031 064 u. Adding 4.002 603 u to this for the mass of ^{4}He, we get 232.033 667 u for the total mass of the decay products. This is less than the mass of ^{232}Th by 0.004 382 u, which multiplied by 931.5 MeV/c^2 gives 4.08 MeV/c^2 for the excess mass of ^{232}Th over that of the decay products. The isotope ^{232}Th is therefore theoretically unstable to α decay. This decay does in fact occur in nature with the emission of an α particle of kinetic energy 4.08 MeV. (The kinetic energy of the α particle is actually somewhat less than 4.08 MeV because some of the decay energy is shared by the recoiling ^{228}Ra nucleus.)

When a nucleus emits an α particle, both N and Z decrease by 2 and A decreases by 4. The daughter of a radioactive nucleus is often itself radioactive and decays by either α decay or β decay or both. If the original nucleus has a mass number A that is 4 times an integer, the daughter nucleus and all those in the chain will also have mass numbers equal to 4 times an integer. Similarly, if the mass number of the original nucleus is $4n + 1$, where n is an integer, all the nuclei in the decay chain will have mass numbers given by $4n + 1$, with n decreasing by one at each decay. We can see, therefore, that there are four possible α-decay chains, depending on whether A equals $4n$, $4n + 1$, $4n + 2$, or $4n + 3$, where n is an integer. All but one of these decay chains are found on the earth. The $4n + 1$ series is not found because its longest-lived member (other than the stable end product ^{209}Bi) is ^{237}Np, which has a half-life of only 2×10^6 y. Because this is much less than the age of the earth, this series has disappeared.

Figure 40-6 shows the thorium series, for which $A = 4n$. It begins with an α decay from ^{232}Th to ^{228}Ra. The daughter nuclide of an α decay is on the left or neutron-rich side of the stability curve (the dashed line in the figure), so it often decays by β^- decay. In the thorium series, ^{228}Ra decays by β^- decay to ^{228}Ac, which in turn decays by β^- decay to ^{228}Th. There are then four α decays to ^{212}Pb, which decays by β^- decay to ^{212}Bi. The series branches at ^{212}Bi, which decays either by α decay to ^{208}Tl or by β^- decay to ^{212}Po. The branches meet at the stable lead isotope ^{208}Pb.

The energies of α particles from natural radioactive sources range from approximately 4 MeV to 7 MeV, and the half-lives of the sources range from approximately 10^{-5} s to 10^{10} y. In general, the smaller the energy of the emitted α particle, the longer the half-life. As we discussed in Section 35-4, the enormous variation in half-lives was explained by George Gamow in 1928. He considered α decay to be a process in which an α particle is first formed inside a nucleus and then tunnels through the Coulomb barrier (Figure 40-7). A slight increase in the energy of the α particle reduces the relative height $U - E$ of the barrier and also the thickness. Because the probability of penetration is so sensitive to the relative height and thickness of the barrier, a small increase in E leads to a large increase in the probability of barrier penetration and therefore to a shorter lifetime. Gamow was able to derive an expression for the half-life as a function of E that is in excellent agreement with experimental results.

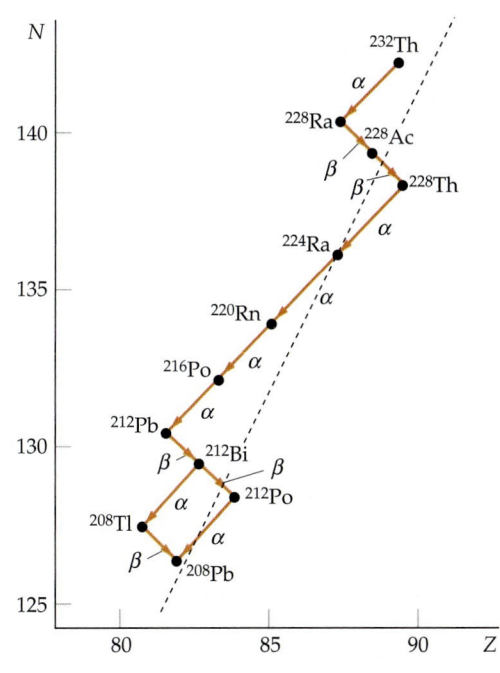

FIGURE 40-6 The thorium ($4n$) α decay series. The dashed line is the curve of stability.

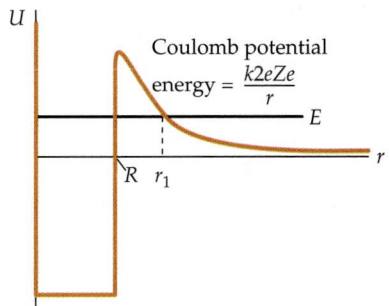

FIGURE 40-7 A model of the potential energy for an α particle and a nucleus. The strong attractive nuclear force that exists for values of r less than the nuclear radius R is indicated by the potential well. Outside the nucleus, the nuclear force is negligible, and the potential energy is given by Coulomb's law $U = +k2eZe/r$, where Ze is the nuclear charge and $2e$ is the charge of the α particle. The energy E is the kinetic energy of the α particle when it is far away from the nucleus. A small increase in E reduces the relative height $U - E$ of the barrier and also reduces its thickness, leading to a much greater chance of penetration. An increase in the energy of the emitted α particles by a factor of 2 results in a reduction of the half-life by a factor of more than 10^{20}.

40-3 Nuclear Reactions

Information about nuclei is typically obtained by bombarding the nuclei with various particles and observing the results. Although the first experiments of this type were limited by the need to use naturally occurring radiation, they produced many important discoveries. In 1932, J. D. Cockcroft and E. T. S. Walton succeeded in producing the reaction

$$p + {}^7\text{Li} \rightarrow {}^8\text{Be} \rightarrow {}^4\text{He} + {}^4\text{He}$$

using artifically accelerated protons. At about the same time, the Van de Graaff electrostatic generator was built (by R. Van de Graaff in 1931) as was the first cyclotron (by E. O. Lawrence and M. S. Livingston in 1932). Since then, enormous advances in the technology for accelerating and detecting particles have been made, and many nuclear reactions have been studied.

When a particle is incident on a nucleus, several different things can happen. The incident particle may be scattered elastically or inelastically, or the incident particle may be absorbed by the nucleus, and another particle or particles may be emitted. In inelastic scattering, the nucleus is left in an excited state and subsequently decays by emitting photons (or other particles).

The amount of energy released or absorbed in a reaction (in the center of mass reference frame) is called the **Q value** of the reaction. The Q value equals c^2 times this mass difference. When energy is released by a nuclear reaction, the reaction is said to be an **exothermic reaction.** In an exothermic reaction, the total mass of the incoming particles is greater than the total mass of the outgoing particles, and the Q value is positive. If the total mass of the incoming particles is less than that of the outgoing particles, energy is required for the reaction to take place, and the reaction is said to be an **endothermic reaction.** The Q value of an endothermic reaction is negative. In general, if Δm is the increase in mass, the Q value is

$$Q = -(\Delta m)c^2 \qquad \text{40-19}$$

Q VALUE

An endothermic reaction cannot take place below a specific threshold energy. In the laboratory reference frame in which stationary particles are bombarded by incoming particles, the threshold energy is somewhat greater than $|Q|$ because the outgoing particles must have some kinetic energy to conserve momentum.

A measure of the effective size of a nucleus for a particular nuclear reaction is the **cross section** σ. If I is the number of incident particles per unit time per unit area (the incident intensity) and R is the number of reactions per unit time per nucleus, the cross section is

$$\sigma = \frac{R}{I} \qquad \text{40-20}$$

The cross section σ has the dimensions of area. Since nuclear cross sections are of the order of the square of the nuclear radius, a convenient unit for them is the **barn,** which is defined as

$$1 \text{ barn} = 10^{-28} \text{ m}^2 \qquad \text{40-21}$$

The cross section for a particular reaction is a function of energy. For an endothermic reaction, it is zero for energies below the threshold energy.

EXOTHERMIC OR ENDOTHERMIC?   **EXAMPLE 40-5**

Find the Q value of the reaction p + ^{7}Li → ^{4}He + ^{4}He and state whether the reaction is exothermic or endothermic.

PICTURE THE PROBLEM We find the masses of the atoms from Table 40-1 and calculate the difference in the total mass of the outgoing particles and the incoming particles. The Q value is $-(\Delta m)c^2$. If we use the mass of hydrogen rather than the mass of the proton, there will be four electrons on each side of the reaction, so the electron masses will cancel.

1. Find the mass of each atom from Table 40-1:	^{1}H	1.007 825 u
	^{7}Li	7.016 004 u
	^{4}He	4.002 603 u

2. Calculate the initial mass m_i of the incoming particles: $m_i = 1.007\,825 \text{ u} + 7.016\,004 \text{ u} = 8.023\,829 \text{ u}$

3. Calculate the final mass m_f: $m_f = 2(4.002\,603 \text{ u}) = 8.005\,206 \text{ u}$

4. Calculate the increase in mass: $\Delta m = m_f - m_i = 8.005\,206 \text{ u} - 8.023\,829 \text{ u}$

$= -0.018\,623 \text{ u}$

5. Calculate the Q value:

$$Q = -(\Delta m)c^2 = (+0.018\,623\,\text{u})c^2 \times \left[931.5\,\text{MeV}/(\text{u}\cdot c^2)\right]$$

$$= \boxed{17.35\,\text{MeV}}$$

$\boxed{Q \text{ is positive, so the reaction is exothermic.}}$

 MASTER the CONCEPT WEB

REMARKS Since the initial mass is greater than the final mass, the initial energy is greater than the final energy and the reaction is exothermic, yielding 17.35 MeV.

Reactions With Neutrons

Nuclear reactions that involve neutrons are important for understanding nuclear reactors. The most likely reaction between a nucleus and a neutron having an energy of more than about 1 MeV is scattering. However, even if the scattering is elastic, the neutron loses some energy to the nucleus because the nucleus recoils. If a neutron is scattered many times in a material, its energy decreases until the neutron is of the order of the energy of thermal motion kT, where k is Boltzmann's constant and T is the absolute temperature. (At ordinary room temperatures, kT is approximately 0.025 eV.) The neutron is then equally likely to gain or lose energy from a nucleus when it is elastically scattered. A neutron with energy of the order of kT is called a **thermal neutron.**

At low energies, a neutron is likely to be captured, with the emission of a γ ray from the excited nucleus. Figure 40-8 shows the neutron-capture cross section for silver as a function of the energy of the neutron. The large peak in this curve is called a **resonance.** Except for the resonance, the cross section varies fairly smoothly with energy, decreasing with increasing energy roughly as $1/v$, where v is the speed of the neutron. We can understand this energy dependence as follows: Consider a neutron moving with speed v near a nucleus of diameter $2R$. The time it takes the neutron to pass the nucleus is $2R/v$. Thus, the neutron-capture cross section is proportional to the time spent by the neutron in the vicinity of the silver nucleus. The dashed line in Figure 40-8 indicates this $1/v$ dependence. At the maximum of the resonance, the value of the cross section is very large ($\sigma > 5000$ barns) compared with a value of only about 10 barns just past the resonance. Many elements show similar resonances in their neutron-capture cross sections. For example, the maximum cross section for ^{113}Cd is approximately 57,000 barns. This material is thus very useful for shielding against low-energy neutrons.

An important nuclear reaction that involves neutrons is fission, which is discussed in the next section.

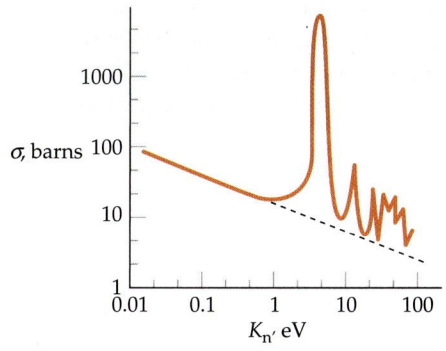

FIGURE 40-8 Neutron-capture cross section for silver as a function of the energy of the neutron. The straight line indicates the $1/v$ dependence of the cross section, which is proportional to the time spent by the neutron in the vicinity of the silver nucleus. Superimposed on this dependence are a large resonance and several smaller resonances.

40-4 Fission and Fusion

Figure 40-9 shows a plot of the nuclear mass difference per nucleon $(M - Zm_\text{p} - Nm_\text{n})/A$ in units of MeV/c^2 versus A. This is just the negative of the binding-energy curve shown in Figure 40-3. From Figure 40-9, we can see that the mass per nucleon for both very heavy ($A \approx 200$) and very light ($A \leq 20$) nuclides is more than that for nuclides of intermediate mass. Thus, energy is released when a very heavy nucleus, such as ^{235}U, breaks up into two lighter nuclei—a process called **fission**—or when two very light nuclei, such as ^2H and ^3H, fuse together to form a nucleus of greater mass—a process called **fusion.**

The application of both fission and fusion to the development of nuclear weapons has had a profound effect on our lives during the past 58 years. The peaceful application of these reactions to the development of energy resources approximately 15 percent of the time and undergoes fission approximately 85 percent of the time. The fission process is somewhat analogous to the oscillation of a liquid drop, as shown in Figure 40-10. If the oscillations are violent enough, the drop splits in two. Using the liquid-drop model, Niels Bohr and John Wheeler calculated the critical energy E_c needed by the ^{236}U nucleus to undergo fission. (^{236}U is the nucleus formed momentarily by the capture of a neutron by ^{235}U.) For this nucleus, the critical energy is 5.3 MeV, which is less than the 6.4 MeV of excitation energy produced when ^{235}U captures a neutron. The capture of a neutron by ^{235}U therefore produces an excited state of the ^{236}U nucleus that has more than enough energy to break apart. On the other hand, the critical energy for fission of the ^{239}U nucleus is 5.9 MeV. The capture of a neutron by a ^{238}U nucleus produces an excitation energy of only 5.2 MeV. Therefore, when a neutron is captured by ^{238}U to form ^{239}U, the excitation energy is not great enough for fission to occur. In this case, the excited ^{239}U nucleus deexcites by γ emission and then decays to $^{239}N_p$ by β decay, and then again to ^{239}Pu by β decay.

A fissioning nucleus can break into two medium-mass fragments in many different ways, as shown in Figure 40-11. Depending on the particular reaction, 1, 2, or 3 neutrons may be emitted. The average number of neutrons emitted in the fission of ^{235}U is approximately 2.5. A typical fission reaction is

$$n + {}^{235}U \rightarrow {}^{141}Ba + {}^{92}Kr + 3n$$

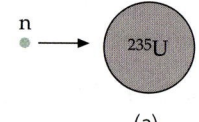

(a)

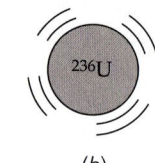

(b)

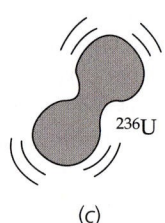

(c)

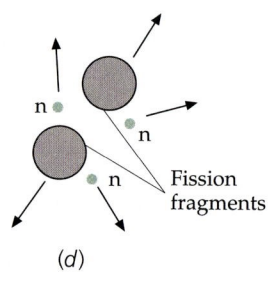

(d)

FIGURE 40-10 Schematic illustration of nuclear fission. (*a*) The absorption of a neutron by ^{235}U leads to (*b*) ^{236}U in an excited state. (*c*) The oscillation of ^{236}U has become unstable. (*d*) The nucleus splits apart into two nuclei of medium mass and emits several neutrons that can produce fission in other nuclei.

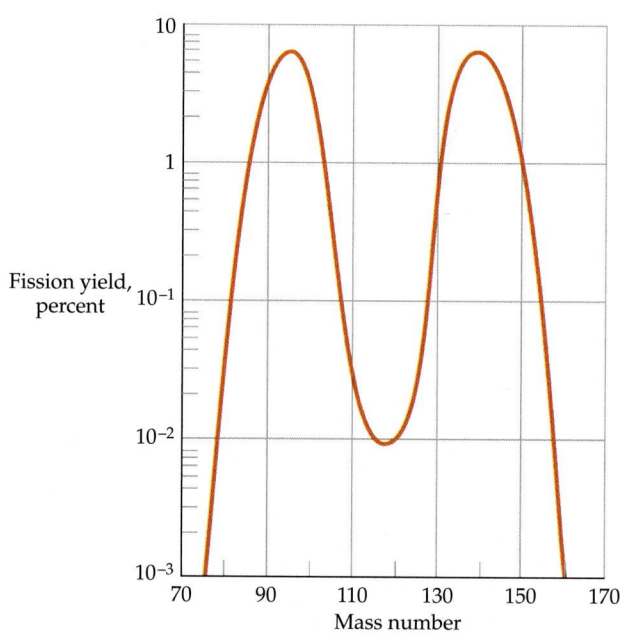

FIGURE 40-11 Distribution of the possible fission fragments of ^{235}U. The splitting of ^{235}U into two fragments of unequal mass is more likely than its splitting into fragments of equal mass.

Nuclear Fission Reactors

To sustain a chain reaction in a fission reactor, one of the neutrons (on the average) emitted in the fission of ^{235}U must be captured by another ^{235}U nucleus and cause it to fission. The **reproduction constant** k of a reactor is defined as the average number of neutrons from each fission that cause a subsequent fission. The maximum possible value of k is 2.5, but it is normally less than this for two important reasons: (1) Some of the neutrons may escape from the region containing fissionable nuclei and (2) some of the neutrons may be captured by nonfissioning nuclei in the reactor. If k is exactly 1, the reaction will be self-sustaining. If k is less than 1, the reaction will die out. If k is significantly greater than 1, the reaction

The inside of a nuclear power plant in Kent, England. A technician is standing on the reactor charge transfer plate, into which uranium fuel rods fit.

rate will increase rapidly and *run away*. In the design of nuclear bombs, such a runaway reaction is desired. In power reactors, the value of k must be kept very nearly equal to 1.

Since the neutrons emitted in fission have energies of the order of 1 MeV, whereas the chance for neutron capture leading to fission in ^{235}U is largest at small energies, the chain reaction can be sustained only if the neutrons are slowed down before they escape from the reactor. At high energies (1 MeV to 2 MeV), neutrons lose energy rapidly by inelastic scattering from ^{238}U, the principal constituent of natural uranium. (Natural uranium contains 99.3 percent ^{238}U and only 0.7 percent fissionable ^{235}U.) Once the neutron energy is below the excitation energies of the nuclei in the reactor (about 1 MeV), the main process of energy loss is by elastic scattering, in which a fast neutron collides with a nucleus at rest and transfers some of its kinetic energy to that nucleus. Such energy transfers are efficient only if the masses of the two bodies are comparable. A neutron will not transfer much energy in an elastic collision with a heavy uranium nucleus. Such a collision is like one between a marble and a billiard ball. The marble will be deflected by the much more massive billiard ball, and very little of its kinetic energy will be transferred to the billiard ball. A **moderator** consisting of material, such as water or carbon, that contains light nuclei is therefore placed around the fissionable material in the core of the reactor to slow down the neutrons. The neutrons are slowed down by elastic collisions with the nuclei of the moderator until they are in thermal equilibrium with the moderator. Because of the relatively large neutron-capture cross section of the hydrogen nucleus in water, reactors that use ordinary water as a moderator cannot easily achieve $k \approx 1$ unless they use enriched uranium, in which the ^{235}U content has been increased from 0.7 percent to between 1 percent and 4 percent. Natural uranium can be used if heavy water (D_2O) is used instead of ordinary (light) water (H_2O) as the moderator. Although heavy water is expensive, most Canadian reactors use heavy water for a moderator to avoid the cost of constructing uranium-enrichment facilities.

Figure 40-12 shows some of the features of a pressurized-water reactor commonly used in the United States to generate electricity. Fission in the core heats the water to a high temperature in the primary loop, which is closed. This water, which also serves as the moderator, is under high pressure to prevent the water from boiling. The hot water is pumped to a heat exchanger, where it heats the water in the secondary loop and converts the water to steam, which is then used to drive the turbines that produce electrical power. Note that the water in the secondary loop is isolated from the water in the primary loop to prevent its contamination by the radioactive nuclei in the reactor core.

the heat-transfer material leads to positive feedback, since it will absorb fewer neutrons than before. Because of these safety considerations, breeder reactors are not yet in commercial use in the United States. There are, however, several in operation in France, Great Britain, and the former Soviet Union.

Fusion

In fusion, two light nuclei, such as deuterium (^{2}H) and tritium (^{3}H), fuse together to form a heavier nucleus. A typical fusion reaction is

$$^2\text{H} + {}^3\text{H} \rightarrow {}^4\text{He} + \text{n} + 17.6\,\text{MeV}$$

The energy released in fusion depends on the particular reaction. For the ^{2}H + ^{3}H reaction, the energy released is 17.6 MeV. Although this is less than the energy released in a fission reaction, it is a greater amount of energy per unit mass. The energy released in this fusion reaction is (17.6 MeV)/(5 nucleons) = 3.52 MeV per nucleon. This is approximately 3.5 times as great as the 1 MeV per nucleon released in fission.

The production of power from the fusion of light nuclei holds great promise because of the relative abundance of the fuel and the absence of some of the dangers inherent in fission reactors. Unfortunately, the technology necessary to make fusion a practical source of energy has not yet been developed. We will consider the ^{2}H + ^{3}H reaction; other reactions present similar problems.

Because of the Coulomb repulsion between the ^{2}H and ^{3}H nuclei, very large kinetic energies, of the order of 1 MeV, are needed to get the nuclei close enough together for the attractive nuclear forces to become effective and to cause fusion. Such energies can be obtained in an accelerator, but since the scattering of one nucleus by the other is much more probable than fusion, the bombardment of one nucleus by another in an accelerator requires the input of more energy than is recovered. To obtain energy from fusion, the particles must be heated to a temperature great enough for the fusion reaction to occur as the result of random thermal collisions. Because a significant number of particles have kinetic energies greater than the mean kinetic energy, $\frac{3}{2}kT$, and because some particles can tunnel through the Coulomb barrier, a temperature T corresponding to $kT \approx 10$ keV is adequate to ensure that a reasonable number of fusion reactions will occur if the density of the particles is sufficiently high. The temperature corresponding to $kT = 10$ keV is of the order of 10^8 K. These temperatures occur in the interiors of stars, where such reactions are common. At these temperatures, a gas consists of positive ions and negative electrons and is called a **plasma**. One of the problems arising in attempts to produce controlled fusion reactions is the problem of confining the plasma long enough for the reactions to take place. In the interior of the sun, the plasma is confined by the enormous gravitational field of the sun. In a laboratory on the earth, confinement is a difficult problem.

The energy required to heat a plasma is proportional to the number density of its ions, n, whereas the collision rate is proportional to n^2, the square of the number density. If τ is the confinement time, the output energy is proportional to $n^2\tau$. If the output energy is to exceed the input energy, we must have

$$C_1 n^2 \tau > C_2 n$$

where C_1 and C_2 are constants. In 1957, the British physicist J. D. Lawson evaluated these constants from estimates of the efficiencies of various hypothetical fusion reactors and derived the following relation between density and confinement time, known as **Lawson's criterion:**

$$n\tau > 10^{20}\ \text{s·particles/m}^3 \qquad \text{40-22}$$

LAWSON'S CRITERION

If Lawson's criterion is met and the thermal energy of the ions is great enough ($kT \sim 10$ keV), the energy released by a fusion reactor will just equal the energy input; that is, the reactor will just break even. For the reactor to be practical, much more energy must be released.

Two schemes for achieving Lawson's criterion are currently under investigation. In one scheme, **magnetic confinement,** a magnetic field is used to confine the plasma (see Section 26-2). In the most common arrangement, first developed in the former USSR and called the Tokamak, the plasma is confined in a large toroid. The magnetic field is a combination of the doughnut-shaped magnetic field due to the windings of the toroid and the self-field due to the current of the circulating plasma. The break-even point has been achieved recently using magnetic confinement, but we are still a long way from building a practical fusion reactor.

In a second scheme, called **inertial confinement,** a pellet of solid deuterium and tritium is bombarded from all sides by intense pulsed laser beams of energies of the order of 10^4 J lasting about 10^{-8} s. (Intense beams of ions are also used.) Computer simulation studies indicate that the pellet should be compressed to approximately 10^4 times its normal density and heated to a temperature greater than 10^8 K. This should produce approximately 10^6 J of fusion energy in 10^{-11} s, which is so brief that confinement is achieved by inertia alone.

Because the break-even point is just barely being achieved in magnetic-confinement fusion, and because the building of a fusion reactor involves many practical problems that have not yet been solved, the availability of fusion to meet our energy needs is not expected for at least several decades. However, fusion holds great promise as an energy source for the future.

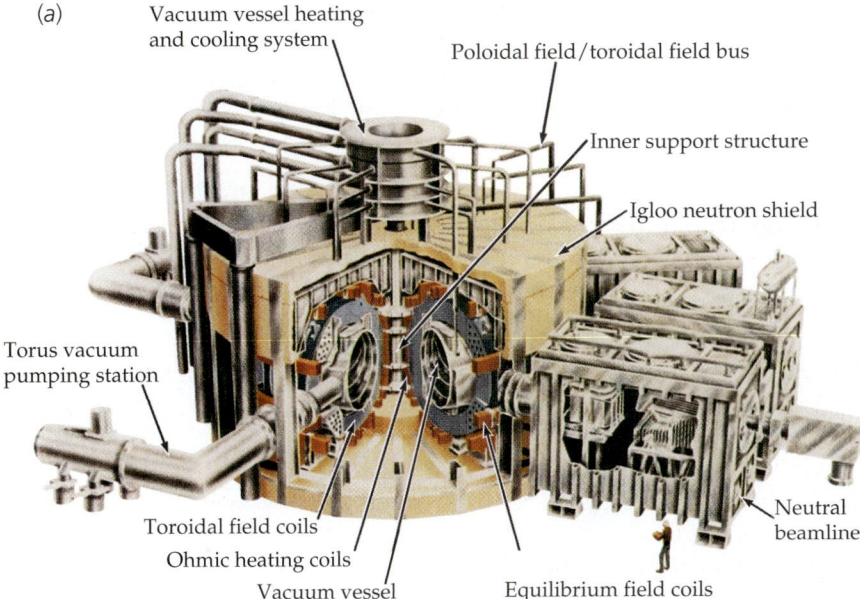

(a)

Vacuum vessel heating and cooling system
Poloidal field/toroidal field bus
Inner support structure
Igloo neutron shield
Torus vacuum pumping station
Neutral beamline
Toroidal field coils
Ohmic heating coils
Vacuum vessel
Equilibrium field coils

(b)

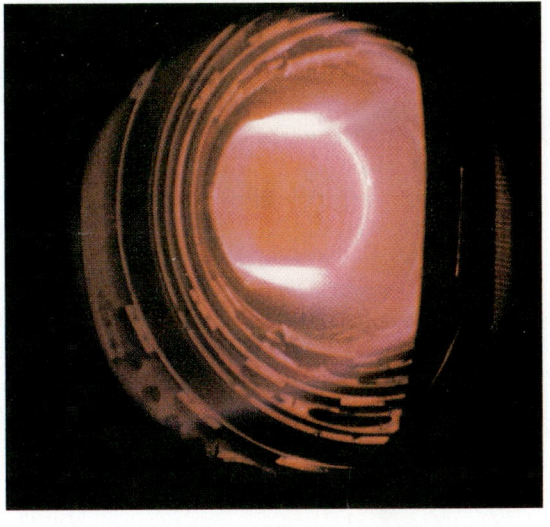

(c)

(*a*) **Schematic of the Tokamak Fusion Test Reactor (TFTR). The toroidal coils, surrounding the doughnut-shaped vacuum vessel, are designed to conduct current for 3-s pulses, separated by waiting times of 5 min. Pulses peak at 73,000 A, producing a magnetic field of 5.2 T. This magnetic field is the principal means of confining the deuterium–tritium plasma that circulates within the vacuum vessel. Current for the pulses is delivered by converting the rotational energy of two 600-ton flywheels. Sets of poloidal coils, perpendicular to the toroidal coils, carry an oscillating current that generates a current through the confined plasma itself, heating it ohmically. Additional poloidal fields help stabilize the confined plasma. Between four and six neutral-beam injection systems (only one of which is shown in the schematic) are used to inject high-energy deuterium atoms into the deuterium–tritium plasma, heating beyond what could be obtained ohmically, ultimately to the point of fusion. (*b*) The TFTR itself. The diameter of the vacuum vessel is 7.7 m. (*c*) An 800-*kA* plasma, lasting 1.6 s, as it discharges within the vacuum vessel.**

(a)

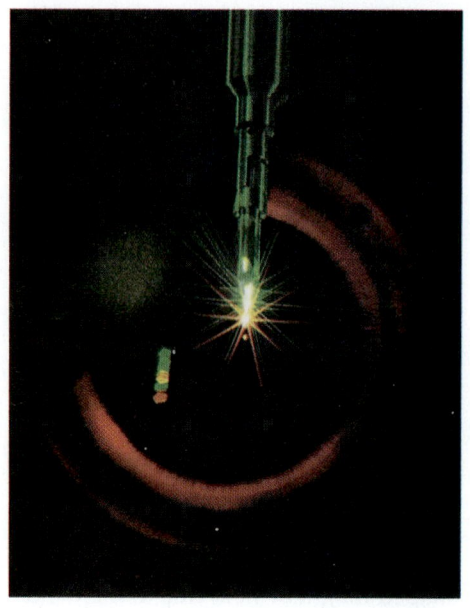

(b)

(*a*) The Nova target chamber, an aluminum sphere approximately 5 m in diameter, inside which 10 beams from the world's most powerful laser converge onto a hydrogen-containing pellet 0.5 mm in diameter. (*b*) The resulting fusion reaction is visible as a tiny star, lasting 10^{-10} s, releasing 10^{13} neutrons.

SUMMARY

Topic	Relevant Equations and Remarks
1. Properties of Nuclei	Nuclei have N neutrons, Z protons, and a mass number $A = N + Z$. For light nuclei, N and Z are approximately equal, whereas for heavy nuclei, N is greater than Z.
Isotopes	Isotopes consist of two or more nuclei with the same atomic number Z but with different values of N and A.
Size and shape	Most nuclei are approximately spherical in shape and have a volume that is proportional to A. Because the mass is proportional to A, nuclear density is independent of A.
Radius	$$R = R_0 A^{1/3} \approx (1.2 \text{ fm}) A^{1/3} \qquad \textbf{40-1}$$
Mass and binding energy	The mass of a stable nucleus is less than the sum of the masses of its nucleons. The mass difference Δm times c^2 equals the binding energy E_b of the nucleus. The binding energy is approximately proportional to the mass number A.
2. Radioactivity	Unstable nuclei are radioactive and decay by emitting α particles (^{4}He nuclei), β particles (electrons or positrons), or γ-rays (photons). All radioactivity is statistical in nature and follows an exponential decay law: $$N = N_0 e^{-\lambda t} \qquad \textbf{40-6}$$

Decay law	$R = \lambda N = R_0 e^{-\lambda t}$	**40-7**
Mean life	$\tau = \dfrac{1}{\lambda}$	**40-9**
Half-life	$t_{1/2} = 0.693\tau$	**40-11**

The half-lives of α decay range from a fraction of a second to millions of years. For β decay, the half-lives range up to hours or days. For γ decay, the half-lives are usually less than a microsecond.

Decay-rate units	The number of decays per second of 1 g of radium is the curie (Ci).

$$1 \text{ Ci} = 3.7 \times 10^{10} \text{ decays/s} = 3.7 \times 10^{10} \text{ Bq}$$

$$(1 \text{ Bq} = 1 \text{ decay/s})$$

3. Nuclear Reactions

Q value	The Q value equals c^2 times the total mass of the incoming particles less the total mass of the outgoing particles in the center of mass reference frame. If the net mass change is Δm, the Q value is	
	$Q = -(\Delta m)c^2$	**40-19**
Exothermic reaction	The mass decreases, Q is positive and measures the energy released.	
Endothermic reaction	The mass increases, Q is negative. Then $|Q|$ is the threshold energy for the reaction in the center of mass reference frame.	

4. Fission

Fission occurs when some heavy elements, such as ^{235}U or ^{239}Pu, capture a neutron and split apart into two medium-mass nuclei. The two nuclei then fly apart because of electrostatic repulsion, releasing a large amount of energy. A chain reaction is possible because several neutrons are emitted by a nucleus when it undergoes fission. A chain reaction can be sustained in a reactor if, on the average, one of the emitted neutrons is slowed down by scattering in the reactor and is then captured by another fissionable nucleus. Very heavy nuclei ($Z > 92$) are subject to spontaneous fission.

5. Fusion

A large amount of energy is released when two light nuclei, such as ^{2}H and ^{3}H, fuse together. Fusion takes place spontaneously inside the sun and other stars, where the temperature is great enough (about 10^8 K) for thermal motion to bring the charged hydrogen ions close enough together to fuse. Although controlled fusion holds great promise as a future energy source, practical difficulties have thus far prevented its development.

Lawson criterion	The minimum product of particle density n and confinement time τ to get more energy out of a fusion reactor than is put in is $n\tau > 10^{20}$ s·particles/m^3.

PROBLEMS

- • Single-concept, single-step, relatively easy
- •• Intermediate-level, may require synthesis of concepts
- ••• Challenging
- SSM Solution is in the *Student Solutions Manual*
- iSOLVE Problems available on iSOLVE online homework service
- iSOLVE ✓ These "Checkpoint" online homework service problems ask students additional questions about their confidence level, and how they arrived at their answer.

In a few problems, you are given more data than you actually need; in a few other problems, you are required to supply data from your general knowledge, outside sources, or informed estimates.

Conceptual Problems

1 • Give the symbols for two other isotopes of (a) ^{14}N, (b) ^{56}Fe, and (c) ^{118}Sn.

2 • Why is the decay series $A = 4n + 1$ not found in nature?

3 • A decay by α emission is often followed by β decay. When this occurs, it is by β^- and not β^+ decay. Why?

4 • SSM The half-life of ^{14}C is much less than the age of the universe, yet ^{14}C is found in nature. Why?

5 • What effect would a long-term variation in cosmic-ray activity have on the accuracy of ^{14}C dating?

6 • Why is there not an element with Z = 130?

7 • Why is a moderator needed in an ordinary nuclear fission reactor?

8 • Explain why water is more effective than lead in slowing down fast neutrons.

9 • The stable isotope of sodium is ^{23}Na. What kind of radioactivity would you expect of (a) ^{22}Na and (b) ^{24}Na?

10 • What is the advantage of a breeder reactor over an ordinary reactor? What are the disadvantages?

11 • True or false:
(a) The atomic nucleus contains protons, neutrons, and electrons.
(b) The mass of ^{2}H is less than the mass of a proton plus a neutron.
(c) After two half-lives, all the radioactive nuclei in a given sample have decayed.
(d) In a breeder reactor, fuel can be produced as fast as it is consumed.

12 • Why do extreme changes in the temperature or the pressure of a radioactive sample have little or no effect on the radioactivity?

13 • SSM Write balanced reactions for each of the following nuclear decays: (a) beta decay of ^{16}N, (b) alpha decay of ^{248}Fm, (c) positron decay of ^{12}N, (d) beta decay of ^{81}Se, (e) positron decay of ^{61}Cu, and (f) alpha decay of ^{228}Th.

14 • SSM Write and balance reaction equations for each of the following: (a) ^{240}Pu undergoes spontaneous fission to form two fission fragments and three neutrons. One of the fission fragments is a ^{90}Sr nucleus. (b) A ^{72}Ge nucleus absorbs an alpha particle and ejects a photon. (c) An ^{127}I nucleus absorbs a deuteron and ejects a neutron. (d) A ^{235}U nucleus absorbs a slow neutron and fissions forming a ^{113}Ag nucleus, two neutrons, and another fission fragment. (e) A ^{55}Mn nucleus is struck with a high-energy ^{7}Li nucleus resulting in a triton, ^{3}H, and a new nucleus. (f) ^{238}U absorbs a slow neutron, resulting in a compound nucleus that emits a beta particle, followed a short time later by a second beta particle. What is the resulting nucleus?

Estimation and Approximation

15 • We found in Chapter 25 that the ratio of the resistivity of the most insulating material to that of the least resistive material (excluding superconductors) is approximately 10^{22}. There are very few properties of materials that show such a wide range of values. Using information in the textbook or other resources, find the ratio of largest to smallest for some nuclear related properties of matter. Some examples might be the range of mass densities found in an atom, the half-life of radioactive nuclei, or the range of nuclear masses.

16 •• According to the United States Department of Energy, the U.S. population consumes approximately 10^{20} joules of energy each year. Estimate the mass (in kg) of (a) uranium that would be needed to produce this much energy using nuclear fission and (b) deuterium and tritium that would be needed to produce this much energy using nuclear fusion.

Properties of Nuclei

17 • SSM Calculate the binding energy and the binding energy per nucleon from the masses given in Table 40-1 for (a) ^{12}C, (b) ^{56}Fe, and (c) ^{238}U.

18 • Repeat Problem 17 for (a) ^{6}Li, (b) ^{39}K, and (c) ^{208}Pb.

19 • Use Equation 40-1 to compute the radii of the following nuclei: (a) ^{16}O, (b) ^{56}Fe, and (c) ^{197}Au.

20 • In a fission process, a ^{239}Pu nucleus splits into two nuclei whose mass number ratio is 3 to 1. Calculate the radii of the nuclei formed in this process.

21 •• SSM The neutron, when isolated from an atomic nucleus, decays into a proton, an electron, and an antineutrino as follows: $^1_0n \rightarrow {}^1_1H + {}^0_{-1}e + {}^0_0\bar{v}$. The thermal energy of a neutron is of the order of kT, where k is the Boltzmann constant. (a) In both joules and electron volts, calculate the energy of a thermal neutron at 25°C. (b) What is the speed of this thermal neutron? (c) A beam of monoenergetic thermal neutrons is produced at 25°C with an intensity I. After traveling 1350 km, the beam has an intensity of $I/2$. Using this information, estimate the half-life of the neutron. Express your answer in minutes.

22 • Use Equation 40-1 for the radius of a spherical nucleus and the approximation that the mass of a nucleus of mass number A is $A \times (1u)$ to calculate the density of nuclear matter in grams per cubic centimeter.

23 •• Consider the following fission process: $^{235}_{92}U + ^{1}_{0}n \rightarrow ^{95}_{42}Mo + ^{139}_{57}La + 2^{1}_{0}n$. Determine the electrostatic potential energy, in MeV, of the reaction products when the ^{95}Mo nucleus and the ^{139}La nucleus are just touching immediately after being formed in the fission process.

24 •• SSM In 1920, 12 years before the discovery of the neutron, Ernest Rutherford argued that proton–electron pairs might exist in the confines of the nucleus in order to explain the mass number, A, being greater than the nuclear charge, Z. He also used this argument to account for the source of beta particles in radioactive decay. Rutherford's scattering experiments in 1910 showed that the nucleus had a diameter of approximately 10 fm. Using this nuclear diameter, the uncertainty principle, and given that beta particles have an energy range of 0.02 MeV to 3.40 MeV, show why electrons cannot be contained within the nucleus.

Radioactivity

25 • ISOLVE Homer enters the visitors' chambers, and his Geiger beeper goes off. He shuts off the beep, removes the device from his shoulder patch, and holds it near the only new object in the room—an orb that is to be presented as a gift from the visiting Cartesians. Pushing a button marked "monitor," Homer reads that the orb is a radioactive source with a counting rate of 4000 counts/s. After 10 min, the counting rate has dropped to 1000 counts/s. The source's half-life appears on the Geiger-beeper display. (a) What is the half-life of the source? (b) What will the counting rate be 20 min after the monitoring device was switched on?

26 • ISOLVE ✔ A certain source gives 2000 counts/s at time $t = 0$. Its half-life is 2 min. What is the counting rate after (a) 4 min, (b) 6 min, and (c) 8 min?

27 • ISOLVE ✔ The counting rate from a radioactive source is 8000 counts/s at time $t = 0$, and 10 min later the rate is 1000 counts/s. (a) What is the half-life? (b) What is the decay constant? (c) What is the counting rate after 20 min?

28 • ISOLVE The half-life of radium is 1620 y. Calculate the number of disintegrations per second of 1 g of radium, and show that the disintegration rate is approximately 1 Ci.

29 • ISOLVE A radioactive silver foil ($t_{1/2} = 2.4$ min) is placed near a Geiger counter and 1000 counts/s are observed at time $t = 0$. (a) What is the counting rate at $t = 2.4$ min and at $t = 4.8$ min? (b) If the counting efficiency is 20 percent, how many radioactive nuclei are there at time $t = 0$? At time

$t = 2.4$ min? (c) At what time will the counting rate be about 30 counts/s?

30 • Use Table 40-1 to calculate the energy in MeV for the α decay of (a) ^{226}Ra and (b) ^{242}Pu.

31 •• SSM Plutonium is a highly hazardous and toxic material to the human body. Once it enters the body it collects primarily in the bones, although it also can be found in other organs. Red blood cells are synthesized within the marrow of the bones. The isotope ^{239}Pu is an alpha emitter with a half-life of 24,360 years. Since alpha particles are an ionizing radiation, the blood-making ability of the marrow is, in time, destroyed by the presence of ^{239}Pu. In addition, many kinds of cancers will also be initiated in the surrounding tissues by the ionizing effects of the alpha particles. (a) If a person accidentally ingested 2.0 μg of ^{239}Pu and it is absorbed by the bones of the victim, how many alpha particles are produced per second within the skeleton? (b) When, in years, will the activity be 1000 alpha particles per second?

32 •• Consider a parent, $^{A}_{Z}X$, alpha-emitting nucleus initially at rest. The nucleus decays into a daughter nucleus, $^{A-4}_{Z-2}Y$, and an alpha particle as follows: $^{A}_{Z}X \rightarrow ^{A-4}_{Z-2}Y + ^{4}_{2}\alpha + Q$. (a) Show that the alpha particle has a kinetic energy of $(A-4)Q/A$. (b) Show that the kinetic energy of the recoiling daughter nucleus is given by $K_Y = 4Q/A$.

33 •• SSM The fissile material ^{239}Pu is an alpha emitter. Write the equation of this reaction. Given that ^{239}Pu, ^{235}U, and an alpha particle have respective masses of 239.052 156 u, 235.043 923 u, and 4.002 603 u, use the relations appearing in Problem 32 to calculate the kinetic energies of the alpha particle and the recoiling daughter nucleus.

34 • Through a friend in the security department at the museum, Angela obtained a sample with 175 g of carbon. The decay rate of ^{14}C was 8.1 Bq. How old is the sample?

35 • ISOLVE A sample of a radioactive isotope is found to have an activity of 115.0 Bq immediately after it is pulled from the reactor that formed the isotope. Its activity 2 h 15 min later is measured to be 85.2 Bq. (a) Calculate the decay constant and the half-life of the sample. (b) How many radioactive nuclei were there in the sample initially?

36 •• SSM Radiation has long been used in medical therapy to control the development and growth of cancer cells. Cobalt-60, a gamma emitter of 1.17 MeV and 1.33 MeV energies, is used to irradiate and destroy deep-seated cancers. Small needles made of ^{60}Co of a specified activity are encased in gold and used as body implants in tumors for time periods that are related to tumor size, tumor cell reproductive rate, and the activity of the needle. (a) A 1.00 μg sample of ^{60}Co, with a half-life of 5.27 y, is prepared in the cyclotron of a medical center to irradiate a small internal tumor with gamma rays. In curies, determine the initial activity of the sample. (b) What is the activity of the sample after 1.75 y?

37 •• ISOLVE Measurements of the activity of a radioactive sample have yielded the results shown in the following table. Plot the activity as a function of time, using semilogarithmic paper, and determine the decay constant and the half-life of the radioisotope.

Time, min	Activity	Time, min	Activity
0	4287	20	880
5	2800	30	412
10	1960	40	188
15	1326	60	42

38 •• (a) Show that if the decay rate is R_0 at time $t = 0$ and R_1 at some later time t_1, the decay constant is given by $\lambda = t_1^{-1} \ln(R_0/R_1)$ and the half-life is given by $t_{1/2} = 0.693 t_1 / \ln(R_0/R_1)$. (b) Use these results to find the decay constant and the half-life if the decay rate is 1200 Bq at $t = 0$ and 800 Bq at $t_1 = 60$ s.

39 •• **ISOLVE** A wooden casket is thought to be 18,000 years old. How much carbon would have to be recovered from this object to yield a ^{14}C counting rate of no less than 5 counts/min?

40 •• **ISOLVE** A 1.00-mg sample of substance of atomic mass 59.934 u emits β particles with an activity of 1.131 Ci. Find the decay constant for this substance in s^{-1} and its half-life in years.

41 •• **SSM** The counting rate from a radioactive source is measured every minute. The resulting counts per second are 1000, 820, 673, 552, 453, 371, 305, and 250. Plot the counting rate versus time on semilogarithmic graph paper, and use your graph to find the half-life of the source.

42 •• **ISOLVE** A sample of radioactive material is initially found to have an activity of 115.0 decays/min. After 4 d 5 h, its activity is measured to be 73.5 decays/min. (a) Calculate the half-life of the material. (b) How long (from the initial time) will it take for the sample to reach an activity level of 10.0 decays/min?

43 •• **ISOLVE** ✓ The rubidium isotope ^{87}Rb is a β emitter with a half-life of 4.9×10^{10} y that decays into ^{87}Sr. It is used to determine the age of rocks and fossils. Rocks containing the fossils of early animals contain a ratio of ^{87}Sr to ^{87}Rb of 0.0100. Assuming that there was no ^{87}Sr present when the rocks were formed, calculate the age of these fossils.

44 ••• If there are N_0 radioactive nuclei at time $t = 0$, the number that decay in some time interval dt at time t is $-dN = \lambda N_0 e^{-\lambda t}\, dt$. If we multiply this number by the lifetime t of these nuclei, sum over all the possible lifetimes from $t = 0$ to $t = \infty$, and divide by the total number of nuclei, we get the mean lifetime τ.

$$\tau = \frac{1}{N_0} \int_0^\infty t|dN| = \int_0^\infty t\lambda e^{-\lambda t}\, dt$$

Show that $\tau = 1/\lambda$.

Nuclear Reactions

45 • Using Table 40-1, find the Q values for the following reactions: (a) ^{1}H + ^{3}H → ^{3}He + n + Q and (b) ^{2}H + ^{2}H → ^{3}He + n + Q.

46 • Using Table 40-1, find the Q values for the following reactions: (a) ^{2}H + ^{2}H → ^{3}H + ^{1}H + Q, (b) ^{2}H + ^{3}He → ^{4}He + ^{1}H + Q, and (c) ^{6}Li + n → ^{3}H + ^{4}He + Q.

47 •• **SSM** (a) Use the atomic masses $m = 14.003\ 242$ u for $^{14}_{6}$C and $m = 14.003\ 074$ u for $^{14}_{7}$N to calculate the Q value (in MeV) for the β decay

$$^{14}_{6}\text{C} \to {}^{14}_{7}\text{N} + \beta^- + \bar{v}_e$$

(b) Explain why you do not need to add the mass of the β^- to that of atomic $^{14}_{7}$N for this calculation.

48 •• (a) Use the atomic masses $m = 13.005\ 738$ u for $^{13}_{7}$N and $m = 13.003\ 354$ u for $^{13}_{6}$C to calculate the Q value (in MeV) for the β decay

$$^{13}_{7}\text{N} \to {}^{13}_{6}\text{C} + \beta^+ + v_e$$

(b) Explain why you need to add two electron masses to the mass of $^{13}_{6}$C in the calculation of the Q value for this reaction.

Fission and Fusion

49 • **SSM** **ISOLVE** ✓ Assuming an average energy of 200 MeV per fission, calculate the number of fissions per second needed for a 500-MW reactor.

50 • **ISOLVE** If the reproduction factor in a reactor is $k = 1.1$, find the number of generations needed for the power level to (a) double, (b) increase by a factor of 10, and (c) increase by a factor of 100. Find the time needed in each case if (d) there are no delayed neutrons, so that the time between generations is 1 ms and (e) there are delayed neutrons that make the average time between generations 100 ms.

51 •• **SSM** Consider the following fission reaction: $^{235}_{92}$U + $^1_0$n → $^{95}_{42}$Mo + $^{139}_{57}$La + 2^1_0n + Q. The masses of the neutron, U, Mo, and La are 1.008 665 u, 235.043 923 u, 94.905 842 u, and 138.906 348 u, respectively. Calculate the Q value, in MeV, for this fission reaction. Compare the result to the result obtained in Problem 23.

52 •• **ISOLVE** In 1989, researchers claimed to have achieved fusion in an electrochemical cell at room temperature. They claimed a power output of 4 W from deuterium fusion reactions in the palladium electrode of their apparatus. If the two most likely reactions are

$$^2\text{H} + {}^2\text{H} \to {}^3\text{He} + \text{n} + 3.27\ \text{MeV}$$

and

$$^2\text{H} + {}^2\text{H} \to {}^3\text{He} + {}^1\text{H} + 4.03\ \text{MeV}$$

with 50 percent of the reactions going by each branch, how many neutrons per second would we expect to be emitted in the generation of 4 W of power?

53 •• **ISOLVE** A fusion reactor that uses only deuterium for fuel would have the two reactions in Problem 52 taking place in the reactor. The ^{3}H produced in the second reaction reacts immediately with another ^{2}H to produce

$$^3\text{H} + {}^2\text{H} \to {}^4\text{He} + \text{n} + 17.6\ \text{MeV}$$

The ratio of ^{2}H to ^{1}H atoms in naturally occurring hydrogen is 1.5×10^{-4}. How much energy would be produced from 4 L of water if all of the ^{2}H nuclei undergo fusion?

54 ••• SSM ISOLVE The fusion reaction between ^{2}H and ^{3}H is

$$^3H + ^2H \rightarrow {}^4He + n + 17.6 \text{ MeV}$$

Using the conservation of momentum and the given Q value, find the final energies of both the ^{4}He nucleus and the neutron, assuming the initial kinetic energy of the system is 1.00 MeV and the initial momentum of the system is zero.

55 ••• Energy is generated in the sun and other stars by fusion. One of the fusion cycles, the proton–proton cycle, consists of the following reactions:

$$^1H + {}^1H \rightarrow {}^2H + \beta^+ + v_e$$

$$^1H + {}^2H \rightarrow {}^3He + \gamma$$

followed by

$$^1H + {}^3H \rightarrow {}^4He + \beta^+ + v_e$$

(a) Show that the net effect of these reactions is

$$4{}^1H \rightarrow {}^4He + 2\beta^+ + 2v_e + \gamma$$

(b) Show that rest energy of 24.7 MeV is released in this cycle (not counting the energy of 1.02 MeV released when each positron meets an electron and the two annihilate). (c) The sun radiates energy at the rate of approximately 4×10^{26} W. Assuming that this is due to the conversion of four protons into helium plus γ-rays and neutrinos, which releases 26.7 MeV, what is the rate of proton consumption in the sun? How long will the sun last if it continues to radiate at its present level? (Assume that protons constitute about half of the total mass, 2×10^{30} kg, of the sun.)

General Problems

56 • (a) Show that $ke^2 = 1.44$ MeV·fm, where k is the Coulomb constant and e is the electron charge. (b) Show that $hc = 1240$ MeV·fm.

57 • SSM ISOLVE ✓ The counting rate from a radioactive source is 6400 counts/s. The half-life of the source is 10 s. Make a plot of the counting rate as a function of time for times up to 1 min. What is the decay constant for this source?

58 • Find the energy needed to remove a neutron from (a) ^{4}He and (b) ^{7}Li.

59 • The isotope ^{14}C decays according to $^{14}C \rightarrow {}^{14}N + e^- + \overline{v}_e$. The atomic mass of ^{14}N is 14.003 074 u. Determine the maximum kinetic energy of the electron. (Neglect recoil of the nitrogen atom.)

60 • ISOLVE A neutron star is an object of nuclear density. If our sun were to collapse to a neutron star, what would be the radius of that object?

61 •• SSM Show that the ^{109}Ag nucleus is stable against alpha decay, $^{109}_{47}Ag \rightarrow {}^4_2He + {}^{105}_{45}Rh + Q$. The mass of the ^{109}Ag nucleus is 108.904 756 u, and the products of the decay are 4.002 603 u and 104.905 250 u, respectively.

62 •• Gamma rays can be used to induce photofission (fission triggered by the absorption of a photon) in nuclei. Calculate the threshold photon wavelength for the following nuclear reaction: $^2H + \gamma \rightarrow {}^1H + {}^1n$. Use Table 40-1 for the masses of the interacting particles.

63 • The relative abundance of ^{40}K (molecular mass 40.0 g/mol) is 1.2×10^{-4}. The isotope ^{40}K is radioactive with a half-life of 1.3×10^9 y. Potassium is an essential element of every living cell. In the human body the mass of potassium constitutes approximately 0.36 percent of the total mass. Determine the activity of this radioactive source in a student whose mass is 60 kg.

64 •• When a positron makes contact with an electron, the electron–positron pair annihilate via the reaction $\beta^+ + \beta^- \rightarrow 2\gamma$. Calculate the minimum total energy, in MeV, of the two photons created when a positron–electron pair annihilate.

65 •• The isotope ^{24}Na is a β emitter with a half-life of 15 h. A saline solution containing this radioactive isotope with an activity of 600 kBq is injected into the bloodstream of a patient. Ten hours later, the activity of 1 mL of blood from this individual yields a counting rate of 60 Bq. Determine the volume of blood in this patient.

66 •• SSM (a) Determine the closest distance of approach of an 8-MeV α particle in a head-on collision with a nucleus of ^{197}Au and a nucleus of ^{10}B, neglecting the recoil of the struck nuclei. (b) Repeat the calculation taking into account the recoil of the struck nuclei.

67 •• Twelve nucleons are in a one-dimensional infinite square well of length $L = 3$ fm. (a) Using the approximation that the mass of a nucleon is 1 u, find the lowest energy of a nucleon in the well. Express your answer in MeV. What is the ground-state energy of the system of 12 nucleons in the well if (b) all the nucleons are neutrons so that there can be only 2 in each state and (c) 6 of the nucleons are neutrons and 6 are protons so that there can be 4 nucleons in each state? (Neglect the energy of Coulomb repulsion of the protons.)

68 •• The helium nucleus or α particle is a very tightly bound system. Nuclei with $N = Z = 2n$, where n is an integer (e.g., ^{12}C, ^{16}O, ^{20}Ne, and ^{24}Mg), may be thought of as agglomerates of α particles. (a) Use this model to estimate the binding energy of a pair of α particles from the atomic masses of ^{4}He and ^{16}O. Assume that the four α particles in ^{16}O form a regular tetrahedron with one α particle at each vertex. (b) From the result obtained in Part (a) determine, on the basis of this model, the binding energy of ^{12}C and compare your result with the result obtained from the atomic mass of ^{12}C.

69 •• Radioactive nuclei with a decay constant of λ are produced in an accelerator at a constant rate R_p. The number of radioactive nuclei N then obeys the equation $dN/dt = R_p - \lambda N$. (a) If N is zero at $t = 0$, sketch N versus t for this situation. (b) The isotope ^{62}Cu is produced at a rate of 100 per second by placing ordinary copper (^{63}Cu) in a beam of high-energy photons. The reaction is

$$\lambda + {}^{63}Cu \rightarrow {}^{62}Cu + n$$

^{62}Cu decays by β decay with a half-life of 10 min. After a time long enough so that $dN/dt \approx 0$, how many ^{62}Cu nuclei are there?

70 •• SSM ISOLVE The total energy consumed in the United States in 1 y is approximately 7.0×10^{19} J. How many kilograms of ^{235}U would be needed to provide this amount of energy if we assume that 200 MeV of energy is released by each fissioning uranium nucleus, that all of the uranium

observed only in nuclear reactions with high-energy accelerators. In addition to the usual particle properties of mass, charge, and spin, new properties have been found and given whimsical names such as strangeness, charm, color, topness, and bottomness.

➤ In this chapter, we will first look at the various ways of classifying the multitude of particles that have been found. We will then describe the current theory of elementary particles, called the *standard model,* in which all matter in nature—from the exotic particles produced in the giant accelerator laboratories to ordinary grains of sand—is considered to be constructed from just two families of elementary particles, leptons and quarks. In the final section, we will use our knowledge of elementary particles to discuss the big bang theory of the origin of the universe.

41-1 Hadrons and Leptons

All the different forces observed in nature, from ordinary friction to the tremendous forces involved in supernova explosions, can be understood in terms of the four basic interactions: (1) the strong nuclear interaction (also called the hadronic interaction), (2) the electromagnetic interaction, (3) the weak (nuclear) interaction, and (4) the gravitational interaction. The four basic interactions provide a convenient structure for the classification of particles. Some particles participate in all four interactions, whereas other particles participate in only some of the interactions. For example, all particles participate in gravity, the weakest of the interactions. All particles that carry electric charge participate in the electromagnetic interaction.

Particles that interact via the strong interaction are called **hadrons.** There are two kinds of hadrons: **baryons,** which have spin $\frac{1}{2}$ (or $\frac{3}{2}$, $\frac{5}{2}$, etc.), and **mesons,** which have zero or integral spin. Baryons, which include nucleons, are the most massive of the elementary particles. Mesons have intermediate masses between the mass of the electron and the mass of the proton. Particles that decay via the strong interaction have very short lifetimes of the order of 10^{-23} s, which is about the time it takes light to travel a distance equal to the diameter of a nucleus. On the other hand, particles that decay via the weak interaction have much longer lifetimes of the order of 10^{-10} s. Table 41-1 lists some of the properties of those hadrons that are stable against decay via the strong interaction.

Hadrons are rather complicated entities with complex structures. If we use the term *elementary particle* to mean a point particle without structure that is not constructed from some more elementary entities, hadrons do not fit the bill. It is now believed that all hadrons are composed of more fundamental entities called *quarks,* which are truly elementary particles.

Particles that participate in the weak interaction but not in the strong interaction are called **leptons.** These include electrons, muons, and neutrinos, which are all less massive than the lightest hadron. The word *lepton,* meaning "light particle," was chosen to reflect the relatively small mass of these particles. However, the most recently discovered lepton, the *tau,* found by Martin Lewis Perl in 1975, has a mass of 1784 MeV/c^2, nearly twice the mass of the proton (938 MeV/c^2), so we now have a "heavy lepton." As far as we know, leptons are point particles with no structure and can be considered to be truly elementary in the sense that they are not composed of other particles.

There are six leptons. They are the electron and the electron neutrino, the muon and the muon neutrino, and the tau and the tau neutrino. (Each of these leptons has an antiparticle.) The masses of the electron, the muon, and the tau are quite different. The mass of the electron is 0.511 MeV/c^2, the mass of the muon is 106 MeV/c^2, and the mass of the tau is 1784 MeV/c^2. The standard model predicts that neutrinos, like photons, are without mass. Neutrinos were originally

The Super-Kamiokande detector, built in Japan in 1996 as a joint Japanese–American experiment, is essentially a water tank the size of a large cathedral installed in a deep zinc mine 1 mile inside a mountain. When neutrinos pass through the tank, one of the nutrinos occasionally collides with an atom, sending blue light through the water to an array of detectors. This photograph shows the detector wall and top with approximately 9000 photomultiplier tubes that help detect the neutrinos. Experimental results reported in June 1998 indicate that the mass of the neutrino cannot be zero.

TABLE 41-1

Hadrons That Are Stable Against Decay via the Strong Nuclear Interaction

Name	Symbol	Mass, MeV/c^2	Spin, $\hbar$	Charge, e	Antiparticle	Mean Lifetime, s	Typical Decay Products[†]
Baryons							
Nucleon	p (proton)	938.3	$\frac{1}{2}$	+1	p^-	Infinite	
	n (neutron)	939.6	$\frac{1}{2}$	0	$\overline{\mathrm{n}}$	930	$\mathrm{p} + \mathrm{e}^- + \overline{\nu}_e$
Lambda	Λ^0	1116	$\frac{1}{2}$	0	$\overline{\Lambda}^0$	2.5×10^{-10}	$\mathrm{p} + \pi^-$
Sigma[‡]	Σ^+	1189	$\frac{1}{2}$	+1	$\overline{\Sigma}^-$	0.8×10^{-10}	$\mathrm{n} + \pi^+$
	Σ^0	1193	$\frac{1}{2}$	0	$\overline{\Sigma}^0$	10^{-20}	$\Lambda^0 + \gamma$
	Σ^-	1197	$\frac{1}{2}$	−1	$\overline{\Sigma}^+$	1.7×10^{-10}	$\mathrm{n} + \pi^-$
Xi	Ξ^0	1315	$\frac{1}{2}$	0	$\overline{\Xi}^0$	3.0×10^{-10}	$\Lambda^0 + \pi^0$
	Ξ^-	1321	$\frac{1}{2}$	−1	$\overline{\Xi}^+$	1.7×10^{-10}	$\Lambda^0 + \pi^-$
Omega	Ω^-	1672	$\frac{3}{2}$	−1	Ω^+	1.3×10^{-10}	$\Xi^0 + \pi^-$
Mesons							
Pion	π^+	139.6	0	+1	π^-	2.6×10^{-8}	$\mu^+ + \nu_\mu$
	π^0	135	0	0	$\overline{\pi}^0$?	0.8×10^{-16}	$\gamma + \gamma$
	π^-	139.6	0	−1	π^+	2.6×10^{-8}	$\mu^- + \overline{\nu}_\mu$
Kaon[§]	K^+	493.7	0	+1	K^-	1.24×10^{-8}	$\pi^+ + \pi^0$
	K^0	497.7	0	0	$\overline{\mathrm{K}}^0$	0.88×10^{-10} and	$\pi^+ + \pi^-$
						5.2×10^{-8}	$\pi^+ + \mathrm{e}^- + \overline{\nu}_e$
Eta	η^0	549	0	0		2×10^{-19}	$\gamma + \gamma$

† Other decay modes also occur for most particles.
‡ The Σ^0 is included here for completeness even though it does decay via the strong interaction.
§ The K^0 has two distinct lifetimes, sometimes referred to as $\mathrm{K}^0_{\text{short}}$ and $\mathrm{K}^0_{\text{long}}$. All other particles have a unique lifetime.

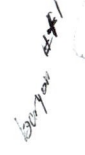

thought to be massless. However, there is now strong evidence that their mass, though very small, is greater than zero. In the late 1990s, experiments using a detector in Japan called the Super-Kamiokande (Super-K) found that neutrinos emitted from the sun arrived on the earth in much smaller numbers than the numbers predicted from the fusion processes in the sun. This can be explained if the mass of the neutrino were not zero.[†] In addition, a neutrino mass as small as a few eV/c^2 would have great cosmological significance. The answer to the question of whether the universe will continue to expand indefinitely or will reach a maximum size and begin to contract depends on the total mass in the universe. Thus, the answer could depend on whether the mass of the neutrino is actually zero, or is merely small, since the cosmic density of each species of neutrino is ~ 100 per cm³. The observation of electron neutrinos from the supernova 1987A puts an upper limit on the mass of these neutrinos. Since the velocity of a particle with mass depends on its energy, the arrival time of a burst of neutrinos with mass from a supernova would be spread out in time. The fact that the electron neutrinos from the 1987 supernova all arrived at the earth within 13 s of one another results in an upper limit of about 16 eV/c^2 for their mass. Note that

† The connection between the shortfall of solar-neutrino detections and the mass of the neutrino is elucidated in "On Morphing Neutrinos and Why They Must Have Mass" by Eugene Hecht, *The Physics Teacher* 41 (2003): 164–168.

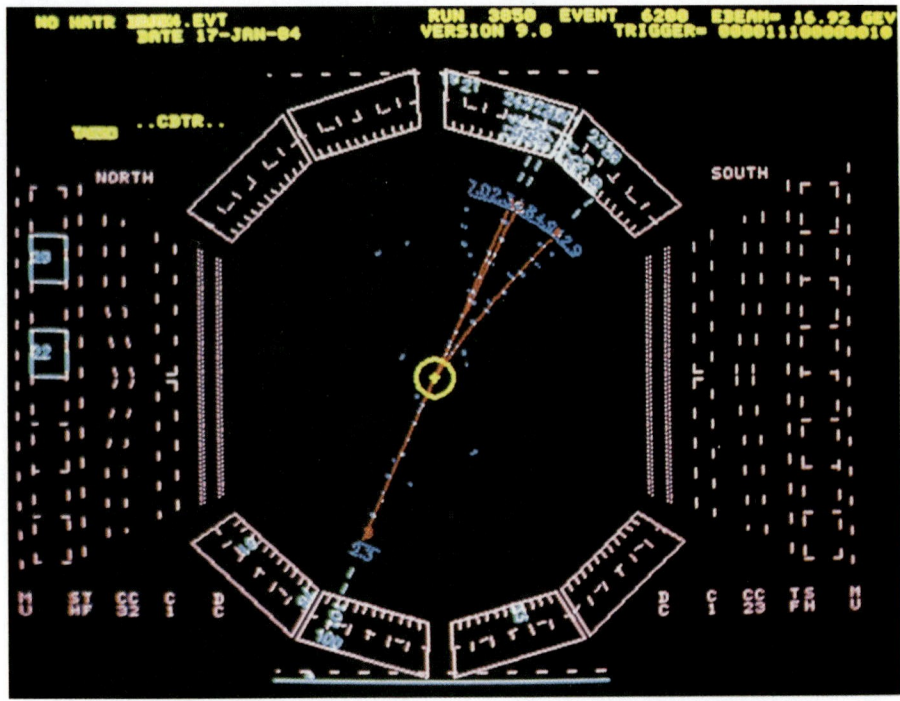

(a)

(b)

(*a*) A computer display of the production and decay of a τ^+ and τ^- pair. An electron and a positron annihilate at the center marked by the yellow cross, producing a τ^+ and τ^- pair, which travel in opposite directions, but quickly decay while still inside the beam pipe (yellow circle). The τ^+ decays into two invisible neutrinos and a μ^+, which travels toward the bottom left. Its track in the drift chamber is calculated by a computer and indicated in red. It penetrates the lead–argon counters outlined in purple and is detected at the blue dot near the bottom blue line that marks the end of a muon detector. The τ^- decays into three charged pions (red tracks moving upward) plus invisible neutrinos. (*b*) The Mark I detector, built by a team from the Stanford Linear Accelerator Center (SLAC) and the Lawrence Berkeley Laboratory, became famous for many discoveries, including the ψ/J meson and the τ lepton. Tracks of particles are recorded by wire spark chambers wrapped in concentric cylinders around the beam pipe extending out to the ring where physicist Carl Friedberg has his right foot. Beyond this are two rings of protruding tubes, housing photomultipliers that view various scintillation counters. The rectangular magnets at the left guide the counterrotating beams that collide in the center of the detector.

an upper limit does not imply that the mass is not zero. Measurements of the relative number of muon neutrinos and electron neutrinos entering the huge, underground Super-K detector suggest that at least one type of neutrino can oscillate between types (e.g., between a mu neutrino and a tau neutrino). Further measurements of antineutrinos from nuclear reactors strongly shows that all three types of neutrinos oscillate between types and thus have mass. Measurements made in Japan, using the *Kamioka Liquid Scintillator Anti-Neutrino Detector* (KamLAND), show that oscillations from one species of neutrino to another species of neutrino can be observed over path lengths as short as 180 km (Figure 41-1).

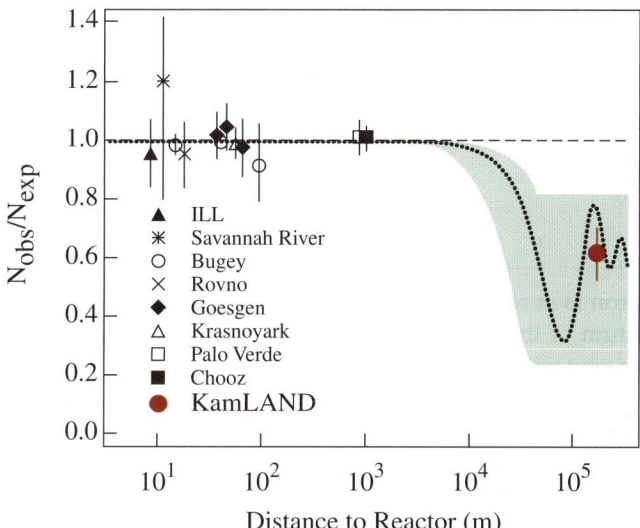

FIGURE 41-1 First evidence for antineutrino disappearance. The ratio of the number of antineutrinos observed N_{obs} to the number that one would expect to observe N_{exp} (assuming no neutrino oscillations) is plotted versus distance to the nearest antineutrino sources. The KamLAND site is 180,000 m (180 km) from nearby antineutrino sources (nuclear reactors), while the other eight detector sites are less than 1000 m from nearby nuclear reactors. For these eight sites, $N_{obs}/N_{exp} = 1.0$, which is what is expected assuming no neutrino oscillations. However, the KamLAND detector found $N_{obs}/N_{exp} = 0.6$. This result is strong evidence that while neutrinos do not oscillate in significant numbers while traveling over path lengths of less than 1.0 km, they do oscillate in significant numbers while traveling over path lengths only a few orders of magnitude longer than that.

41-2 Spin and Antiparticles

One important characteristic of a particle is its intrinsic spin angular momentum. We have already discussed the fact that the electron has a quantum number m_s that corresponds to the z component of its intrinsic spin characterized by the quantum number $s = \frac{1}{2}$. Protons, neutrons, neutrinos, and the various other particles that also have an intrinsic spin characterized by the quantum number $s = \frac{1}{2}$ are called **spin-$\frac{1}{2}$ particles.** Particles that have spin $\frac{1}{2}$ (or $\frac{3}{2}$, $\frac{5}{2}$, etc.) are called fermions and obey the Pauli exclusion principle. Particles such as pions and other mesons have zero spin or integral spin ($s = 0, 1, 2$, etc.). These particles are called bosons and do not obey the Pauli exclusion principle. That is, any number of these particles can be in the same quantum state.

Spin-$\frac{1}{2}$ particles are described by the Dirac equation, which is an extension of the Schrödinger equation that includes special relativity. One feature of Paul Dirac's theory, proposed in 1927, is the prediction of the existence of antiparticles. In special relativity, the energy of a particle is related to the mass and the momentum of the particle by $E = \pm\sqrt{p^2c^2 + m^2c^4}$ (Equation 39-28). We usually choose the positive solution and dismiss the negative-energy solution with a physical argument. However, the Dirac equation requires the existence of wave functions that correspond to the negative-energy states. Dirac got around this difficulty by postulating that all the negative-energy states were filled and would therefore not be observable. Only holes in the "infinite sea" of negative-energy states would be observed. For example, a hole in the negative sea of electron energy states would appear as a particle identical to the electron except with positive charge. When such a particle came in the vicinity of an electron the two particles would annihilate, releasing two photons with a total energy of $2m_ec^2$, where m_e is the mass of the electron. This interpretation received little attention until a particle with just these properties, called the positron, was discovered in 1932 by Carl Anderson.

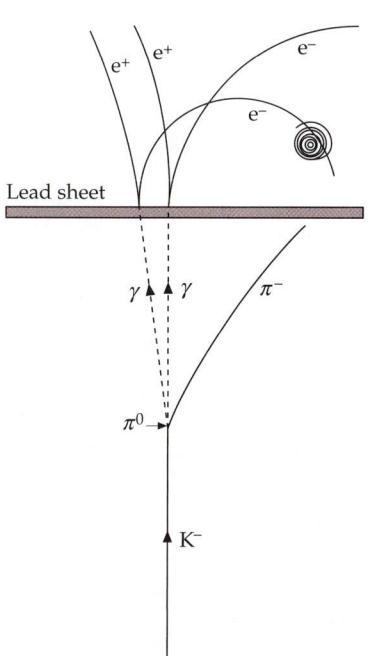

A negative kaon (K^-) enters a bubble chamber from the bottom and decays into a π^-, which moves off to the right, and a π^0, which immediately decays into two photons whose paths are indicated by the dashed lines in the drawing. Each photon interacts in the lead sheet, producing an electron–positron pair. The spiral at the right is another electron that has been knocked out of an atom in the chamber. (Other extraneous tracks have been removed from the photograph.)

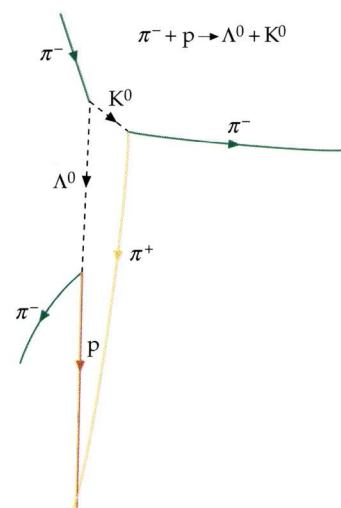

$$\pi^- + p \rightarrow \Lambda^0 + K^0$$

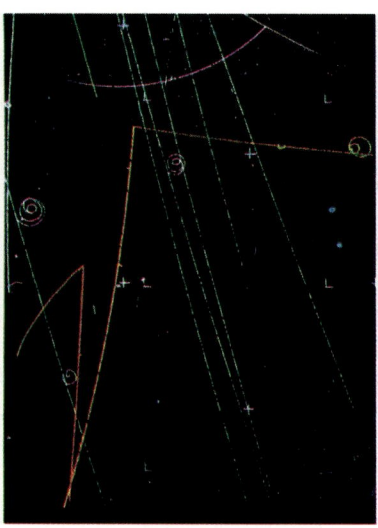

An early photograph of bubble-chamber tracks at the Lawrence Berkeley Laboratory, showing the production and the decay of two strange particles, the K^0 and the Λ^0. These neutral particles are identified by the tracks of their decay particles. The lambda particle was named because of the similarity of the tracks of its decay particles to the Greek letter Λ. (The blue tracks are particles not involved in the reaction of Equation 41-6.)

and

$$K^0 \rightarrow \pi^+ + \pi^- \tag{41-8}$$

However, the decay times for both the Λ^0 and K^0 are of the order of 10^{-10} s, which is characteristic of the weak interaction, rather than 10^{-23} s, which would be expected for the strong interaction. Other particles showing similar behavior were called **strange particles.** These particles are always produced in pairs and never singly, even when all other conservation laws are met. This behavior is described by assigning a new property called strangeness to these particles. In reactions and decays that occur via the strong interaction, strangeness is conserved. In reactions and decays that occur via the weak interaction, the strangeness can change by ± 1. The strangeness of the ordinary hadrons—the nucleons and pions—was arbitrarily taken to be zero. The strangeness of the K^0 was arbitrarily chosen to be $+1$. The strangeness of the Λ^0 particle must then be -1 so that strangeness is conserved in the reaction of Equation 41-6. The strangeness of other particles could then be assigned by looking at their various reactions and decays. In reactions and decays that occur via the weak interaction, the strangeness can change by ± 1.

Figure 41-3 shows the masses of the baryons and the mesons that are stable against decay via the strong interaction versus strangeness. We can see from this figure that these particles cluster in multiplets of one, two, or three particles of approximately equal mass, and that the strangeness of a multiplet of particles is related to the *center of charge* of the multiplet.

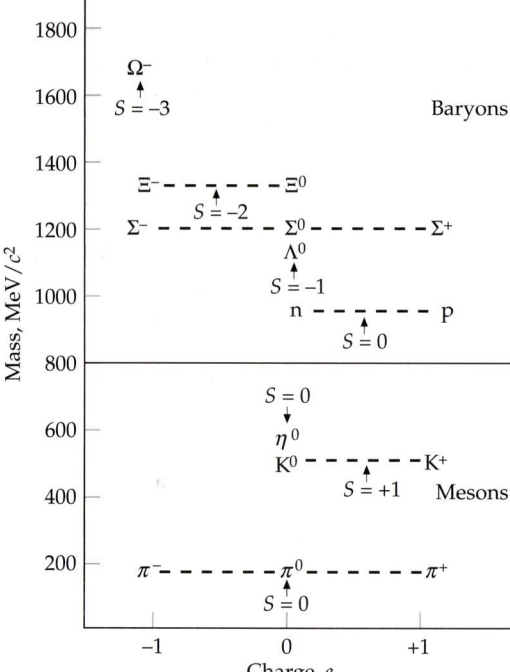

FIGURE 41-3 The strangeness of hadrons shown on a plot of mass versus charge. The strangeness of a baryon-charge multiplet is related to the number of places the center of charge of the multiplet is displaced from that of the nucleon doublet. For each displacement of e, the strangeness changes by ± 1. For mesons, the strangeness is related to the number of places the center of charge is displaced from that of the pion triplet. Because of the unfortunate original assignment of $+1$ for the strangeness of kaons, all of the baryons that are stable against decay via the strong interaction have negative or zero strangeness.

EXAMPLE 41-3

STRONG INTERACTION, WEAK INTERACTION, OR NO INTERACTION

State whether the following decays can occur via the strong interaction, via the weak interaction, or not at all: (*a*) $\Sigma^+ \rightarrow p + \pi^0$, (*b*) $\Sigma^0 \rightarrow \Lambda^0 + \gamma$, and (*c*) $\Xi^0 \rightarrow n + \pi^0$.

PICTURE THE PROBLEM We first note that the mass of each decaying particle is greater than the mass of the decay products, so there is no problem with energy conservation in any of the decays. In addition, there are no leptons involved in any of the decays, and charge and baryon number are both conserved in all the decays. The decay will occur via the strong interaction if strangeness is conserved. If $\Delta S = \pm 1$, the decay will occur via the weak interaction. If $|S|$ changes by more than 1, the decay will not occur.

(*a*) From Figure 41-3, we can see that the strangeness of the Σ^+ is -1, whereas the strangeness of both the proton and the pion is zero. This decay is possible via the weak interaction but not the strong interaction. It is, in fact, one of the decay modes of the Σ^+ particle with a lifetime of the order of 10^{-10} s.

(*b*) Since the strangeness of both the Σ^0 and Λ^0 is -1, this decay can proceed via the strong interaction. It is, in fact, the dominant mode of decay of the Σ^0 particle with a lifetime of approximately 10^{-20} s.

(*c*) The strangeness of the Ξ^0 is -2, whereas the strangeness of both the neutron and the pion is zero. Since strangeness cannot change by 2 in a decay or in a reaction, this decay cannot occur.

41-4 Quarks

Leptons appear to be truly elementary particles in that they do not break down into smaller entities and they seem to have no measurable size or structure. Hadrons, on the other hand, are complex particles with size and structure, and they decay into other hadrons. Furthermore, at the present time, there are only six known leptons, whereas there are many more hadrons. Except for the Σ^0 particle, Table 41-1 includes only hadrons that are stable against decay via the strong interaction. Hundreds of other hadrons have been discovered; their properties, such as charge, spin, mass, strangeness, and decay schemes, have been measured.

The most important advance in our understanding of elementary particles was the quark model proposed by M. Gell-Mann and G. Zweig in 1963 in which all hadrons consist of combinations of two or three truly elementary particles called **quarks.**[†] In the original model, quarks came in three types, called **flavors,** labeled u, d, and s (for *up*, *down*, and *strange*). An unusual property of quarks is that they carry fractional electron charges. The charge of the u quark is $+\frac{2}{3}e$ and the charge of the d and s quarks is $\frac{1}{3}e$. Each quark has spin $\frac{1}{2}$ and a baryon number of $\frac{1}{3}$. The strangeness of the u and d quark is 0, and the strangeness of the s quark is -1. Each quark has an antiquark with the opposite electric charge, baryon number, and strangeness. Baryons consist of three quarks (or three antiquarks for antiparticles), whereas mesons consist of a quark and an antiquark, giving mesons a baryon number $B = 0$, as required. The proton consists of the combination uud and the neutron consists of the combination udd. Baryons with a strangeness $S = -1$ contain one s quark. All the particles listed in Table 41-1 can be constructed from these three quarks and three antiquarks.[‡] The great strength of the quark model is that all the allowed combinations of three quarks or

[†] The name *quark* was chosen by M. Gell-Mann from a quotation from *Finnegan's Wake* by James Joyce.

[‡] The correct quark combinations of hadrons are not always obvious, because of the symmetry requirements on the total wave function. For example, the π^0 meson is represented by a linear combination of $u\bar{u}$ and $d\bar{d}$.

quark–antiquark pairs result in known hadrons. Strong evidence for the existence of quarks inside a nucleon is provided by high-energy scattering experiments called *deep inelastic scattering*. In these experiments, a nucleon is bombarded with electrons, muons, or neutrinos of energies from 15 GeV to 200 GeV. Analyses of particles scattered at large angles indicate that inside the nucleon are three spin-$\frac{1}{2}$ particles of sizes much smaller than that of the nucleon. These experiments are analogous to Rutherford's scattering of α particles by atoms in which the presence of a tiny nucleus in the atom was inferred from the large-angle scattering of the α particles.

Given the Constituent Quark Species, Identify the Particle

EXAMPLE 41-4

What are the properties of the particles made up of the following quarks: (a) $u\bar{d}$, (b) $\bar{u}d$, (c) dds, and (d) uss?

PICTURE THE PROBLEM Baryons are made up of three quarks, whereas mesons consist of a quark and an antiquark. We add the electric charges of the quarks to find the total charge of the hadron. We also find the strangeness of the hadron by adding the strangeness of the quarks.

(a) Because $u\bar{d}$ is a quark–antiquark combination, it has baryon number 0 and is therefore a meson. There is no strange quark here, so the strangeness of the meson is zero. The charge of the up quark is $+\frac{2}{3}e$ and the charge of the antidown quark is $+\frac{1}{3}e$, so the charge of the meson is $+1e$. This is the quark combination of the π^+ meson.

(b) The particle $\bar{u}d$ is also a meson with zero strangeness. Its electric charge is $-\frac{2}{3}e + (-\frac{1}{3}e) = -1e$. This is the quark combination of the π^- meson.

(c) The particle dds is a baryon with strangeness -1 because it contains one strange quark. Its electric charge is $-\frac{1}{3}e - \frac{1}{3}e - \frac{1}{3}e = -1e$. This is the quark combination for the Σ^- particle.

(d) The particle uss is a baryon with strangeness -2. Its electric charge is $+\frac{2}{3}e - \frac{1}{3}e - \frac{1}{3}e = 0$. This is the quark combination for the Ξ^0 particle.

MASTER the CONCEPT WEB

In 1967, a fourth quark was proposed to explain some discrepancies between experimental determinations of certain decay rates and calculations based on the quark model. The fourth quark is labeled c for a new property called **charm.** Like strangeness, charm is conserved in strong interactions but changes by ± 1 in weak interactions. In 1975, a new heavy meson called the ψ/J particle (or simply the ψ particle) was discovered that has the properties expected of a $c\bar{c}$ combination. Since then, other mesons with combinations such as $c\bar{d}$ and $\bar{c}d$ as well as baryons containing the charmed quark, have been discovered. Two more quarks labeled t and b (for *top* and *bottom*) were proposed in the 1970s. In 1977, a massive new meson called the Υ meson or **bottomonium,** which is considered to have the quark combination $b\bar{b}$, was discovered. The top quark was observed in 1995. The properties of the six quarks are listed in Table 41-2.

The six quarks and six leptons (and their antiparticles) are thought to be the fundamental elementary particles of which all matter is composed. Table 41-3 lists the masses of the fundamental particles. In this table, the masses given for neutrinos are upper limits. The masses given for quarks are educated guesses. There is experimental evidence for the existence of each of these particles.

TABLE 41-2

Properties of Quarks and Antiquarks

Flavor	Spin	Charge	Baryon Number	Strangeness	Charm	Topness	Bottomness
Quarks							
u (up)	$\frac{1}{2}\hbar$	$+\frac{2}{3}e$	$+\frac{1}{3}$	0	0	0	0
d (down)	$\frac{1}{2}\hbar$	$-\frac{1}{3}e$	$+\frac{1}{3}$	0	0	0	0
s (strange)	$\frac{1}{2}\hbar$	$-\frac{1}{3}e$	$+\frac{1}{3}$	-1	0	0	0
c (charmed)	$\frac{1}{2}\hbar$	$+\frac{2}{3}e$	$+\frac{1}{3}$	0	$+1$	0	0
t (top)	$\frac{1}{2}\hbar$	$+\frac{2}{3}e$	$+\frac{1}{3}$	0	0	$+1$	0
b (bottom)	$\frac{1}{2}\hbar$	$-\frac{1}{3}e$	$+\frac{1}{3}$	0	0	0	$+1$
Antiquarks							
$\bar{u}$	$\frac{1}{2}\hbar$	$-\frac{2}{3}e$	$-\frac{1}{3}$	0	0	0	0
$\bar{d}$	$\frac{1}{2}\hbar$	$+\frac{1}{3}e$	$-\frac{1}{3}$	0	0	0	0
$\bar{s}$	$\frac{1}{2}\hbar$	$+\frac{1}{3}e$	$-\frac{1}{3}$	$+1$	0	0	0
$\bar{c}$	$\frac{1}{2}\hbar$	$-\frac{2}{3}e$	$-\frac{1}{3}$	0	-1	0	0
$\bar{t}$	$\frac{1}{2}\hbar$	$-\frac{2}{3}e$	$-\frac{1}{3}$	0	0	-1	0
$\bar{b}$	$\frac{1}{2}\hbar$	$+\frac{1}{3}e$	$-\frac{1}{3}$	0	0	0	-1

Quark Confinement

Despite considerable experimental effort, no isolated quark has ever been observed. It is now believed that it is impossible to obtain an isolated quark. Although the force between quarks is not known, it is believed that the potential energy of two quarks increases with increasing separation distance so that an infinite amount of energy would be needed to separate the quarks completely. This would be true, for example, if the force of attraction between two quarks remains constant or increases with separation distance, rather than decreasing with increasing separation distance as is the case for other fundamental forces, such as the electric force between two charges, the gravitational force between two masses, and the strong nuclear force between two hadrons.

When a large amount of energy is added to a quark system, such as a nucleon, a quark–antiquark pair is created and the original quarks remain confined within the original system. Because quarks cannot be isolated, but are always bound in a baryon or a meson, the mass of a quark cannot be accurately known, which is why the masses listed in Table 41-3 are merely educated guesses.

TABLE 41-3

Masses of Fundamental Particles

Particle	Mass
Quarks	
u (up)	336 MeV/c^2
d (down)	338 MeV/c^2
s (strange)	540 MeV/c^2
c (charmed)	1,500 MeV/c^2
t (top)	174,000 MeV/c^2
b (bottom)	4,500 MeV/c^2
Leptons	
e^- (electron)	0.511 MeV/c^2
ν_e (electron neutrino)	< 7 eV/c^2
μ^- (muon)	105.659 MeV/c^2
ν_μ (muon neutrino)	< 0.27 MeV/c^2
τ^- (tau)	1,784 MeV/c^2
ν_τ (tau neutrino)	< 31 MeV/c^2

41-5 Field Particles

In addition to the six fundamental leptons and six fundamental quarks, there are other particles, called *field particles* or *field quanta,* that are associated with the forces exerted by one elementary particle on another. In **quantum electrodynamics,** the electromagnetic field of a single charged particle is described by **virtual photons** that are continuously being emitted and reabsorbed by the particle. If we put energy into the system by accelerating the charge, some of these virtual photons are shaken off and

Chapter 34

1. *(c)*

3. *(a)* True
   (b) False
   (c) True
   (d) True

5. *(a)*

7. *(a)* True
   (b) True
   (c) True
   (d) False

9. *(c)*

11. In the photoelectric effect, an electron absorbs the energy of a single photon. Therefore, $K_{max} = hf - \phi$, independent of the number of photons incident on the surface. However, the number of photons incident on the surface determines the number of electrons that are emitted.

13. According to quantum theory, the average value of many measurements of the same quantity will yield the expectation value of that quantity. However, any single measurement may differ from the expectation value.

15. *(a)*

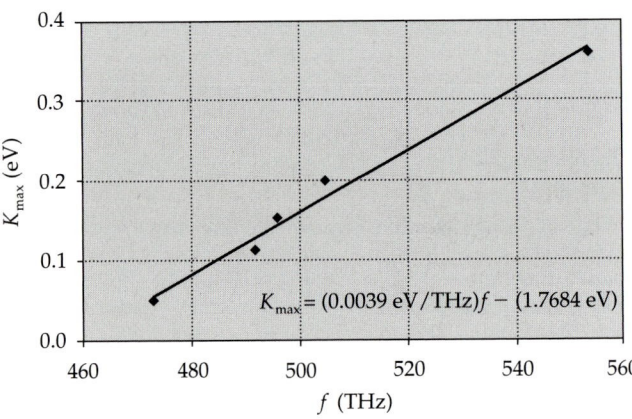

$K_{max} = (0.0039 \text{ eV/THz})f - (1.7684 \text{ eV})$

(b) 1.77 eV
(c) Cesium

17. Soccer ball

19. *(a)* 2.42×10^{14} Hz
    (b) 2.42×10^{17} Hz
    (c) 2.42×10^{20} Hz

21. *(a)* 12.4 keV
    (b) 1.24 GeV

23. 1.95×10^{16} s^{-1}

25. *(a)* 4.13 eV
    (b) 2.10 eV
    (c) 0.784 eV
    (d) 590 nm

27. *(a)* 653 nm; 4.59×10^{14} Hz
    (b) 3.06 eV
    (c) 1.64 eV

29. 1.21 pm

31. 180 pm

33. $p_1 = 9.32 \times 10^{-24}$ kg·m/s; $p_e = 1.80 \times 10^{-23}$ kg·m/s

35. 42

37. 2.91 nm

39. *(a)* $p_e = 2.09 \times 10^{-22}$ N·s; $p_p = 8.95 \times 10^{-21}$ N·s; $p_\alpha = 1.79 \times 10^{-20}$ N·s
    (b) $\lambda_p = 7.41 \times 10^{-14}$ m; $\lambda_e = 3.17 \times 10^{-12}$ m; $\lambda_\alpha = 3.70 \times 10^{-14}$ m

41. 20.2 fm

43. *(a)* 0.820 meV
    (b) 820 MeV

45. 0.167 nm

47. 4.65 pm

49. 1.66×10^{-33} m; This is many orders of magnitude smaller than even the diameter of a proton.

51. 0.0872 nm; This distance is of the order of the size of an atom.

53. *(a)* $E_1 = 206$ MeV; $E_2 = 824$ MeV; $E_3 = 1.85$ GeV

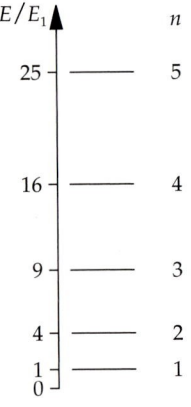

(b) 2.01 fm
    (c) 1.20 fm
    (d) 0.752 fm

55. *(a)* 0.004
    (b) 0.003
    (c) 0

57. *(a)* $\dfrac{L}{2}$
    (b) $0.321L^2$

59. *(a)* $\dfrac{1}{\sqrt{a}}$
    (b) 0.865

61. *(a)* 0.500
    (b) 0.402
    (c) 0.750

65. $<x> = 0; <x^2> = L^2 \left[\dfrac{1}{12} - \dfrac{1}{2\pi^2} \right]$

67. *(a)* 3.10 eV
    (b) 6.25×10^{16} eV
    (c) 2.02×10^{16}

69. *(a)* 1.00 μm; 10^{-16} kg·m/s
    (b) 0.949×10^{12}

71. 121 eV

73. 6.80×10^3 km

75. (a) 3.18 W/m^2
     (b) 1.04×10^{15}
79. 1.28 MeV
81. 1.04 eV; 554 nm
83. (b)   0.2 percent
     (c)   Classically, the energy is continuous. For very large values of n, the energy difference between adjacent levels is infinitesimal.
85. (a) $6.25 \times 10^{-4} \text{ eV/s}$
     (b)   53.3 min

Chapter 35

1.   True
3.   (a)   (b)

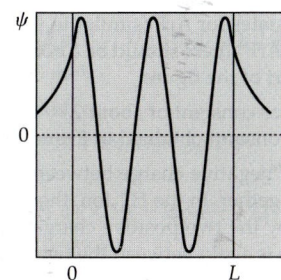 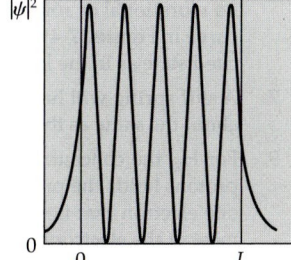

11. (a) 9.49×10^{-9} m
     (b)   4.19 MeV

13. $\Delta x \, \Delta p = \dfrac{\hbar}{2}$

15. (a) $\sqrt{\dfrac{3}{2}} k_1$
     (b)   0.0102
     (c)   0.990
     (d) 9.90×10^5

17. 0.341

19. (a) $r_{1,4\,\text{MeV}} = 6.62 \times 10^{-14}$ m; $r_{1,7\,\text{MeV}} = 3.78 \times 10^{-14}$ m
     (b) $T_{4\,\text{MeV}} = 3.27 \times 10^{-51}$; $T_{7\,\text{MeV}} = 1.55 \times 10^{-4}$

21. $\psi = A \sin\left(\dfrac{n_1 \pi}{L_1} x\right) \sin\left(\dfrac{n_2 \pi}{2L_1} y\right) \sin\left(\dfrac{n_3 \pi}{3L_1} z\right)$

23. $\psi = A \sin\left(\dfrac{n_1 \pi}{L_1} x\right) \sin\left(\dfrac{n_2 \pi}{2L_1} y\right) \sin\left(\dfrac{n_3 \pi}{4L_1} z\right)$

25. (a) $\psi(x, y) = A \sin\dfrac{n\pi}{L} x \sin\dfrac{m\pi}{L} y$
     (b) $E_{n,m} = \dfrac{h^2}{8mL^2} (n^2 + m^2)$
     (c) $E_{1,2} = E_{2,1} = \dfrac{5h^2}{8mL^2}$
     (d) $(1, 7), (7, 1), (5, 5)$; $E = \dfrac{25h^2}{4mL^2}$

27. $E_{0,10\,\text{bosons}} = \dfrac{5h^2}{4mL^2}$

33. $E_0 = \dfrac{5h^2}{mL^2}$; $E_1 = E_2 = \dfrac{21h^2}{4mL^2}$

39. $A_2 = 4\sqrt{\dfrac{8m\omega_0}{h}}$

Chapter 36

1.   Examination of Figure 36-4 indicates that as n increases, the spacing of adjacent energy levels decreases.
3.   (a)
5.   (d)
7.   (a)
9.   (c)
11. In conformity with the exclusion principle, the total number of electrons that can be accommodated in states of quantum number n is n^2 (see Problem 37). The fact that closed shells correspond to $2n^2$ electrons indicates that there is another quantum number that can have two possible values.
13. (a)   phosphorus
     (b)   chromium
15. (d)
17. The optical spectrum of any atom is due to the configuration of its outer-shell electrons. Ionizing the next atom in the periodic table gives you an ion with the same number of outer-shell electrons and almost the same nuclear charge. Hence, the spectra should be very similar.
19. (a)   allowed
     (b)   not allowed
     (c)   not allowed
     (d)   allowed
     (e)   allowed
21. (b)   75.2 nK
23. (a)   103 nm
     (b)   97.3 nm
25. (a)   1.51 eV; 821 nm
     (b)   0.661 eV; 1876 nm
           0.967 eV; 1282 nm
           1.13 eV; 1097 nm

	$6 \to 3$	$5 \to 3$		$4 \to 3$
	1097 nm	1282 nm		1876 nm

27. (a)   657.8 nm
     (b) $1.0945 \times 10^7 \text{ m}^{-1}$
29. (b) $1.096776 \times 10^7 \text{ m}^{-1}$; $1.097374 \times 10^7 \text{ m}^{-1}$; 0.0546 percent
31. (a) 1.49×10^{-34} J·s
     (b) $-1, 0, +1$
     (c) z

33. (a)   0, 1, 2
     (b) 0, $-1, 0, +1$, $-2, -1, 0, +1, +2$
     (c)   18

49.

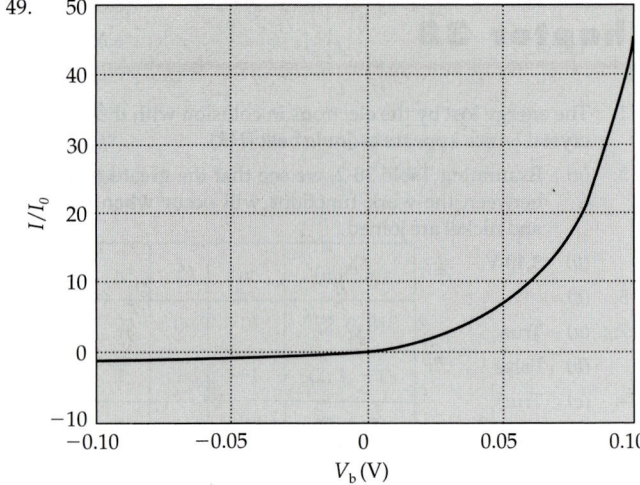

51.   250

53.   *(a)*   *(b)*

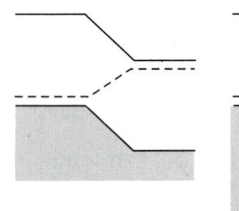 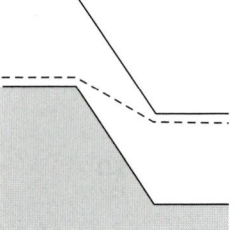

55. $n = 1.00 \times 10^{23}$ m^{-3}; the semiconductor is *p*-type.

57.   *(a)*   2.17 meV; $0.8E_{g,\text{measured}}$

     (b)   0.454 mm

59. 1.97×10^{18}

63.   1

67.   0.596

73.   1.07

75.   *(a)* 5.51×10^{-3}

     (b) 1.84×10^{-2}

77.   689 nm

Chapter 39

1.   *(a)*

3.   *(a)*   True

     (b)   True

     (c)   False

     (d)   True

     (e)   False

     (f)   False

     (g)   True

$\Delta y = \Delta y'$

$\Delta t \neq \Delta t'$

5.   Although $\Delta y = \Delta y'$; $\Delta t \neq \Delta t'$. Consequently, $u_y = \Delta y / \Delta t' \neq \Delta y' / \Delta t' = u_y'$.

7.   *(a)*   0.946

     (b)   12.3 Gc·y

9.   *(a)* $L_p = 978.5$ m; the width of the beam is unchanged

     (b) 9.57×10^7 m

     (c)   0.102 μm

11.   *(a)*   0.914*c*

     (b)   22.5*c*·y

     (c)   101 y

15. 1.85×10^4 y

17.   *(a)*   1.76 μs

     (b)   6.32 μs

     (c)   3.09 μs

     (d)   1.70 km

19.   4.39 μs

21.   *(a)*   2.10 μs

     (b)   2.59 μs

     (c)   0.49 μs

     (d)   2.59 μs

     (e)   4.36 h

     (f)   18.8 h

25. 2.22×10^7 m/s

29.   10.5 ms

31.   *(a)* $u_x = v; u_y = \dfrac{c}{\gamma}$

33.   *(a)*   0.976*c*

     (b)   0.997*c*

35.   66.7 percent

39.   *(a)*   290 MeV

     (b)   629 MeV

41.   *(a)*   0.943*c*

     (b)   3.00 MeV

     (c)   2.83 MeV/*c*

     (d)   0.878 MeV

     (e)   4.12 MeV/*c*^{2}

43.   In a freely falling reference frame, both cannonballs travel along straight lines, so they must hit each other, as they were pointed at each other when they were fired.

45.   0.999*c*

47.   *(a)* S' moves in the negative *x* direction

     (b)   1.73 y

49.   281 MeV

51.   *(a)* $v = -\dfrac{E}{Mc}$

     (b) $d = -\dfrac{LE}{Mc^2}$

53. $a_x' = \dfrac{a_x}{\gamma^3 \delta^3}; a_y' = \dfrac{a_y}{\gamma^2 \delta^2} + \dfrac{v u_y}{\gamma^3 \delta^3 c^2} a_x$

Chapter 40

1.   *(a)* ^{15}N, ^{16}N

     (b) ^{54}Fe, ^{55}Fe

     (c) ^{54}Fe, ^{55}Fe

3. Generally, β-decay leaves the daughter nucleus neutron rich, that is, above the line of stability. The daughter nucleus therefore tends to decay via β^- emission, which converts a nuclear neutron to a proton.

5. It would make the dating unreliable because the current concentration of ^{14}C is not equal to that at some earlier time.

7. The probability for neutron capture by the fissionable nucleus is large only for slow (thermal) neutrons. The neutrons emitted in the fission process are fast (high energy) neutrons and must be slowed to thermal neutrons before they are likely to be captured by another fissionable nucleus.

9. (a) β^-

   (b) β^+

11. (a) False

    (b) True

    (c) False

    (d) True

13. (a) $^{16}_{7}N \rightarrow {}^{16}_{8}O + {}^{0}_{-1}\beta + {}^{0}_{0}\overline{v} + Q$

    (b) $^{248}_{100}Fm \rightarrow {}^{244}_{98}Cf + {}^{4}_{2}He + Q$

    (c) $^{12}_{7}N \rightarrow {}^{12}_{6}C + {}^{0}_{+1}\beta + {}^{0}_{0}v + Q$

    (d) $^{81}_{34}Se \rightarrow {}^{81}_{35}Br + {}^{0}_{-1}\beta + {}^{0}_{0}\overline{v} + Q$

    (e) $^{61}_{29}Cu \rightarrow {}^{61}_{28}Ni + {}^{0}_{+1}\beta + {}^{0}_{0}v + Q$

    (f) $^{228}_{90}Th \rightarrow {}^{224}_{88}Ra + {}^{4}_{2}He + Q$

15. Mass density $= 10^{15}$, half life $= 10^{15}$, and nuclear masses $= 2$

17. (a) 92.2 MeV; 7.68 MeV

    (b) 492 MeV; 8.79 MeV

    (c) 1802 MeV; 7.57 MeV

19. (a) 3.02 fm

    (b) 4.59 fm

    (c) 6.98 fm

21. (a) 4.11×10^{-21} J; 25.7 MeV

    (b) 2.22 km/s

    (c) 10.1 min

23. 295 MeV

25. (a) 5 min

    (b) 250 Bq

27. (a) 200 s

    (b) 3.47×10^{-3} s^{-1}

    (c) 125 Bq

29. (a) 500 Bq; 250 Bq

    (b) 1.04×10^6; 5.19×10^5

    (c) 12.1 min

31. (a) 4.55×10^3 α/s

    (b) 5.32×10^4 y

33. $^{239}_{94}Pu \rightarrow {}^{235}_{92}U + {}^{4}_{2}\alpha + Q$; $Q = 5.25$ MeV; $K_\alpha = 5.16$ MeV; $K_U = 87.9$ keV

35. (a) 0.133 h^{-1}; 5.20 h

    (b) 3.11×10^6

37.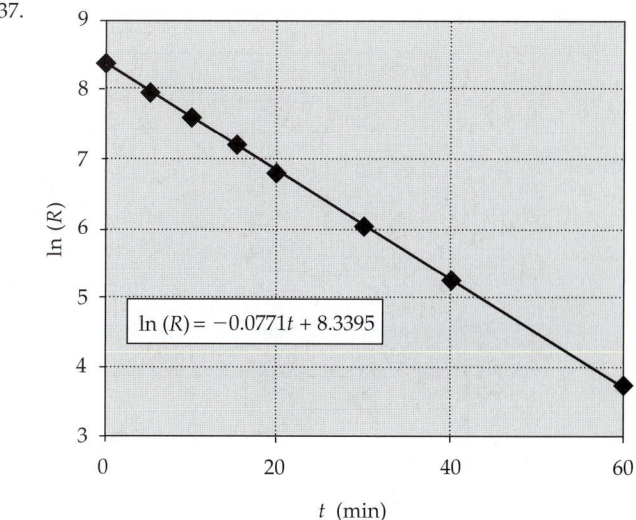

    $\lambda = 0.0771$ min^{-1}

    $t_{1/2} = 8.99$ min

39. 2.94 g

41. 3.50 min

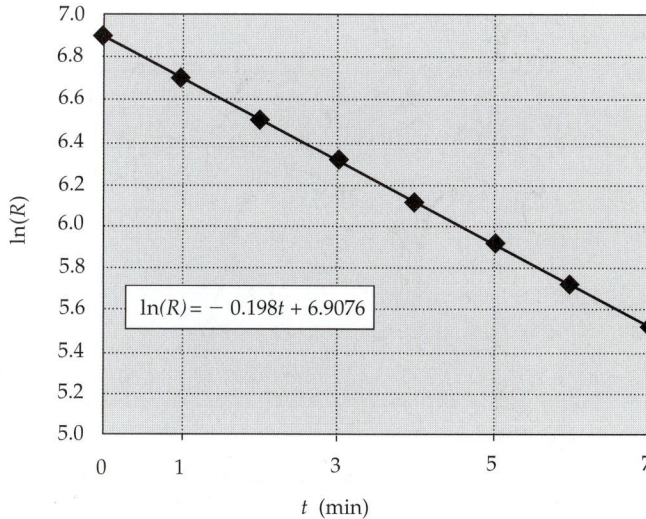

43. 7.03×10^8 y

45. (a) 0.156 MeV

    (b) 3.27 MeV

47. (a) 0.158 MeV

    (b) The masses given are for atoms, not nuclei, so for nuclear masses the masses are too large by the atomic number times the mass of an electron. For the given nuclear reaction, the mass of the carbon atom is too large by $6m_e$ and the mass of the nitrogen atom is too large by $7m_e$. Subtracting $6m_e$ from both sides of the reaction equation leaves an extra electron mass on the right. Not including the mass of the beta particle (electron) is mathematically equivalent to explicitly subtracting $1m_e$ from the right side of the equation.

49. 1.56×10^{19} s^{-1}

51. 208 MeV; 88.1 percent

53. 3.20×10^{10} J

55. (a) $4{}^{1}H \rightarrow {}^{4}He + 2\beta^+ + 2v_e + \gamma$

    (b) 24.7 eV

    (c) 3.74×10^{38} s^{-1}; 5.04×10^{10} y

57. 0.693 s^{-1}

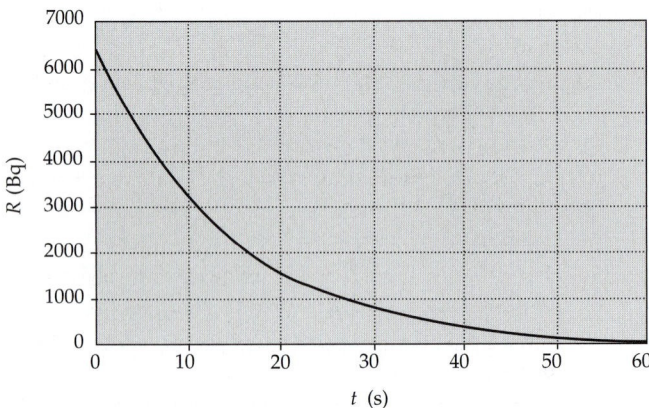

59. 156 keV

61. -2.88 MeV

63. 6.75×10^3 Bq

65. 6.30 L

67. (a) 22.9 MeV

    (b) 4.17 GeV

    (c) 1.28 GeV

69. (a)

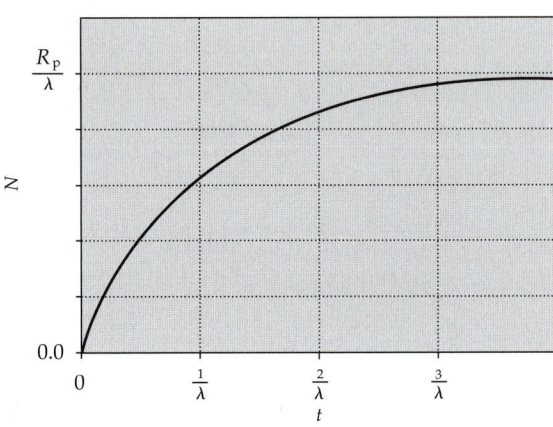

    (b) 8.66×10^4

71. (a) 4.00 fm

    (b) 310 MeV/c

    (d) 310 MeV

75. (a) 1.188 MeV/c

    (b) 752 eV

    (c) 0.0962 percent

77. (b) 55

79. (d)

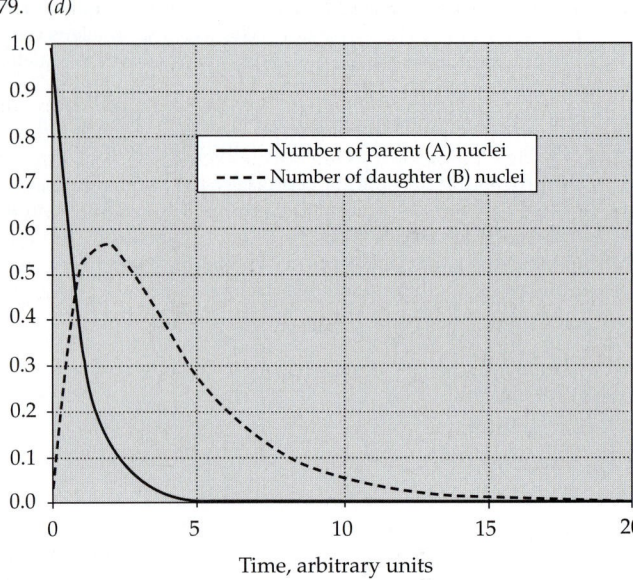

Chapter 41

1. *Similarities:* Baryons and mesons are hadrons; that is, they participate in the strong interaction. Both are composed of quarks.

   Differences: Baryons consist of three quarks and are fermions. Mesons consist of two quarks and are bosons. Baryons have baryon number $+1$ or -1. Mesons have baryon number 0.

3. A decay process involving the strong interaction has a very short lifetime ($\sim 10^{-23}$ s), whereas decay processes that proceed via the weak interaction have lifetimes of order 10^{-10} s.

5. False

7. No. From Table 41-2, it is evident that any quark–antiquark combination always results in an integral or zero charge.

9. No. Such a reaction is impossible. A proton requires three quarks. Three quarks are not available because a pion is made of a quark and an antiquark and the antiproton consists of three antiquarks.

11. (a) 279.2 MeV

    (b) 1877 MeV

    (c) 211.3 MeV

13. (a) $+1$; Because $\Delta S = +1$, the reaction can proceed via the weak interaction.

    (b) $+2$; Because $\Delta S = +2$, the reaction is not allowed.

    (c) $+1$; Because $\Delta S = +1$, the reaction can proceed via the weak interaction.

15. (a) $+2$; Because $\Delta S = +2$, the reaction is not allowed.

    (b) $+1$; Because $\Delta S = +1$, the reaction can proceed via the weak interaction.

17. *(a)* No. The neutron is not stable. $n \rightarrow p^+ + e^- + \bar{v}_e$

    (b) $\Omega^- \rightarrow p^+ + e^+ + 3e^- + v_e + 3\bar{v}_e + 2\bar{v}_\mu + 2v_\mu$

    (c) Because $Q = -1$ before and after the decay, charge is conserved. Because $B = 1$ before and after the decay, the baryon number is conserved. Because $L_e = 0$ before and after the decay, the lepton number for electrons is conserved. Strangeness is not conserved. However, in each baryon decay $\Delta S = +1$, and each decay is allowed via the weak interaction.

19.

Combination		B	Q	S	Hadron
(a)	*uud*	1	+1	0	p^+
(b)	*udd*	1	0	0	n
(c)	*uus*	1	+1	−1	Σ^+
(d)	*dds*	1	−1	−1	Σ^-
(e)	*uss*	1	0	−2	Ξ^0
(f)	*dss*	1	−1	−2	Ξ^-

21. From Table 41-2, we see that to satisfy the conditions of charge $= +2$ and zero strangeness, charm, topness, and bottomness, the quark combination must be *uuu*.

23. *(a)* $c\bar{d}$

    (b) $\bar{c}d$

25. *(a)* *uds*

    (b) $\bar{u}\bar{u}\bar{d}$

    (c) *dds*

27. *(a)* *sss*

    (b) *ssd*

29. $3.26 \times 10^8 \ c\cdot y$

33. *(a)* It must be a meson, and it must consist of a quark and its antiquark.

    (b) The π^0 is its own antiparticle.

    (c) The Ξ^0 is a baryon; it cannot be its own antiparticle. The antiparticle is the $\Xi^0 = \overline{uss}$.

35. *(a)* The u and $\bar{u}$ annihilate, resulting in the photons.

    (b) Two photons are created to conserve linear momentum.

37. *(a)* These properties indicate that the particle is the kaon, K^0.

    (b) These properties indicate that the particle is either the Σ^0 or the Λ^0 baryon.

    (c) These properties indicate that the particle is the kaon, K^+.

39. *(a)* $1.99 \times 10^5 \ km/s$

    (b) $8.65 \times 10^9 \ c\cdot y$

41. *(a)* 1193 MeV

    (b) 77.0 MeV/c

    (c) 2.66 MeV

    (d) 74.3 MeV; 74.3 MeV/c

INDEX